THE WINES OF CALIFORNIA

THE CLASSIC WINE LIBRARY

There is something uniquely satisfying about a good wine book, preferably read with a glass of the said wine in hand. The Classic Wine Library is a series of wine books written by authors who are both knowledgeable and passionate about their subject. Each title in The Classic Wine Library covers a wine region, country or type and together the books are designed to form a comprehensive guide to the world of wine as well as an enjoyable read, appealing to wine professionals, wine lovers, tourists, armchair travellers and wine trade students alike.

Other titles in the series include:
Côte d'Or: The wines and winemakers of the heart of Burgundy, Raymond Blake
Madeira: The islands and their wines, Richard Mayson
Rosé: Understanding the pink wine revolution, Elizabeth Gabay MW
Sherry, Julian Jeffs
The wines of Australia, Mark Davidson
The wines of Austria, Stephen Brook
The wines of Bulgaria, Romania and Moldova, Caroline Gilby MW
The wines of Chablis and the Grand Auxerrois, Rosemary George MW
The wines of Georgia, Lisa Granik MW
The wines of Germany, Anne Krebiehl MW
The wines of Greece, Konstantinos Lazarakis MW
The wines of New Zealand, Rebecca Gibb MW
The wines of northern Spain, Sarah Jane Evans MW
The wines of Piemonte, David Way
The wines of Portugal, Richard Mayson
The wines of South Africa, Jim Clarke
The wines of Southwest USA, Jessica Dupuy
Wines of the Languedoc, Rosemary George MW
Wines of the Loire Valley, Beverley Blanning MW
Wines of the Rhône, Matt Walls

THE WINES OF
CALIFORNIA

ELAINE CHUKAN BROWN

ACADEMIE DU VIN LIBRARY

Elaine Chukan Brown is an award-winning writer, speaker, and global wine educator, as well as a consultant on mentorship and diversity programs worldwide. Brown reviews Napa Valley wine and wines of northeastern Spain for *Wine Enthusiast*. Previously, they served as the Executive Editor US for JancisRobinson.com, a columnist for *Decanter*, and a contributing writer to *Wine & Spirits*. They contributed to both the fourth and fifth editions of *The Oxford Companion to Wine*, the eighth and upcoming ninth editions of the *World Atlas of Wine*, and the compilations *On Burgundy* and *On California* from Académie du Vin Library. Indigenous (Inupiaq and Unangan-Sugpiaq) from what is now Alaska, Brown specializes in American wine and has dedicated their career to the intersection of sustainability, climate action, and reducing gatekeeping in the wine industry. Brown has served as a cultural advisor and diversity consultant for book projects, university courses, organizations, events, and television shows. They co-founded the Diversity in Wine Leadership Forum, serve as chair of judges for the 67 Pall Mall Global Wine Communicator Awards, and as a board member of the Wine Writer Symposium. Brown lives in Sonoma, California with their pet guinea pigs, rabbit, hamster, and Rosie bird.

For Dottie, who did it first,

And, as in all things,
for Ellaita

First published in 2025 by Académie du Vin Library Ltd
academieduvinlibrary.com

A CIP catalogue record for this book is available from the British Library
ISBN 978–1–913141–87–5

Publisher: Hermione Ireland
Editor: Rebecca Clare
Indexer: Catherine Hall
Illustrator: Darren Lingard

Printed and bound in the U.S.A.

CONTENTS

Acknowledgments ix

Introduction: How California 1

Part 1: How we got here **11**

1. Indigenous peoples and the founding of California wine 13
2. Prohibition, war, and a culture of curiosity 23
3. California farming evolves 41
4. Vietnam, Nixon, and the rise of California wine 55
5. Social media, recessions, and weather 73

Part 2: Where we grow **91**

6. California growing conditions 93
7. The North Coast 112
8. The Santa Cruz Mountains 188
9. The Central Coast 202
10. The Central Valley 251
11. The Sierra Foothills 273
12. The far north and southern California 285

Part 3: What we're facing **297**

13. The cultural significance of wine 298

14. Social considerations 304

15. Climate change 322

Epilogue: Regeneration and innovation 350

Appendix I: What is an AVA? 361

Appendix II: Wine label requirements 363

Appendix III: Sustainability certifications 365

Appendix IV: Farmworker protections 370

Appendix V: The three-tier system 373

Appendix VI: 154 AVAs of California 375

Glossary 390

Notes 415

References 429

Index 454

Maps

Map 1: California wine regions and topography 92

Map 2: California climate types 100

Map 3: Major wine regions of the North Coast 114

Map 4: The Santa Cruz Mountains and the Central Coast 201

Map 5: The Sierra Foothills and the Central Valley 272

Map 6: The Far North 287

Map 7: Southern California 292

ACKNOWLEDGMENTS

Rebecca Clare and Hermione Ireland cannot be thanked enough. Their patience, support, and trust as I struggled through a series of family health challenges that delayed my writing made this book possible. Thank you for your efforts, and your guidance. Thank you for entrusting me with the subject of California wine. It is a pleasure to work with both of you.

Thank you to James Tidwell who recommended me, Sarah Jane Evans and Richard Mayson for supporting his recommendation, and Richard Burton for initially signing me to this project.

The greatest gift rests in friends who can be trusted through uncertainty, fear, confidence, pride, and persistence alike. Kelli White generously read or listened to parts of this book as I was writing and provided feedback as well as endless encouragement. Katie Canfield reminded me of what's important, like the simple idea of enjoying the project as I worked on it. She also helped me whittle down lists and ideas. Their friendships helped get me through the family challenges too. Thank you.

Drew DiMatteo helped through everything. He listened to sections, talked through ideas, stood up for the risks I was taking, told me plainly when he disagreed, laughed with me, drank with me, challenged me when it was needed, and helped me zone out through watching movies. He also fed me, drove me to the airport, and introduced me to some of the best croissants I've had in my life. Drew loves wine more than anybody. Through so much of this book I was motivated by remembering his joy in it.

Sydney Love listened and shared ideas and resources when I was wrapping my head around how to approach recovering erased histories.

Vivianne Kennedy listened to portions of the history section and brightened my spirit with cute animal videos and by sharing photos of our silly pets. Sam Cole-Johnson talked through producers and regions with me.

Tahiirah Habibi, Yannick Benjamin, Davon Hatchett, Paul Johnson, Annemarie Succop, Susan Collicott, Jude Rose, Drew Herman, Håkon Skurtveit, Danielle Shehab, Tye Thompson, Mick Descamps, Todd Trzaskos, Eric Crane, Chris McLloyd, Tonya Pitts, Pascaline Lepeltier, Andrew Caillard, Jasper Morris, Angus Hughson, Denise Matthews, Diana Hawkins, Maryam Ahmed, Elaine Heide, Aidan Kennedy, Vivian Gay, Sam Coturri, James Joiner, Erin Drain, Paul Pujol, Nina Marfell, Stu Marfell, Scott Kirkpatrick, Jim Sligh, Sue McNulty, Gaïa Orain, Cameron and Janet Douglas, and Christy Frank all kept reminding me I have support while sending me hilarious internet memes, posting videos of donkeys being their goofy selves, and pouring me gin and tonics.

Jirka Jireh and Jahde Marley, you each helped me write this more than you know.

Emily Ticasuk Ivanoff Brown, Paul and Anna Chukan, Anisha McCormick, Leda Garside, Jessica Hutchings, Santiago De Marco, Olga Barbosa, Stacey Abrams, Frantz Fanon, James Baldwin, Toni Morrison, bell hooks, Shane Burcaw, Dolores Huerte, Larry Itiong, Billy Mills, Levar Burton, Haider Ackermann, Daphne Guinness, Ursula K. Le Guin, and S'chn T'gai Spock, your work got me here.

More vignerons, sommeliers, wine writers and wine experts have helped me than I could ever name. Thank you.

Fred Wojtkielewicz, you helped me get started. I love you.

Everything I do is in gratitude to my elders, ancestors and family, as well as Josef Rivers and Gita Jahn, and with love for Jason Harris-Talley, Louis Angeles, and most of all, in all things, Ellaita.

INTRODUCTION: HOW CALIFORNIA

California revels in discovery. Its wine and its identity depend as much on the state's ethos as its climate or geology. Almost as soon as the United States occupied the California territory, it was adopted as a state. Territories typically suffered a trial period to determine if they were state worthy, but in California, gold was found. It was an accidental discovery with economic promise that quickly changed the world.

The California gold rush became a founding narrative for the region that continues today. *We might find riches. Everywhere, if we try, we could create a new opportunity from what was thought impossible.* The gold rush. Hollywood. Disneyland. Silicon Valley. Apple computers. Tesla. Monarch tractor. Each an innovation, founded in California. And each a reminder for people around the world: in California you can create riches.

Even wine made in the state has created something new. What is surprising about California wine is less how successful it has been (and how quickly!), and more that it managed to become one of the world's leading wine regions based entirely on imported grapes. Like, essentially, all the so-called new world, California depends on varieties native to Europe. But so many cultivars of Europe brought to California became not merely derivations of their original homeland, but novel and satisfying expressions that reveal new possibilities in wine. An innovation as much as electric cars, early internet companies, or the first talking films.

CALIFORNIA'S VARIETAL CONTRIBUTIONS

Amidst many successes, three varieties are among California's unique contributions to the world of wine: Zinfandel, Chardonnay, and Cabernet Sauvignon. Each grows elsewhere, but in California they created new genres.

Zinfandel was almost completely lost from the vineyards of Croatia, its original home, and while Primitivo (as the grape was known in Italy) grew in Puglia, it was the success of Zinfandel in California that bolstered Italy's interest in it. Genetic innovations at UC Davis played a key role in recognizing Zinfandel and Primitivo as the same grape. Today, Italy has increased sales by renaming wines made from Primitivo to Zinfandel, and the rosé version, known as white Zinfandel, has gained popularity through southeast Asia.

Chardonnay unquestionably comes from Burgundy. But it was California, in the 1930s, that named it by variety rather than region first. Then, through a combination of varietal labeling and a series of market revolutions, California established the variety's fame *as Chardonnay* (instead of white burgundy, or Chablis, or Montrachet, or Puligny) and helped make it one of the most planted grapes in the world. Even wine, it would seem, can fulfill the American dream in California.

Cabernet Sauvignon enjoys royal status in the left bank of Bordeaux. But with its growth in California, it created two new categories in the world of wine. Its move into the steep slopes and mountain ridges of the state brought recognition of mountain tannins. While there are high elevation Cabernet vineyards elsewhere in the world, nowhere else is there a high enough concentration of them to create the notion of mountain Cabernet. Then, stepping into the lust and abundance of the late twentieth century, Robert Parker ushered in another new expression of the variety seemingly native to California: cult Cabernet. Harlan, Screaming Eagle, Dalla Valle, Colgin, Bryant, each one a Napa Cabernet reaching icon status and collectors' cellars through the praise of Parker for their vivacity and size.

This is the ethos of California: that when we do something, we can change the world. California is about believing luck, innovation, and hard work can convert impossibility into the next great industry. (You probably also need capital, but with California luck, you might not!)

It shows up today in the continued willingness of California's winegrowers and vintners to experiment. We've established what grows well here, and yet we keep trying varieties that shouldn't work where we plant them if we are to believe merely the growing conditions of their original country.

The obscure grape Valdiguié originated in southwestern France, where it was first planted in the 1870s. It was imported to California in the twentieth century and labeled Napa Gamay. In France, it has all but disappeared with only small plantings remaining around the Languedoc. Today, it is comparatively more widely planted in California. Its wines initially were meant to emulate the charm and freshness of Beaujolais Nouveau, but today we know the grape is not Gamay at all but another mid-weight, delicious red. It's still a niche variety, but our work with it has helped instigate curiosity for it again in France.

Back in California, we find Albariño, but not grown along the ocean shores like we see in Spain. Instead, it's been planted inland at the intersection of California's Central Valley and its interior delta region through Lodi, or perched on the high elevation and forested slopes of Calaveras County in the Sierra Foothills. Each location is far from the ocean, yet both deliver delicious, thirst-quenching Albariño enjoyed by in-the-know wine lovers.

For another example, take our willingness to give Petite Sirah a chance after it was rejected as a climate mismatch in southern France, where it was intentionally bred (and known as Durif). In California, it became the backbone for the historic field-blend vineyards of the late 1800s and early 1900s, bringing depth and structure to the exuberance of Zinfandel. Then, decades later, with the desire for full-bodied, tannic, grandiose reds of the late twentieth and early twenty-first centuries, Petite Sirah became an impeccable blender for more impoverished varieties, and its own varietal wine. Concannon of Livermore created the first varietal bottling of Petite Sirah in 1961. Others soon followed. While it is mostly regarded as a full-bodied wine of intensity, some producers refused to believe brawn and thrust was all Petite Sirah had to offer. Older bottles from the likes of Freemark Abbey, York Creek, Souverain, and Ridge, as well as more recent examples from (just shuttered) Carlisle, (the now defunct) Mountain Tides, (the also finished) Dirty & Rowdy, or the deservedly still successful Tres Sabores, Theopolis, and Turley have asserted the variety's aromatic, more finessed side. Altogether, here in California, Petite Sirah has done so well that a few growers in France have brought the variety home to plant there again.

GROWING CONDITIONS

Innovation proves to be as much a part of the growing conditions of California as the San Andreas fault, the Pacific Ocean, the state's Mediterranean climate, its temperature range, or exposure to sunlight. It is the influence of the human element brought into harmony with the textured landscape.

Understanding how California came to be the fourth largest producer of wine in the world, as well as the largest in the United States by a factor of more than 16, depends on recognizing that human element. This includes more than mere planting or farming choices made by a few individuals. Who can make these choices in the vineyard depends on the bigger picture.

National – and global – political, economic, and social dynamics and events influence who has access: the history that established wine in the state; the political and market forces brought to bear on the wine industry; the socio-political decisions that led to legal changes, immigration policies, and even intentional genocide. Each of these shaped the wine industry California has today.

But understanding the growing potential of California wine also depends on knowing the state's unique topography; the relationship between the cold ocean currents of the Pacific Ocean with the rising air off the hot inland valleys; the marine layer; seasonal fog; and the important role of wind. These determine which grape types can succeed in particular areas.

How California wine came to be is due to multiple factors. The intersection of the state's environment with broader global economic and political forces channels through the decisions of local communities and individual wine growers, investors, and winemakers to make the state of wine. The Spanish missions planted vines along the coast of Monterey Bay, but the grapes couldn't ripen there. It was too cold. Later, new venture seekers tried other varieties that would.

GLOBAL FORCES

Wine first came to California in the late 1700s, via Spain. The empire sought to protect its investment in New Spain, or modern-day Mexico, by expanding up the coast along the Pacific. Over time, the settlers also brought grapes native to Europe to colonize the landscape and help convert the Indigenous of the region. Those who were converted became the viticulturists of the first hundred years. Only a few Spanish monks were involved in the push to plant grapes. It was enslaved Indigenous peoples that conquered winegrowing in America. After Mexico gained independence, and in the early years of the United States in California, it was the Indigenous population of the area that created the region's wine.

Vines were established in each of the 21 missions built by Spain between San Diego and San Francisco (the twenty-second, Sonoma, was added by Mexico), but four of them never produced wine. The temperatures were too low and ocean influence too great in San Francisco, Santa Cruz, Monterey, and Carmel to sufficiently ripen the fruit.

Unlike Spain, Mexico fostered open trade policies and offered citizenship incentives to anyone who moved into Alta California with hard work and an entrepreneurial spirit. The population grew. Under Mexico, Alta California included an enormous expanse across not only what is California today but also Nevada, Utah, and parts of Colorado, Arizona, Wyoming, and New Mexico. After it was ceded to the United States, Alta California was partitioned into smaller states. The United States also furthered the population boom with a Homestead Act. Land was given to new occupants and farmers, the cost pro-rated to how long they lived on it. Stay long enough, you could own it for free, time served.

New immigrants brought new vines. Spain planted only what was eventually named the Mission grape, known as Pais or Criollo in South America, and Listán Prieto in its native Spain. While it served its purpose, its potential for quality was limited. In young vines, the texture of the resulting wine tends to be unpleasant. But as the population of California expanded, so did its range of vines. Investors from the east coast brought vine cuttings gathered throughout the Austro-Hungarian empire. Second waves of missionaries carried in new varieties from northwestern Spain. Wine aficionados traveled around France capturing selections of vines to plant in California. These became the nurseries and expanding vineyards of a new industry.

But a wave of economic recessions in the late 1800s, as well as phylloxera and Pierce's disease, depressed its development. Then social mores did. In the early 1900s, when the United States finally joined World War I, though its loss of life was significant, the impact on the wine industry was almost irrelevant. Prohibition had already stopped the sale of not only wine but essentially all alcohol.

World War II, on the other hand, helped build the wine industry. It was then supported by the prospering of the fledgling American middle class. But racial segregation maintained throughout the country well into the second half of the 1900s meant interest in wine and investment in its growth was primarily a phenomenon of the white population.

Only recently did new efforts to include people of non-white ethnic and racial groups begin. Today, California celebrates the Association of African American Vintners, the longest-standing Black wine festival in the country – Black Vines, the nation's first Native American winemaker, several of the country's few Arab-American winemakers, a handful of Japanese-American vintners, and wineries all over the state are owned by people who originate from India, China, Vietnam, Chile, Mexico, Iran, Lebanon, Palestine, and others. Even so, many of these examples feel like exceptions. The socio-economic history of the United States means white populations, broadly speaking, were historically advantaged by the nation's policies so had the money to invest in wine.

But the innovative spirit of California persists. Solutions to farming amid climate change develop swiftly here. No other wine industry in the world has such proximity to or availability of technological ability, in the form of Silicon Valley, or scientific research acumen, from UC Davis, combined with the investment capital of the state's wealth. Here in California, a robotics expert joined forces with a manufacturing logistics specialist, a mobility engineer, and an organic farmer to create Monarch, the world's first fully electric, automated, smart tractor. They then also convinced the state to offer financial incentives to farming companies to make owning it easier.

In Napa, a plant biologist and climate scientist from UC Davis teamed up with a viticulturist and vineyard consultant, and convinced growers throughout Napa Valley (including some of the most proprietary and tight-lipped in the region) to share weather data and farming choices to create a more up-to-date climate model. It will help growers better understand the relationship between climate, farming, and fruit quality.

In San Francisco, a community organizer and urban farmer partnered with a viticulturist from the north coast to create a fully funded, yearlong and immersive apprenticeship for Black, Indigenous, and people of color to learn vineyard work and management from some of the top producers in California. Together, they're slowly expanding access to leadership in the wine industry for people who historically have been excluded. These are only a few examples of innovation happening today.

NEW CHALLENGES

Climate change has come to California. As overall temperatures increase, the heat is perhaps less a concern than how it simultaneously increases the vapor pressure deficit, making the air effectively drier. As air becomes drier, the wildfire season that has always existed grows longer. It's pushed California to collaborate with researchers in Bordeaux and Australia at a level beyond what's happened before. As a result, the world of wine has a better understanding of how smoke impacts vines, when it bonds with fruit sugars, and how it shows up in wine. They're still working on how to mitigate it.

Climate change has already shifted where people plant. Over time, vineyards have pushed closer to the coast and higher up the mountain slopes in search of more even temperatures. Mendocino used to be regarded as the northern boundary of the state's growing regions. Today, vines grow around Mount Shasta and close to the border of Oregon. It's not merely an evolving climate that's helped these newer areas develop. It's also the deepening of our understanding of how vines grow. As producers have gained a more global perspective, they've expanded their views of what's possible. Newer plantings take inspiration from inland Galicia and the Jura as much as the French alps.

Shifts in planting choices, grape varieties, and regions have steadily altered the state map of tasting experience. Writing a digest of California wine is no longer a straightforward endeavor of listing the grapes and describing the regions, outlining their founding and the most inspired producers.

WRITING CALIFORNIA

Today, a comprehensive book on California wine depends on considering the complicated (and often painful) history of how the state's industry became a world leader, what producers are doing to face new challenges, and how its regions continue to evolve. That's what this book hopes to do, offering readers not merely an accounting of California wine today, but an in-depth look at its broader context.

The book is written in three parts. It begins with the state's history. Brilliant work has already been done outlining the major moments in the evolution of California wine. The writings of Thomas Pinney and Charles Sullivan are recommended for in-depth, wine-specific history. Rather than recount the work that's already been done, this book seeks to examine wine's development within the broader state, national, and international changes occurring at the time.

People often forget that the triumphs of history arise from problem solving great challenges. As one of the most successful and interesting wine regions in the world, California has grown from communities of people responding to a series of difficulties with solutions. At times changes in wine were catalysed by forces that at first would seem to have little connection to wine.

In looking at the state's wine history within a larger context, I have tried to demonstrate the connections between its successes and pivotal challenges, its triumphs and the costs that came from them. This includes acknowledging that its history of winegrowing has depended on the labor of people of color. Other histories on the wine industry have tended to overlook or erase this reality. By incorporating this important part of wine's development, we can better understand the history of California and perhaps also what solutions for today's labor issues might be possible.

My hope is that providing this bigger picture will make California's story more memorable – and perhaps also open additional room for unique conversations in the endless potential for wine to be interesting. In truth, I have also done this to show respect for so many people whose work I admire, not only the laborers who have built our communities, but numerous people now actively evolving the future of wine in America.

Each history chapter begins with a list of key dates relevant to the time period. In short, the history section considers the question of *how we got here*.

Part 2 looks at *where we grow*. It examines the overall growing conditions of the state before delving into the particulars of the larger regions, such as the North Coast, the Central Coast, or inland California, and the nuances of more specific appellations. Brief consideration of the history important to understanding the formation of that area, and maps to illustrate are also included. At the back of the book, there are also appendices to help understand the technicalities of some of these topics, for example, the US legal definition of an appellation, the AVA.

Finally, Part 3 considers *what we're facing*. The challenges of today are considered in series, looking at the reality of climate action through farming, the impact of wildfires, the challenges of a changing market, and more.

Altogether, these materials aim to provide an inclusive and far-reaching look at California wine. The book provides information for wine students and wine lovers, and, hopefully, also compelling perspectives on the wines of California in a larger context.

PART 1

HOW WE GOT HERE

The first hundred years
Key dates
1769
First Spanish settlers arrive, the missions start
1779
First Mission vines are planted, at San Juan Capistrano
1781 or 1782
First wine made in Alta California
1797
First brandy made in Alta California
1810–22
Mexican war with Spain
1822
Mexico wins independence, and Alta California
1833
Mexico secularizes the missions
1847
United States claims Alta California as territory
1848
Discovery of gold is announced
1850
California becomes a state
1850
California passes the Indian Indenture Act
1866
Thirteenth Amendment passes, abolishing slavery and indentured servitude
1869
Transcontinental railroad arrives in California
1875
Page Act is passed
1882
Chinese Exclusion Act is passed

1

INDIGENOUS PEOPLES AND THE FOUNDING OF CALIFORNIA WINE

The first hundred years

Indigenous peoples started California wine. Franciscan priests brought the first grape cuttings to make sacramental wines for the native population they intended to convert, but the priests depended on those same Indigenous people to grow the vines. Decades later, when Mexico succeeded in gaining independence and wrestling Alta California from Spain, it was still the Indigenous vineyard workers who had the skills needed for winegrowing. Their knowledge perpetuated wine under the United States as well.

After Mexico secularized the missions and established international trade, commercial winegrowing took hold in Alta California. The pueblo of Los Angeles became the epicenter of a new wine industry led by the ambition of a mix of wealthy Mexican families and immigrants newly welcomed to the region by the country's more open trade policies. The expansion of vineyards depended on the labor and know-how of the Native peoples already trained in viticulture. Under the new economy, Indigenous peoples were the primary resource for winegrowing.

Mission era California is often credited with introducing the region's first *Vitis vinifera* grapes. But other aspects of winegrowing also arose during this time that became standard throughout the area for subsequent decades. The first vineyards were universally head-pruned. This approach was adopted from Spanish viticultural practice and proliferated by Indigenous vineyard workers taught the technique by Franciscans. It remained the dominant form well into the twentieth century and continues at smaller scale today.

The success of the first hundred years of winegrowing depended on deeply anti-Indigenous sentiment, established by the Spanish colonizers, perpetuated by Mexico, and escalated by the United States. Anti-Native racism became both the motivation and the means to grow wine.

Historical evidence suggests that prior to the first explorers, Indigenous populations in the region exceeded several hundred thousand residents, with plausibly as many as a million people, and certainly over one hundred unique languages as well as additional dialects.[1] Tribes enjoyed robust trade relationships with other Indigenous societies not only throughout California but into what today are regarded as neighboring states.

Europeans began exploring California's coastline as early as the 1540s.[2] The trails and trade routes established by the area's First Peoples served as lifelines for explorers to traverse the landscape, and later for the first settlers to establish themselves in the region. But even so, the first Spanish colonizers saw the Indigenous people of California as primitive, childish, and in need of outside instruction on how to live.

When the Spanish established the colony in what came to be known as Alta California, their law did not recognize Indigenous people as citizens, but neither did it allow them freedom. Instead, the goal of the first settlements of Alta California was to "civilize" Indigenous people into Spanish ways of life, first through conversion, and then by enslavement. This attitude of the first colonizers laid the foundation for similar attitudes practiced by Mexico and then more extreme efforts to eradicate Native peoples entirely by the United States.[3] Indigenous peoples under Mexican rule were treated as indentured servants. When California joined the United States, the first legislature created a law that made forced labor of Native peoples to non-Natives legal and denied Natives citizenship; soon after, laws were added that paid for their extermination.[4]

Even so, into the 1860s, Indigenous workers served as the primary means for planting and growing wine in California. Knowledge gained from work within the missions was carried into subsequent plantings for private landowners. The know-how Indigenous peoples provided made the first hundred years of California wine possible, and under Mexico and the United States, helped build the region's economy.

THE SPANISH ARRIVAL

Franciscan missionaries arrived in 1769 alongside Spanish soldiers on what was called the Sacred Expedition. Though numerous European explorers had charted the coast over the two centuries before, this was the first non-Native group to settle in the area. Spain was protecting its venture in Mexico. Meanwhile, Russians were steadily migrating down the coastline from the north.[5] English ships had landed in Canada and led at least one expedition through the waters off California.[6] So Spain moved north from Baja to stop other countries from getting too close.[7] With enormous struggle, and the loss of at least half their party, the Spanish established themselves from San Diego to San Francisco,[8] beginning the mission era.

The missions were meant to be a temporary gamble, and part of a three-part approach to settling the region. The military would establish *presidios,* to serve as troop garrisons and protection. Officials would build pueblos, or townships for territorial citizens. And the churches would create missions to convert, house, and enslave the Indigenous peoples of California. Once California was brought to the Catholic faith, the literal mission of the Church would be complete. Mission lands, it was said, would be divided and distributed to the converted,[9] and the missionaries could return to other parts of the Spanish empire.

As settlements were established, surrounding forests were cut down and livestock were let loose into areas to feed themselves and be caught when needed. The lack of trees, plus the arrival of new settlers and their animals, scared off native wildlife, while livestock grazing cleared the land for invasive, imported plant species to proliferate.[10] The landscape Indigenous people relied on for food was changed.[11] As the immigrants moved north through Alta California, they transformed not only the landscape but Indigenous peoples' access to their ways of life.[12]

Over time, the destruction of traditional ways of accessing food became a means of control, as soldiers and missionaries brought Native peoples to the missions to offer food and housing. Franciscan missionaries did not venture out into Native communities to proselytize. Instead, the altered landscape pushed Natives into new areas, and histories of trade meant they sometimes sought contact with the new arrivals. This allowed some Natives with more intact resources or greater distance from settlements to subsist on their own. Those who did travel to missions were offered food in exchange for work. If they stayed, priests would strive to convert them.

Missionaries called converted Natives neophytes, children of God needing parental control and supervision sanctioned by the priest. According to the Spanish Church, the Indigenous peoples of North America were savages needing to be saved and civilized through conversion.[13] Under Spanish law, they had no rights to citizenship or independence. Conversion began with taking Communion, but once taken, neophytes were no longer allowed to leave. Those who tried were recaptured by Spanish soldiers and returned to the strict control of the Franciscans. Families and tribal groups were separated, sent to different missions to ensure they remained dependent on their Spanish masters. Breaking family and tribal ties in this way aided in erasing cultural practices to speed the longer process of conversion under Christianity. Now residents, neophytes were required to work to fulfill the needs of the mission and its neighboring *presidio*. By accepting Communion, Indigenous individuals were unknowingly entering enslavement.[14]

Although the Franciscans wielded control over the converted, they depended on soldiers to enforce it. Few Franciscans ever resided in Alta California. Each mission housed less than a handful, sometimes only one. As a result, Franciscans relied on the labor of the captured neophytes to perform whatever work was needed for survival and the building of a new civilization. Indigenous peoples built the infrastructure for early California, from buildings and forts to farms and irrigation. Though Franciscans and soldiers forced the change, Indigenous peoples established the foundation of the California we live in today. Eventually, neophytes were also planting and maintaining vineyards.

BRINGING OF THE VINE

Conversion of neophytes depended on sacramental wine. Communion could not be performed without it. For at least the first ten years, priests relied on barrels of wine imported by ship. We know from letters written by Priest Junipero Serra, who led the Franciscans building the missions until his death in 1784, that getting enough wine from Mexico or Spain to adequately serve Communion was difficult.[15] The prevalence of wild grapevines throughout California assured Spanish settlers that cultivated vines would proliferate.[16] *Vitis girdiana* in the south and *Vitis californica* in the north prospered in streambeds,[17] often alongside wild roses and other blooms. However, survival of the new settlements depended on growing food, so other crops and livestock were established first.[18] Once the newcomers escaped their early scarcity, they began to work on growing vines.

In 1777, Serra wrote to the viceroy in Mexico requesting cuttings from Baja to establish vineyards in Alta California. By 1779, vines were planted at mission San Juan Capistrano, establishing what became known as the Mission grape.[19] From these vines we can say that 1781 or 1782 was likely California's first vintage of wine.[20] Eventually, the missions began distilling as well, creating their first brandy in 1797.[21]

The Franciscans brought with them a book on viticultural practice in Spain, *Agricultura General.*[22] The priests translated its recommendations to the California landscape and instructed Indigenous converts on how to plant and maintain the vines. Although trellising techniques were included in the text, that approach required additional materials like wire and posts for attaching vines. So a trellis was considered a disadvantage. Mission vineyards were established using a head-pruned technique,[23] thus beginning what would become the standard for California vineyards for almost the next two hundred years.[24]

For irrigation purposes, vineyards were planted close to streambeds or waterways. Indigenous vineyard workers dug a series of ditches to channel water between vine rows, to allow for irrigation once or twice a season. This was done to service other crops as well. The wine and food produced through Indigenous ingenuity fed both the missions and at least some of the nearby *presidios.*

Distillation meant both still wines and fortified wines could be made. A little north of San Juan Capistrano in the Los Angeles basin, Mission San Gabriel eventually became the largest producer of wine in the mission era. There, four types of wine were made: a dry red table wine, a sweet red table wine, a dry white table wine, and a fortified white.[25] The whites relied on juice from the Mission grape separated quickly enough from the skins to avoid color. The fortified white was likely Angelica, made by adding brandy to freshly pressed grape juice, or partially fermented wine so that it remained sweet.

It is likely that the earliest wines in the mission era were red. After harvesting the fruit, Indigenous workers spread the grapes on a flat platform and smashed them with their feet. The released juice was captured in bag-shaped skins and then fermented in skins or clay jars. Barrels were difficult to come by, but when available they were used as well.

Early records of winemaking are uncommon, but in at least some cases the grape skins were added to the top of fermenting juice to protect it from spoilage. As infrastructure developed, larger missions established areas permanently for this purpose, either a floor designed to capture and funnel the juice, or *lagares*-like troughs, resembling those found today in Portugal.[26]

Vineyards and wine were controlled by the missions. Under Spain, international trade was not allowed within the territory, but missions often used wine or brandy for good favor with local officials or the neighboring *presidio*.[27] In some cases they were able to sell wine or brandy locally.[28] For those with large enough production, it became part of their economy. Although missions were regarded as religious centers, they operated as businesses as well, selling and trading the food, wine, brandy, or other goods produced by the neophytes for additional income.[29] By 1810, the mission system for wine production was stable enough that it continued during the Mexican war with Spain. In 1822, Mexico won independence, including control of the Alta California territory, and the mission system slowly began to change.

Under Mexico, Franciscans lost their annual government stipend for service. Without the additional income, labor requirements for farming and saleable goods increased and the demand on neophytes did too.[30] Many Natives living in missions had no communities to return to, and certainly no resources to take with them, so they continued to live and work at the missions. Those who tried to leave were often tricked into indentured servitude.

In 1833, Mexico secularized the missions. The transition process began the following year, though in different ways and at different speeds at each site.[31] Since their founding, the plan had been for mission lands to be distributed among converted Natives, which Mexico also intended. But the law did not propose a standardized approach, and so regional officials were left in charge of the process. Though some small parcels were deeded to a few former neophytes, they were not large enough for anyone to subsist on. In the end mission lands, often considered the most agriculturally rich in their area, were bestowed upon wealthy Mexican families.[32]

By the end of the mission era in 1834, the Franciscans had built a chain of 21 missions from San Diego to San Francisco under Spain, and, under Mexico, a twenty-second in Sonoma. All but four eventually produced wine. The maritime climates of San Francisco, Santa Cruz, Monterey, and Carmel meant the vineyards were unsuccessful.[33]

Many of the vineyards were incredibly small and made very little wine, though it is impossible to determine volumes accurately. San Gabriel had the greatest impact on the future of wine production, followed by San Fernando. Both were situated near enough the Los Angeles pueblo to enjoy a reliable demand for wine. The volume and quality of wine from San Gabriel was substantial enough for it to be known in both Spain and Mexico, while San Fernando was known for brandy.[34] In 1834, a census taken of San Gabriel recorded 160 acres of vineyard, certainly the largest producer in California at the time.[35] When Mexico secularized the missions, these vines became the source for cuttings to plant commercial vineyards. San Gabriel provided source material for so many new plantings it became known affectionately as *madre* (mother).

WINE OF ALTA CALIFORNIA UNDER MEXICO

Almost as quickly as Mexico gained independence, the pueblo of Los Angeles (founded in 1781) began expanding its vineyards. In the early 1800s, the first secular vineyards were established within the Los Angeles pueblo under Spain on a family land grant in what today is the city of Glendale.

Records indicate that by 1805, the pueblo had begun making wine.[36] Cuttings were taken from San Gabriel, then planted and farmed with help from Indigenous workers trained under the missions. Considering the economic restrictions in the territory, the wine could only have been intended for family or local use.[37] Trade restrictions under Spain limited economic growth, as well as any influx of foreigners. But Mexico opened international trade and offered a citizenship incentive to anyone outside the region who moved to Alta California. The change encouraged new residents to establish farming and trade relationships.[38]

By the 1830s, Los Angeles had only 764 residents[39] but more than 100,000 vines, most if not all of them planted from San Gabriel cuttings. Winegrowing became one of the territory's great hopes for economic growth.[40] The success of wine as a commercial industry depended on local laborers. Unlike Spain, Mexico made the Indigenous of Alta California citizens. No longer living within the mission system, some former neophytes were adopted by tribal groups that survived inland in eastern California. Others migrated to Los Angeles and remained the best resource for winegrowing. Paying skilled labor within a new economy is challenging, but even so, landowners needed workers for the economy to grow. So Mexico created a sort of legal loophole. Like Spain, Mexico placed a legal restriction on access to alcohol for Native people only. The stereotype of 'the drunken Indian' started under the prejudices of the mission system and continued into secular California.

Citizens and territorial officials of Alta California both protested the Native-only drinking restriction. Letters of the time argue that non-Natives have bigger problems with drunkenness. Even so, the ban persisted, and similar laws continued to be practiced after the region became part of the United States.[41] The law against Native drinking created a labor solution for territorial landowners. If caught drinking, Indigenous people could be arrested and sold at auction to the highest bidder. Landowners in need of workers took advantage of the law. They gave Indigenous workers alcohol instead of wages, then had them arrested for drunkenness. The following Monday morning, those arrested were available for sale by auction overseen by the pueblo. Anyone purchased at auction was obligated to work for the buyer unpaid.[42] The practice allowed Mexico to escape overt enslavement as practiced under Spain, while retaining the benefits of unpaid labor. And through the auctions, local pueblos earned extra money.[43]

The benefits of free labor supported the growth of Mexico's territorial wine industry with Los Angeles as its epicenter. Outside the Los Angeles basin very few vineyards continued beyond the mission era. Its Indigenous workers followed the flow of labor into the pueblo.

CALIFORNIA BECOMES A STATE

Commercial vineyards continued to expand throughout Los Angeles and its surrounding areas into the 1880s, reliant primarily on Indigenous labor. After secularization of the missions, Indigenous workers left those vineyards and began working for commercial ventures. By the time American troops arrived, in 1847, Los Angeles was already being referred to as the city of vines, dominated by vineyards in both its town limits and surroundings.[44]

During the first two years as a US territory, policy remained largely the same as under Mexican occupation. Indigenous workers suffered under a separate set of laws from non-Natives, including restricted drinking, though under the United States they were also denied citizenship.

Native peoples continued to work primarily as indentured servants to the area's landowners.[45] In December 1848, it was announced that gold had been discovered in the mountains of eastern California. Residents petitioned the US government for statehood, which was granted in September 1850. Officially, California joined the country as a slave-free state. Reality was starkly different.

Los Angeles and its surrounding areas remained the epicenter of commercial wine production for California into the 1880s. By the 1850s, the area's wines were winning awards for quality in the eastern United States.[46] In the 1860s, they were being written about in New York City.[47] As the commercial industry grew, mission vineyards without clear ownership became irrelevant. By the 1880s, even those that had been the most celebrated were largely abandoned.[48]

With the advent of the gold rush, wine sales moved north from Los Angeles to San Francisco. It became the gateway to the gold fields of the Sierra Foothills directly to the east. The increase in outsiders created a new regulation problem. Indigenous groups that survived the mission era and Mexican rule did so largely thanks to the expanse of Alta California. Many followed waterways to the east or relocated into unexplored mountains and hidden valleys along the coast. As the gold hunters pushed into these parts of California, these newcomers encountered Indigenous groups with established relationships to the land they now wanted.

In April 1850, the first session of the California State Legislature passed the Act for the Government and Protection of Indians, more commonly called the Indian Indenture Act. The law continued the practice of selling Native adults for forced labor at auction to non-Native landowners, removed tribal groups from their traditional lands, and separated Native children from their families by giving them to non-Natives for unpaid labor, described as apprenticeship.[49] American journals well into the late 1850s describe the promise of economic wealth possible from California, the abundance and promise of its agriculture (including wine), and how most of the work to produce these crops was done by Indigenous laborers.[50] Yet, faced with the desire for land and its potential for gold, California chose to eradicate its laborers.

From the 1850s until the 1870s, the California legislature allocated funds to pay not only the bounty for successful killing of Natives, but also to cover the expenses incurred by those bounty hunters. Between 1851 and 1859, the claims for reimbursement of expenses alone (bounty for each person killed was additional) submitted to the State of California totaled $1,293,179.20.[51] Massacres of entire villages occurred frequently. Those who escaped often did so with poor access to food and survived in hiding. When the United States took control of California, the known population of Indigenous people in the region was more than 150,000. The actual numbers are disputed by historians and are likely to have been much higher.[52] By 1873, the known Indigenous population in California was less than 30,000.[53] As if that was not enough, newcomers to the state of California brought with them smallpox. An epidemic in 1868 most virulently struck the poorer populations in town centers. The Indigenous of Los Angeles all but disappeared.[54] The city had lost its vineyard workers.

In 1869, the completion of the transcontinental railroad from the east helped California's state population increase substantially. But immigrants drawn by the hope for gold were rarely farmers. Agricultural labor shortages became a problem.

A wave of Chinese labor arrived in California thanks to their work completing the transcontinental railroad line to Los Angeles. For a little more than a decade they helped solve the labor shortage. They established some of the earliest commercial vineyards in Sonoma and then Napa and built significant infrastructure including tunnels and wine caves, winery buildings and homes, roads, vineyard terraces, some of northern California's most famous structures, and more that helped found both wine regions. Then, in 1875 the Page Act, and (even more impactfully) in 1882, the Chinese Exclusion Act was passed. These were the first Acts of Congress in United States history to broadly limit immigration.[55] Over the next two decades, the Chinese population of California sharply declined. California farms again faced labor shortages.

Into the 1880s, Los Angeles and southern California remained the epicenter of California wine growing. In the mid-1880s, at the region's peak of wine production, area vineyards were hit by Pierce's disease. More than half of the vines quickly died from this bacterial infection.[56] Although some vineyards persisted, attention for the region's wines declined. The epicenter of California wine shifted to the north.

2

PROHIBITION, WAR, AND A CULTURE OF CURIOSITY

The next 100 years

When Prohibition was enshrined into the US Constitution, the collection of activists (affectionately known as the dry party) that agitated it into national law rejoiced. It took decades of effort. Increased drinking was not what they expected.

The Eighteenth Amendment, ratified in 1919 and put into effect in 1920, made the "manufacture, sale, or distribution of intoxicating liquors" illegal throughout the United States. The expectation was it would result in less drinking and eradicate most social ills. The removal of the demon liquor from the marketplace would surely spawn an idyllic future. It did not. Instead, across the nation, drinking per capita increased.[1] In California, planted vineyard acres escalated. The price of grapes, though volatile, climbed. Railway stations across the eastern seaboard became trading hubs and exchanges for a new grape economy, and "grape broker" became a new career.

While the Eighteenth Amendment made Prohibition part of the US Constitution, the terms of the law were defined by the Volstead Act. Almost as an afterthought, an exemption was included to protect the rights of farmers to make juice from their own produce. But the exception was written in vague terms, and it had the effect of protecting the ability of heads of households to make wine at home. To make wine, they needed grapes.

Grapes grown in California were sent by railcar over the Rocky Mountains and across the country, as well as all over the state. Vineyard plantings expanded in already established regions as well as moving into underdeveloped parts of California. Thanks to its expanse of open land, the Central Valley became one of California's swiftest growing, and most important, growing areas.[2] Its planted acres doubled during Prohibition.[3]

Prohibition created a generation of home winemakers, but it also collapsed the California wine industry. Grape growers experienced a boom, while most professional winemakers lost their careers. When commercial winemaking returned in the 1930s, winemaking know-how had drained from the state – and the American marketplace had changed. After the repeal of Prohibition, the sale of alcohol moved from federal control to state governments. For repeal was not only a matter of the legality of booze. It was also about strengthening the rights of states, limiting the size of federal oversight, and returning self-determination to the individual. Each state became responsible for overseeing alcohol sales, and each had its own set of taxes, laws, and practices. In some states, individual towns or counties added their own taxes and policies too. The effect was an increase in alcohol-related regulations across the country, thus creating a web of policies to sell wine.

Additionally, public tastes, and what consumers were willing to pay for had changed. During Prohibition, with alcohol missing from store shelves, consumers got used to buying soda as their quick drink option. Soda sales jumped 200 percent in the first six weeks of Prohibition and kept climbing from there.[4] Though dry table wine had dominated sales before 1920, during Prohibition the American public developed a taste for sweet beverages. By 1933, consumer interest was in sweet wines. Anything else could be made at home.

Wineries that returned after Repeal had to rebuild not only their own production, but also the interest of potential wine buyers, and deal with far greater government regulation. For the next several decades, the industry would work to create a new culture for California wine.

The second hundred years

Key dates

- **1873–1879** Long Depression
- **1882–1885** Depression of 1882 to 1885
- **1875** Page Act, the United States's first expansive restriction of immigration, is passed
- **1882** Chinese Exclusion Act is passed
- **1894** California Wine Association is formed
- **1914–18** World War I
- **1919** Eighteenth Amendment ratified, to instigate Prohibition
- **1920** Prohibition begins
- **1929–39** Great Depression
- **1933** Twenty-first Amendment passed, ending Prohibition
- **1934** Three-tier system of alcohol sales begins
- **1934** Formation of the Wine Institute
- **1939** Formation of the Wine Advisory Board
- **1939–45** World War II
- **1944** GI Bill passes, helping to launch the American middle class
- **1954** *Brown v. the Board of Education* makes school segregation unconstitutional
- **1964** Civil Rights Act passed
- **1967** Table wine sales exceed sweet wine sales in the United States for the first time since before Prohibition
- **1968** Fair Housing Act becomes law

THE STATE OF THE INDUSTRY PRIOR TO PROHIBITION

Throughout the 1800s, the California wine industry struggled to find a steady footing. By the early 1900s, it had started to succeed. The entrepreneurs, merchants, and winemakers behind it learned from decades of challenges.

Beginning in the 1850s, vineyards were emerging from Sonoma, the Sierra Foothills, and Santa Clara Valley. Soon after, Napa and Monterey began planting as well. Vines were also creeping into parts of the Central Valley. Los Angeles and southern California remained the epicenter of California wine into the 1880s, but the reputation of northern California continued to grow. In the second half of the nineteenth century, the wine industry was struck with a series of challenges.

In the 1860s, vineyards in Europe were hit by the grape louse phylloxera.[5] Its ability to kill vines and destroy vineyards led to a wine shortage even in US markets that investors hoped to help fill by creating a California alternative. By the time they realized the louse had struck in northern California, significant investments had already expanded plantings throughout the region. Into the 1880s, replanting with European varieties grafted onto phylloxera-resistant rootstock was required to continue making wine. As phylloxera ravaged the vineyards of northern California, vines in southern California were hit by an unknown malady later found to be the bacterially caused Pierce's disease. In the 1880s, more than half the region's vines were struck down. The epicenter of California wine was toppled as southern growers switched from vines to citrus.

Replanting efforts in the north were complicated by labor shortages throughout the farming sector. Native peoples and Chinese immigrants had been essential contributors to building vineyards until legal reforms largely forced them from the state. The Indian Indenture Act of California in 1850 helped eradicate Indigenous peoples by the late 1860s. Chinese laborers then became the primary vineyard workers until the federal Page Act of 1875 followed by the Chinese Exclusion Act of 1882. Immigration was broadly limited and led to a large decrease in the Chinese population throughout California (see page 22). These were also the first expansive restrictions to immigration established in the country.

Vineyard losses were happening as the country stepped into its second and third longest recessions in history: the Long Depression from 1873–1879, and the Depression of 1882–1885. Lack of laborers made it difficult to correct economic woes. Recessions meant capital for rebuilding was hard to come by.

Although California wine was besieged by challenges, it kept creating new solutions. Settlers imported a host of new grape cultivars, including as many as several hundred European selections (though a much smaller portion became part of the California repertoire), raisin and table grape varieties, as well as American hybrids such as Catawba and Concord that helped establish winegrowing in the eastern United States.[6, 7]

Vine growers partnered with scientists at the University of California. Researchers worked to match newly imported grape varieties to growing regions. This could help increase the quality of California wine while expanding the winegrowing potential throughout the state. They also studied wine aging and the biochemistry of fermentation, identifying the role of yeasts and isolating unique strains from different parts of wine country.[8]

By the late 1800s, California vintners had learned the power of joint capital to conquer market woes. As producers were forced to solve issues in the vineyards, they failed to adequately build consumer interest in their greatest potential market, the eastern seaboard. Even after the phylloxera crisis, the eastern American marketplace depended on European rather than Californian wine. As production in California grew, its survival depended on increasing market share outside the state. California wineries started to flounder. Then in 1894, seven of the state's largest producers combined assets and incorporated as the California Wine Association (CWA),[9] arguably American wine's first monopoly.

Over time, the CWA captured ownership of other wineries, both large and small, as well as eventually leading around 80 percent of California wine production.[10] Part of the genius of the enterprise was in how it handled marketing the various brands that it consolidated. While CWA oversaw all aspects of business, the wineries continued to produce the same host of wines under the same labels as before incorporation. Consumers had no idea management had changed, but their preferred brands became increasingly prevalent and more reliable in the marketplace. The effort stabilized the industry and helped build consumer demand for California wine across the country.[11]

By the 1900s, sales were growing in major cities and town centers throughout the United States, eventually leading to recognition in

Europe. In 1915, when San Francisco hosted the world's fair[12] it included a California wine exhibit nicknamed the Grape Temple. The temple was decorated with grape garlands, wine casks, and vines. Forty-two booths showcased California wineries, while a special room displayed its history, and another showed seven hours of film on loop. It highlighted the vineyards and wineries of the state.[13] The fair also included several days of presentations and research exhibitions from leading producers, as well as experts in viticulture and enology who shared and celebrated the success of wine in the United States, especially California.[14] California wine had successfully built an industry.

Just five years later, the sale of wine across the country ended. Research in viticulture and winemaking stopped countrywide. Government support ended. The US Department of Agriculture no longer recognized wine grapes as part of its efforts to improve farming in America. Scientists moved into other areas of study. Wine apprentices, and indeed those who would have trained them, moved to other industries. The know-how of American wine moved on, and so did its market.

DURING PROHIBITION

Though the Volstead Act limited home production to 200 gallons per head of household per year, the government deemed what happened inside the home a private matter. The exact amount was never monitored, so there was plenty of room for making more than 200 gallons. Demand grew.

Although commercial winemaking under Prohibition was mostly illegal, the thirst for home wine made grapes more valuable. Pre-Prohibition, the average price per ton hovered at around $25. During Prohibition, the average increased to $50, with prices at times as high as $82 per ton.[15] Vineyard acres in California expanded from 300,000 in 1920, to 577,000 productive acres by 1927. Tonnage doubled in that same time from 1.25 million in the state to 2.5 million.[16] Newly planted vineyards tended towards thicker skinned grapes that would survive the east-bound trip by railway. The heartiest varieties, particularly Alicante Bouschet, as well as Zinfandel, Carignan, and Petite Sirah, were shipped over the Rocky Mountains to cities in the American east: New York, Chicago, and Boston especially. Napa Valley, the Sierra Foothills, and in particular the Central Valley together sent as many as 72,000 train cars over the Rockies in 1927 alone.[17]

However, thanks to the local marketplace, some comparatively thinner-skinned varieties survived Prohibition, albeit at small scale. San Francisco became one of California's most important grape markets, buying fruit especially from Sonoma County, and most particularly its Zinfandel.[18] White wine grapes did not disappear but did decline.

While home winemaking grew, professional winemaking dwindled, although it did not entirely disappear. Under the Volstead Act, commercial winemaking was allowed in only a few exceptional cases: for sacramental or medicinal wine, to turn into vinegar, and as flavoring for food or for soaking cigars. Each required a permit, which came with regular inspection and oversight. It was impossible to make more than was allowed.

At the start of Prohibition, wine businesses were desperate to come up with solutions to survive. The CWA considered pressing grapes for juice in San Francisco, then loading the juice onto ships bound for Japan so it could ferment at sea beyond the borders of the United States and be sold abroad. They quickly recognized the logistical complications and permanently closed instead.[19]

Paul Masson managed to get a permit for making medicinal champagne from his Santa Cruz Mountain vineyards.[20] He sold very little, and after a routine inspection was required to explain to the federal government the great reduction in volume from the juice he'd harvested to the bottled sparkling wine. They didn't understand how much can be lost during disgorgement or from bottles exploding under pressure.[21] A small market for regulated grape products emerged. The Colonial Grape Products Company produced sherry for the Campbell's Soup Company to use as flavoring in their canned soups and provided wine to cure tobacco leaves rolled into cigars.[22]

During the thirteen years of Prohibition, a few wineries made just enough wine to survive until repeal in 1933, but none increased their business, save one notable exception. Georges de la Tour, owner of Beaulieu Vineyards (BV), took a national contract with the Catholic Church to make sacramental wine. The demand was so great, he expanded BV multiple times.[23] But the reality for most was somewhat different. By the time business owners realized Prohibition would not end quickly, it was too late to sell equipment. Most of it rusted from neglect. Wineries closed. The resources from these businesses rarely transferred to new industries, and instead created a sector of economic loss.

Such losses also extended to both federal and state governments. Prior to 1920, many states depended on excise taxes from liquor sales, and alcohol taxes played a significant role in funding the activities of Congress as well. This important source of revenue ended with the passing of Prohibition, and arguably strengthened the reliance on income tax at both state and federal level that continues in the United States today.[24] It is estimated that the federal government alone lost $11 billion in tax revenue during Prohibition, in addition to the $300 million it spent trying to enforce it.[25] State losses were additional. The reduction in state and federal revenue had an impact on government spending and rippled back into the private sector.

Before 1920, more than 700 wineries existed in California. By 1932, the year before Prohibition ended, 140 wineries remained. Most of those were in poor condition, and many of them had a winery licence but no wine.[26]

THE NEW AMERICAN MARKET FOLLOWING REPEAL

On December 5, 1933, when the Twenty-first Amendment repealed Prohibition, the United States had more wine drinkers than ever before thanks to that almost-accidental clause in the Volstead Act. But the market for table wine was non-existent. Consumers could make that at home. People were more willing to buy what they couldn't make themselves. Although table wine led wine sales prior to 1920, fortified wines became the primary market sector after repeal. Before Prohibition, California wineries sold 2 gallons of table wine to every gallon of sweet. After, wineries sold 1 gallon of dry table wine to every 4 of sweet.[27] Distilling spirits to make fortified wine was a specialist venture, not one for the home winemaker. It would take until 1967 for dry wine sales to again surpass those of sweet wines.

The new system of regulations around the country also posed a challenge. The Twenty-first Amendment placed the oversight of alcohol sales in the hands of individual states. Each state created its own answer on how to tax and control alcohol after the demonization of Prohibition. That meant wine sales would have to meet a unique set of laws for each state, as well as having that state's own taxes applied. Many states also established controls to ensure individual producers could not unduly dominate the market. Laws blocked overlapping ownerships of a producer, distributor, or wholesaler, thus spawning the beginning of the country's inimitable three-tier system. The new model created a sales bottleneck. With the three-tier system, a producer couldn't control how wine was sold once it left their winery.

Once businesses began working on wine sales again, the quality of California wine proved part of the problem. The impact of Prohibition created an almost complete break between the accomplishments of the industry before 1920, and its ability to rebuild after 1933. The struggles faced in the 1800s were significant, but their impact varied in degree from region to region and business to business, never creating a total collapse of the industry. Someone always remained to problem solve and help their neighbor. Prohibition, on the other hand, severed the industry's past accomplishments from its future. Vineyards expanded but commercial wine disappeared, and with it a culture of fine wine. Winemaking know-how was lost with the virtual disappearance of commercial wine.

As if vineyard and cellar challenges weren't enough, the rebirth of wine sales arrived during the longest economic recession in US history, the Great Depression, which started in 1929. It would take a decade for the American economy to recover. In the meantime, a determined few began redeveloping technical knowledge in winemaking while working to establish a new method for selling wine.

REBUILDING IN THE GREAT DEPRESSION

Excitement for relaunching California wine following Prohibition was immediate. Enthusiasts eager to become part of the wine revolution rushed to create new wineries, most of them with little or no experience. In 1934, a year after repeal, California celebrated 804 bonded wineries, an increase from 140 the year before Prohibition ended.[28] But rebuilding the industry, especially during the Great Depression, would not be easy. The differing regulations from state to state led to added complexity, increased costs, and limited sales, and the quality challenges facing California wine itself needed to be addressed. Within a few years most of the newly bonded wineries were no longer in business.

But the industry wasn't just wineries. Under Prohibition, the planted vineyard acres in the state had more than doubled. Grapes for home winemakers continued to be shipped over the Rocky Mountains after

1933, but not at the same volume. What were vineyard owners to do with so much extra fruit? California wine had already learned the benefits of collaboration and vineyard owners soon followed. Cooperative wineries formed in regions throughout California in the 1930s. Together, the investment in building a winery was manageable, and it guaranteed fruit purchases for growers. A cooperative was obligated to make wine from the harvest of any of its members. Almost every region in California sported at least one cooperative winery during the 1930s.

Lodi, at the intersection of the California Delta with the Central Valley of inland California, became one of California wine's most important regions in terms of volume and number of co-ops during the co-op era. Lodi's Bear Creek Vineyard Association included 103 growers who could produce upwards of 1.1 million gallons annually. Lodi's East-Side Cooperative included 60 growers making as much as 1.5 million gallons per year. Over the next few years, the region gained six additional co-ops, adding another 4.5 million gallons of production.[29]

The growth of cooperatives helped stabilize the market for California wine post-Prohibition. It built reliable sales for grape growers while ensuring the availability of wine for the fledgling marketplace. But it also normalized the quality associated with bulk wine. Given their enormity, most cooperatives made wine in huge quantities, and from whatever grapes their growers provided. Those vineyards that might have the ability to produce exceptional wine were lost to the blending tank. The job of the co-op was to ensure its growers had a market, not to advance quality or niche wine styles.

The security provided by cooperatives proved necessary for the industry to transition from the damage of Prohibition through the poverty of the Great Depression. But their presence maintained the quality problem that began in the vineyards. Most vineyard acreage established during Prohibition went to varieties not suited to fine wine, but the economic recession meant replanting was untenable. It would take several decades for significant changes in California vineyards to occur.

And vineyards weren't the only issue facing California wineries. The complexity of regulations across the country and the need to rebuild winemaking know-how within California were tasks too big for any one business to solve alone. So, in October 1934, winery owners came together to outline the challenges faced by the California wine industry and consider solutions. By the end of the meeting, twenty-four wineries (a little less than half who attended) founded the Wine Institute.

Included among them were leaders of the CWA, who had helped stabilize the industry following the Depression of 1882–5.

The plan was to advocate for policies as well as scientific research that would foster success for the wine industry, while offering education and resources to support the success of individual wineries. For it to work, producers willing to pay membership dues had to be recruited. The need for such collaboration was so apparent that by the time of the first annual meeting, in August of the following year, the Wine Institute boasted 188 member wineries, which together oversaw 80 percent of the state's production.[30] Soon after, the Wine Institute joined forces with the California Department of Public Health to establish new quality guidelines for the industry. Among these was banning the addition of sugar to fermenting wine, a practice known as chaptalization,[31] which continues to be illegal in California winemaking today.

As quality standards were steadily addressed, wineries recognized they would also have to work together to rebuild market share among the American public. In 1938, the Wine Advisory Board was formed within the California State Department of Agriculture. The goal was to create marketing programs that would draw public interest to California wine. Unlike the Wine Institute, the Wine Advisory Board was state-run, which meant every winery in California was required to pay dues to fund the effort.

The Wine Advisory Board launched a national marketing campaign, as well as a wine appreciation correspondence course complete with certificate and gold seal for those who passed. Both efforts garnered interest from throughout the country and helped increase sales of California wine. The Wine Advisory Board and Wine Institute also worked together to initiate viticultural research efforts overseen by the University of California, and to rebuild educational resources on winemaking, both of which had essentially disappeared during Prohibition.[32]

Eventually, individual regions in California also created their own collaborative research groups to improve quality. In Napa Valley, André Tchelistcheff, who became winemaker at Beaulieu Vineyards post-Prohibition, played an important role in furthering the effort. He'd been trained in viticulture and enology in Bordeaux before arriving in Napa Valley. It was technical training the state sorely needed: few enologists existed in California. As more wineries formed in Napa, Tchelistcheff realized he could support their success by offering his services. In 1947, he founded the Napa Valley Enological Research Laboratory. The lab helped wineries throughout the valley, checking the chemistry and stability of fermenting and aged wines, something wineries struggled to manage alone.[33]

At the same time, the Napa Valley Technical Group was formed. It was a partnership between some of the region's founding winemakers, including Tchelistcheff. Others involved included both Peter and Robert Mondavi, who revived Napa's oldest winery, Charles Krug, before Robert went off on his own in 1966; Louis P. Martini, winemaker and son of Louis M. Martini, who opened his eponymous winery in St. Helena immediately following Prohibition; John Daniel, the winemaker at Inglenook who made some of its most inspiring Cabernets between the 1930s and 1960s; and Lee Stewart, founder of Chateau Souverain on Howell Mountain, who helped bring attention to mountain Cabernet.

Together, they fostered a drive to continually improve quality while also elevating the reputation of Napa Valley. They recognized that by working together they could create success for the entire region and ease the road for themselves. They also shared their efforts with faculty at the University of California. Professors there helped solve technical riddles and used the Technical Group's work as part of their scientific studies and research papers, eventually founding the American Society of Enologists to foster similar efforts nationwide.[34]

By 1940, the Great Depression had subsided. The challenges of the recession had eliminated wineries formed just after Prohibition without the wherewithal to face the vagaries of market. By 1944, there were 465 wineries bonded in the state of California, just over half the number that existed immediately following Prohibition.[35] Though the 1930s brought incredible challenges for the industry, the combined efforts of the cooperatives, the Wine Institute and the Wine Advisory Board brought stability by the end of the decade.[36] Together, they helped those that did survive build a steadily growing future for the California industry. Surprisingly, their efforts would be bolstered by the advent of World War II.

THE IMPACT OF WORLD WAR II

In 1939, with the advent of World War II, it became necessary for the federal government to consolidate the nation's resources and channel them towards the war effort. Railway tank cars used to transport juice, wine, and other liquid goods began exclusively transporting oil. Fruit that could be dried for stable transport was bought in bulk for rations. Cartons, bottle caps, and packaging materials came in short supply. Men from their twenties to forties went overseas to join the war effort. The government implemented fixed pricing on a range of consumer goods to protect the economy. The new restrictions changed how California wine did business and helped speed the drive to improve quality in the industry.[37]

The sheer volume of grapes in California's vineyards thanks to the expansion during Prohibition and the varieties planted meant a serious quality problem. California wines reaching the eastern seaboard did little to demonstrate that the wine was worth paying for. But growers couldn't afford to replant, and improving local winemaking knowledge took time.

In the 1940s, the situation started to change. Thanks to World War II, the federal government contracted with growers to mandate that all table grapes grown were dried as raisins for war rations. Before, most were still funneled into bulk wine. The raisin contract amounted to 44 percent of California's previous wine harvest and perhaps more importantly, removed grapes unsuitable for fine wine from the state's production. With the drop in supply for the wine market, the price of wine grapes also increased.[38] Soon, the loss of table grapes from the mix demonstrated that the type of grape used really did affect the character of the wine. The Wine Institute took note. After the end of World War II, and the revival of the American economy, they began promoting replanting to fine wine varieties.[39]

The transport of finished wine also changed. Prior to the war, California wine was largely made in bulk then shipped in railway tank car to be bottled on the east coast. This meant wine producers had essentially no control over what happened to their wine or how it was sold. Consumers recognized merchant names on labels, but not wineries. With the advent of World War II, the government commandeered tank cars for the exclusive purpose of transporting oil across the country. Wine lost its vehicle for moving bulk wine to bottling.[40] As a result, wineries began purchasing equipment to bottle onsite. For the first time in its history, most of the state's wine was bottled where it was made, by the people who made it. Watchful consumers began to regard the note on wine labels "Bottled at the winery" as a sign of higher quality, and so also more valuable (read: worth paying for).[41] By the end of World

War II, wineries realized that by bottling their own wines they had not only increased control over their product but also improved their margins. When war ended, few wineries wished to return to the practice of shipping bulk back east.[42] The change meant it also became easier to consider bottling smaller lots of specific wines, creating the early steps in moving from large tanks to bespoke lots.

Amid the many wartime changes, the federal government also commandeered the nation's distilleries. In 1942, Congress halted the production of all whiskey, not for the sake of Prohibition but because it needed to increase production of industrial alcohol for cleaners, solvents, and fuels of the wartime effort. The government contracts provided distilleries with plenty of income but left them with two problems. They wanted to avoid losing market share in the nation's alcohol sales, and to keep employed their team of salesmen who could no longer work in Europe because of the war. So distilleries began buying California wineries.[43]

During World War II, distilleries Seagram's, Hiram Walker, Fleishmann, Schenley, and National bought wineries that all together totaled more than 25 percent of the state's wine production, making a volume of more than 200 million gallons annually.[44] They owned wineries from regions throughout California including St. Helena, Healdsburg and Santa Rosa, Livermore, Lodi, San Francisco, Saratoga, Hollister, and Fresno.[45]

The impact of the distilleries was in more than just ownership. They sold alcohol at a scale beyond what the wine industry had experienced, and with professional salesmen. Distilleries also began promoting wine with a variety of newspaper, pamphlet, and magazine advertising. Some also made short radio programs (and later television segments) celebrating the enjoyment of wine. The increased attention focused on wine as part of everyday enjoyment brought by the distillery promotions benefited more than just their own market share. Most California wineries saw the increased advertising as positive for the entire industry.[46]

Soon after, the sales model for wines also changed. Ernest and Julio Gallo founded their wine business in 1933. By the late 1930s, they were bottling their own wines under their own name. Ernest recognized that for their business to grow they would have to develop sales. Unable to afford advertising, Ernest instead hired his own dedicated salesforce of wine specialists, a first for the industry. In any city where Gallo wines were sold, his representative would not only negotiate special deals for buying more Gallo products, but also stock shelves and ensure Gallo wines commanded prime shelf space at eye level. The effort improved Gallo sales but it also increased the visibility of wine in stores across the nation. Prior to Gallo's efforts, spirits dominated that space.[47]

After the end of World War II, distilleries returned their attention to the production of whiskey, and other liquors. By the 1950s, they started selling off winery investments and by the 1970s, all had essentially left California wine.

A NEW CURIOSITY FOR WHITE AMERICANS

The impact of World War II on the nation's infrastructure would continue to change California wine. The necessity of expedited distribution to support the production of goods for the war effort laid the groundwork for transportation channels that remained following it. Bolstered distribution made the nationwide movement of everything from appliances to clothing, to food, as well as wine, easier and more affordable.

The almost universal increase in employment nationwide during the war broke the country out of the Great Depression and created a period of economic prosperity that continued. Seizing on the opportunity to stabilize the economy and support returning troops, in June 1944 President Franklin D. Roosevelt decided to guarantee financial support for war veterans. He signed into law the Servicemen's Readjustment Act, more commonly known as the GI Bill, providing World War II veterans with money for education, job training, and housing, as well as unemployment insurance. By 1951, almost 8 million veterans received funding for higher education or vocational training. By 1955, 4.3 million home loans had been granted, and hundreds of thousands of new businesses created.[48] The GI Bill ensured that millions who had served in the armed forces would be employed after the war, and fostered a sense of security that led to a period affectionately known as the baby boom, a literal increase in the American population. The GI Bill was arguably the greatest (though not sole) contributor to the creation of the American middle class.

The creation of the middle class was not only an economic boost, but also the building of a new way of life. By the mid-1950s, plane travel for pleasure was becoming possible. Americans began to travel to Europe as tourists and return with newfound interests. Italian and French restaurants took hold throughout the eastern United States, and home cooks soon followed suit, making similar dishes on their own. Alongside these cultural cuisines, wine was treated as a normal part of a meal, reintroducing an elevation of fine wine to the American palate.

Improved access to electricity as well as appliances[49] (built in the same factories that had supplied the war effort) made cooking at home easier, which in turn made it easier for Americans to explore their newly diversified palates. New technology and distribution networks made frozen, dried, and packaged foods commonplace in the country as well, and produce could be distributed nationwide for the first time (see chapter 4). With ready-made food, meals could be prepared at home more quickly, allowing additional time for other pleasures. Wine began to re-emerge as not just drunkenness but refined enjoyment.

Only a portion of the population benefited from newly found middle class luxuries, however. As written, the GI Bill included all veterans of the war, but in practice Black Americans struggled to access the benefits offered by the bill.[50] Into the late 1960s, the United States was still a segregated country. State and local laws included white-only towns and neighborhoods, and schools across the country were segregated.[51] Black Americans were often denied newly built homes and forced to live away from where they worked. Many who sought funding for higher education were pushed to vocational training or offered more menial jobs instead.

In 1954, the Supreme Court case *Brown v. the Board of Education* declared segregation of schools based on race unconstitutional. The Civil Rights Act of 1964 outlawed employment discrimination as well as segregation of public accommodations. In 1968, the federal government signed The Fair Housing Act into law. Although the Thirteenth Amendment in 1866 had already made discrimination based on race unconstitutional, there had been no means to enforce it. Additional laws in the 1950s and 1960s provided modest (and arguably still inadequate) enforcement,[52] and removed some of the barriers that restricted the ability of Black Americans to join the middle class. Together, these and other civil rights legislation instigated a slow and painful process of increasing opportunity to not only Black Americans but other non-white citizens as well. Even so, the decades between the end of the war and the beginning of these changes advantaged white Americans and gave them a head start into not only the prosperity of the middle class but also a range of associated luxuries, including fine wine.

And yet, the United States was experiencing a booming economy along with the convenience of new technologies, and exposure to other cultures simultaneously. White working-class Americans were starting to consider what it meant to live the good life. The definition was beginning to include wine at the dinner table, though it would take longer, and for many would not be a part of home life until into the twenty-first century. Interest in wine for Black Americans would slowly increase as well.

In 1967, the sale of table wine exceeded that of sweet or fortified wines for the first time since before Prohibition. California wine was on the verge of a new expansion.

California farming evolves

Key dates

1920s and 1930s	First wave of migrant workers arrives from the Philippines
1935	National Labor Relations Act
1938	Start of the Central Valley Water Project, and the State Water Project
1942–1964	Bracero program
1945	End of World War II
1958	First citizen lawsuit against the United States regarding environmental concerns
1962	Rachel Carson's *Silent Spring* is published
1965	Filipino farmworkers start the largest grape strike in US history; Mexican American farmworkers join them a few weeks later
1966	Filipino farmworkers and Mexican American farmworkers together form what becomes the United Farm Workers union (UFW)
1967	Several wineries in California sign contracts with the UFW
1969	US Senate special subcommittee forms to assess pesticides and farms in California
1970	Occupational Safety and Health Act (OSHA) formed
1970	Environmental Protection Agency (EPA) formed
1970	Growers in the Central Valley sign contracts with the UFW
1972	State of California bans DDT; the United States follows shortly after
1975	California Agricultural Relations Act is passed

3

CALIFORNIA FARMING EVOLVES

Immigration, environmentalism, and the farmworker movement, 1950–1980

The end of the war fed new growth in the United States. World War II improved the national economy and created a boom of new babies across the country. The GI Bill and access to more jobs made the rise of the American middle class possible. The distribution and production systems that supplied the war effort were turned towards sending goods and food all over the continent. No region had to rely any longer on only what it could get locally. Appliances made or produce grown in one part of the country could be moved to urban centers elsewhere. A larger population and its greater efficiency also increased the nation's demand for food, consumer goods, and other comforts. The rise of the middle class created a new period of luxury spending. Americans started drinking and buying more wine.

The move of the distillers to wineries during World War II meant an increased availability in primarily fortified wine after the war. Distilling was, after all, what they knew best. In the mid-1940s, as the war ended, smaller family wineries were poised to move back to winegrowing. But as the decade continued, the larger scale vineyards, especially those in the San Joaquin Valley, and big-volume businesses commanded the market share.

While distilleries took hold of the fortified wine business, other larger wineries producing a combination of sweet wines and bulk table wine also emerged, almost all in the San Joaquin Valley. Sunnyside in Fresno grew. Gallo remodeled and launched new brands. Roma expanded. New operations opened as well: Bisceglia Brothers in Fresno, Di Giorgio Wine Company in Kern County, Liberty Winery in Acampo.[1]

These larger farms helped change the direction of California's economy. New technologies, farming practices, and increased sources of water were ushered into the state. Together, they benefited not only grapes, but most California crops. Within a few years, the region became the nation's top supplier of produce and nuts, as well as farming the greatest diversity of vegetables.

Today, the San Joaquin Valley continues to provide half of all fruits and vegetables grown in the United States, as well as 99 percent of its almonds, walnuts, and pistachios. It leads the country on milk production, is the second-largest cheese producer in the nation, and is the fifth-largest supplier of food in the world. Yet only 4 percent of the nation's farmland is in California.[2]

California became the agricultural center for the country, and the Central Valley became its salad bowl. The Central Valley produces a wider range of vegetables and fruits than any other part of the nation. It also provides the highest volume of wine in California, the biggest producer of wine in the United States. While much of it is disregarded as bulk wine, fine wines are made in the region as well – and bulk wines move the market.

As the success of California agriculture increased, so too did its wealth of vineyards, but harvest industries require workers.[3] After the end of the war, growers needed more farmhands, but citizens could find other jobs for more money. The solution had to be found through immigrant labor. Even so, the need for farm labor was met with a series of anti-immigration laws. The combination perpetuated a cycle of importing workers, then excluding them and thus creating labor shortages. It's a cycle that has been repeated since the late 1800s and continues today.

Growers relied on immigrant labor to produce their crops. Non-citizens did not require health or retirement benefits, consistently worked for lower wages, often lived in buildings deemed untenable for resident workers, and were less likely to use social resources in the area. In every way, immigrant labor was less expensive.[4] But in most cases, such employees were also unprotected and many worked as an exploited lower class. In the 1960s, farmworkers decided to address their lower wages and working conditions directly. The post-war wineries of the

San Joaquin Valley and their reliance on vineyards of the Central Valley put them directly in the path of the upcoming farmworker movement. Farmworkers sought economic gain but did so by educating the public about the realities of chemical farming, the impact of its food on American families, labor exploitation, and the social, moral, and political power of consumer spending. Inspired by the successful boycotts of the civil rights movement, farmworkers employed similar strategies while also collaborating with civil rights groups. Historically, the labor movement and the pursuit of civil liberties had not combined forces. The farmworker movement stands as one of the first cases of labor and civil rights groups working together.

The farmworker movement helped change state and national legislation on chemical farming, and raised international awareness of its detrimental impacts. It also helped create health and safety protections, bathroom facilities and breaks on site, community agencies, and the ability to negotiate with employers – none of which existed for agricultural workers before their efforts. These changes continue to shape how food and wine are grown in California today.

GROWING INFRASTRUCTURE AND AGRICULTURAL EXPANSION

The wide-open tracts of land, water from winter rain, and dry conditions through the growing season, made California ideal for farming. Almost from the state's founding up until World War I, the state primarily grew grain. Fields of wheat and other cereals dominated California agriculture.[5]

The warmth of the Central Valley drew the attention of wine growers, but also brought challenges. The northern portion, the Sacramento Valley, was cooler with more fog, and championed far more water. The San Joaquin Valley in its south, enjoys more than 300 days of sun per year but is comparatively arid.

As early as 1938, and continuing into the late 1960s, the state began a comprehensive water diversion system known as the Central Valley Project (CVP) and a collaborative one known as the State Water Project (SWP). Together, these projects brought water from as far north as Mount Shasta, to the southern point of the San Joaquin Valley and the length of the region between.[6] The CVP and SWP also managed flooding, making more land farmable. The overall system also created power.[7]

The introduction of additional farmland and the water to grow, ushered new wealth into the San Joaquin Valley. Its sun exposure, warmer temperatures, and fertile soils made it ideal for agriculture. Smaller, family-run farms largely left the region. Wealth was required to remain.

Following World War II, migrant laborers, alongside the arrival of mechanization,[8] the increase of water from the north for irrigation, and the advent of chemical farming expanded California agriculture. Farms shifted from grains to more intensive (and profitable) crops[9] such as orchard fruits, citrus, and wine grapes.

IMMIGRATION AND FARM LABOR

As agriculture expanded, so too did the need for farm labor. Immigrant labor made the economic growth of California agriculture possible.[10] The tension between the search for farmworkers, and the country's resistance to immigration kept the labor shortage an ongoing issue. The Indian Indenture Act of 1850, the Chinese Exclusion Act of 1882,[11] the Gentlemen's Agreement of 1908 (meant to reduce Japanese immigration and then sanction Japanese internment in World War II[12]), the forced repatriation of Mexicans to Mexico and the attempt at repatriating Filipino workers, both during the Great Depression, led to a repeating cycle of labor shortages.

When shortages threatened business, growers would lobby the government to allow new ethnic groups to enter and provide inexpensive, temporary, farm labor. The Manongs, the first wave of Filipino farm laborers, arrived in the 1920s and 1930s. They were followed in 1942–64 by migrant laborers from Mexico, through the Bracero program.[13] In between, immigrants from southern Europe, smaller groups of Sikh workers, and some Black laborers from the southern United States also entered California. Together, farm laborers formed a lower class of the state's economy[14] with the least power for invoking change and capacity for avoiding exploitation.[15]

THE START OF COLLABORATION

Until the mid-1960s, attempts to organize farm labor largely failed. The promise of the Bracero program for importing temporary, non-citizen labor guaranteed protesting farm hands were replaceable. Without citizenship, Braceros rarely complained of conditions, for fear of deportation. Knowing they would return to Mexico after their stint as a Bracero kept them less invested in provoking change. When the Bracero program ended in 1964, growers returned to a reliance on local labor. Manongs and Mexican-American workers filled the industry. Both suffered mistreatment and discrimination by white Americans who saw them as outsiders and a threat. In his semi-autobiographical book, *America is in the Heart,* Carlos Bulosan famously said, "In many ways it was a crime to be Filipino in California. I feel like a criminal running away from a crime I did not commit."[16] Darker skinned Mexican Americans described similar experiences.

Due to cultural and language differences, and deliberately by growers, the two ethnic groups were kept largely separated. Lacking proximity to each other, there was little opportunity for the two to join forces. If one group protested, they could simply be replaced with the other.[17] In 1965, that changed. Filipino workers had already created a farmworker union, the Agricultural Farm Workers Organizing Committee (AWOC). It included not only Manongs but also Sikhs, Japanese, and Black laborers. In 1965, discontented with low pay and no protections, Larry Itliong, Pete Manuel, Ben Gines, and Pete Valesco, leaders of AWOC, alongside member Bob Armington,[18] called for a strike against growers in the San Joaquin Valley.

Itliong had learned about community organizing with a commitment to non-violence from Ernesto Mangaoang,[19] one of the leaders of the Filipino cannery workers union in Alaska, where Itliong had also worked. Those principles, along with Itliong's charisma, built the union. The collaboration of its members initiated the strike. It started on September 8, 1965. Itliong had little contact with Cesar Chavez, head of the National Farm Workers Association (NFWA), a union specifically for Mexican-American laborers. But Dolores Huerta, who led alongside Chavez, grew up with Filipino farmworkers thanks to the hotel and restaurant business owned by her mother.[20] Through Huerta, Itliong spoke to Chavez asking NFWA to join the strike. Chavez refused. Then Itliong pointed out that if AWOC failed in their attempts in 1965, a few years later it would be NFWA that failed in theirs. The only chance for success was to work together.[21]

In 1966, AWOC and NFWA joined forces and became the United Farm Workers (UFW) led by Itliong, Chavez, Huerta, AWOC leaders

Philip Vera Cruz and Andy Imutan, and NFWA's Gilbert Padilla.[22] The strike against growers in California quickly evolved into a nationwide boycott of grapes and wines grown in the San Joaquin Valley. Today, the boycott is largely remembered as impacting table grapes, but all grapes were included, and wine production of both fortified and table wine was slowed in the area as well. People all over the country joined a secondary boycott against businesses that sold boycotted goods.

When the farmworker movement started in the mid-1960s, civil rights efforts of the early 1960s had already gained public attention. Initially, the farmworker movement appeared to be another labor rights effort. From this perspective, civil rights activists avoided the movement. The history of labor efforts excluded Black workers. However, as the exploitation of farmworkers became better known, multiple organizations in the Black freedom struggle joined with the UFW. Activists in the south recognized a parallel between the treatment of farmworkers and that of Black sharecroppers. Others seeking rent support and access to basic resources such as food found commonalities in poverty.

The Black Panther party joined the boycott in poor neighborhoods with people of color.[23] In white neighborhoods, UFW allies turned to garnering support from religious organizations. That helped legitimize the farmworker movement and made it easier to reach white communities.[24] Stationed outside grocery stores, the boycott helped educate middle class consumers, especially mothers, who tended to be the grocery buyers for their families, on the harmful effects of pesticides in food.[25] Berkeley students protested the university serving boycotted grapes on campus. It stopped buying grapes.[26]

The success of the grape boycott exceeded all expectations and reached national levels. In New York City, Mayor John Lindsay stopped the annual purchase of fifteen tons of grapes for hospitals and prisons.[27] The *U.S. News and World Report* declared grape sales were down 50 percent in New York, 46 percent in Boston, and reduced in other metropolitan centers around the country.[28]

The support of civil rights and Black freedom groups played an instrumental role in the success of the UFW effort. Some collaborations proved short, but the increase in national attention and allies was essential to the success of the farmworker boycott.[29] The UFW did not have the numbers alone to reach into communities across the country. Collaboration with Black organizations, religious groups, and student bodies made the boycott's reach possible.

President Nixon attempted to help growers[30] by sending grapes to troops in Vietnam and international markets in Northern Europe. The boycott had so successfully garnered media attention that soldiers refused the grapes[31] and port workers in the UK and Scandinavia created a blockade refusing to allow ships carrying California produce into their countries.[32]

LABOR NEGOTIATIONS

While the strikes and boycotts by the United Farmworkers in the 1960s are well remembered events in California's history, they were preceded by far larger labor strikes in the 1930s, during the Great Depression. Across the nation, laborers of multiple industries walked out of their jobs to demand the right to organize and collectively negotiate for better wages, safer working conditions, and reasonable breaks and resources throughout the day. In 1933, 50,000 agricultural workers left the fields during harvest. They were met by physical violence and threats to force them back to work.[33]

In response to the nationwide strikes, President Franklin D. Roosevelt created the National Labor Relations Act of 1935. It gave workers across all industries the right to collective negotiations, or union building, to give them more equal power in relation to the owners of industry. All industries, except one: growers petitioned the United States government to exclude farm laborers from the bill, emphasizing that the growth of food and farming in the nation depended on an ongoing stream of inexpensive labor.[34]

The growers' pleas worked. Part of their strategy proved dependent on the seasonal nature of farmwork. While industries such as manufacturing or mining depended on reliable labor, any shortage from strikes would pause progress but not dismantle it. In the case of agriculture, and especially high commodity crops like grapes or strawberries, strikes during harvest could decimate an entire year's profit.[35]

As a result of the economic growth produced by World War II, citizens had access to better paying jobs outside agriculture. It was only immigrants who would farm food for less money. As farmworkers continued to attempt strikes to force better conditions, growers met them with violence. The enormity and wealth behind agribusiness meant the persuasive power of growers reached into local police forces[36] and government.[37] Policies and enforcement benefited growers. It was not until forty years after the creation of the National Labor Relations Act that the state of California provided protections for farm laborers.[38]

DDT AND ORGANOPHOSPHATES

During and after World War II, the chemical DDT became a commonly used pesticide. It was found effective against insects that damage crops as well as those that might spread disease. People joining the Bracero program were mandatorily sprayed with DDT before entering the United States as a precaution against possible disease they may have carried with them.[39] Wildlife areas were doused from above to protect flora and fauna from damaging insects.[40] Later, soldiers fighting in Vietnam or Cambodia were sprayed to reduce mosquitos and other biting bugs. DDT was an apparent wonder chemical.[41]

The presence of DDT grew throughout the United States. But in the mid-1960s, it was discovered that some insects had developed immunity to it, so growers began relying on other chemicals as well. Organophosphates became a solution. A different class of pesticides than DDT, organophosphates appealed partly for having a shorter lifespan but more powerful initial impact. It was believed they would swiftly damage insect colonies, then gradually disappear from the environment.[42] DDT could last years in soils and remain in foods after harvest. The use of both types of pesticides increased across the United States and elsewhere. Between 1950 and 1969, national spending on chemical pesticides alone increased by 15 percent per year.[43] In agriculture, it was common practice to spray agrichemicals over fields as farmhands worked in them below.

In the 1950s, researchers across the country started to notice harms to wildlife areas. Eggshell thinning decimated bird populations.[44] Plankton in waterways died, depriving fish and birds of a food source.[45] A series of books on the matter, culminating in Rachel Carson's infamous *Silent Spring* in 1962 linked DDT to the problem.[46] In 1958, Marjorie Spock of Long Island launched the first citizen lawsuit against the United States government over the use of DDT. Spock was an organic gardener (who also practiced biodynamics). Her property was sprayed repeatedly by plane with DDT through a government initiative, including 14 times on one day. Fish in nearby waterways died within hours. On testing, area vegetables were found to contain three times the human tolerance limit for DDT. On Spock's property, her cows' milk was also contaminated.[47]

After several years of effort, Spock's lawsuit failed. She received no compensation from the government, and the use of DDT remained unchanged. But the court case and research accumulated for it became substantial research material for Carson's book. The case also set legal precedent and is regarded today as the first step in developing environmental legislation.[48] Researchers in Long Island had found links between DDT use and its presence in foods. Even so, environmentalists at the time tended to focus on protecting wildlife and wilderness areas from the chemical. Carson's *Silent Spring* became a popular focal point for a generation of disenchanted youth, and a call to the white middle and upper class to respond. Large-scale environmental groups, the Audubon Society being the most prominent, campaigned against the chemical and facilitated Carson's writing.

The US Department of Agriculture (USDA) responded in support of the agrichemical industry. Agribusiness leaders from both the grower and chemical industries retaliated. Hearings held with the Federal Department of Agriculture (FDA) to consider revisions to pesticide regulations were dominated by erroneous witnesses sent by the chemical industries.[49] Regulations remained unchanged.

Two years after the release of *Silent Spring*, Carson died of breast cancer at the age of 56. Studies have since confirmed the link between exposure to pesticides and multiple forms of malignancy, including breast cancer.

THE FIGHT AGAINST PESTICIDES

Resistance from the agrichemical industry continued for another decade. Governor Ronald Reagan and President Nixon both publicly sided with the agricultural industry. They organized publicity-seeking photo shoots of themselves eating California grapes, and used their political acumen to increase grower protections.

Several years of increasing public demands finally began to change government protocols. In 1972, the state of California banned the use of DDT. A few months later it was banned nationally as well. With the anti-DDT campaign secure, environmentalists moved on to other matters. The farmworkers continued to fight to raise awareness of other harmful pesticides.

For many, the fight against DDT is still considered a golden age for environmentalism. The campaign's eventual success inspired a wave of other environmental efforts that have continued since. At the same time, most environmentalists of the time treated the issue of DDT as about the natural environment. With a focus on wildlife, it was easy to overlook that chemical harm was also a human issue. Additionally, DDT had a disproportionate impact on racial and ethnic minorities living in poverty precisely because they were forced to work within DDT spray zones.[50] When farmworkers also took up the cause, the issue of agrichemical impact was framed as a consumer rights issue and created far-reaching impact.[51] In 1964, the Bracero program ended. California farming returned to a reliance on local labor. The change gave farmworkers the necessary leverage to fight for change.

Environmentalists had framed their fight against DDT as a moral issue to protect nature. Carson argued protecting nature was needed as that is how we as humans protect ourselves. The UFW boycott went further. Through leaflets at grocery stores and via national campaigns, the UFW boycott confronted consumers with the notion of poisons in food.[52] The approach framed pesticides as more than an environmental choice. It became a personal one. It was in consumers' own best interest, in protecting the safety of their families, to support the fight against harmful chemicals in farming and therefore in their food.[53] A new phenomenon, food co-ops, formed all over the country. They arose as a way of bringing more affordable and healthier food back to the people. Co-ops could support local farmers and keep costs down for members. Co-op newsletters throughout the United States supported the UFW boycott and spread the message of poisons in foods from big farms.[54]

The UFW effort helped instigate a larger consumer awareness of chemicals in farming, opening the way to more deeply considering the origins of food. Over time, it also exposed the government's complicity in the problem. At the same time as instigating a national boycott, the UFW launched a series of lawsuits and initiatives against the government designed to challenge the legal status of pesticides. There were no laws protecting farmworkers. There were also essentially no enforced regulations limiting the use of chemicals in agriculture. The regulations that did exist protected the rights of growers and made it difficult for farmworkers to act.[55] In 1969, UFW actions against pesticides led to a special Senate subcommittee being called in Washington, D.C., led by Walter Mondale. Growers from California, and senators that supported them, appeared and testified.

Tests on California produce showed direct links between pesticide use in the fields and toxicity levels in food. Growers in turn denied using the relevant pesticides, to call the accuracy of the tests into question. A few senators supported them. Grocery stores that had been selling California produce responded by having the fruit professionally tested. Their results confirmed those brought before the Senate subcommittee.[56]

The FDA was called to testify. The contradiction between test results and growers' claims put a spotlight on government enforcement of the issue. By the end of the hearings, it was clear that growers, and their supporting senators, had lied. Not only were the agrichemicals found in the tested food harmful, but growers also used them when they said they didn't. Additionally, the FDA admitted to not supervising the agricultural industry's use of these chemicals. Any existing regulations were irrelevant without enforcement. It appeared the agricultural industry of California was misleading consumers. At the same time, the government was not doing enough to protect them.[57]

In addition to DDT, it was shown that the newer class of pesticides, organophosphates, might have less long-term impact on food, but were even more harmful for farmworkers.[58] The production of these chemicals turned out to be linked to the creation of highly toxic nerve gas by German scientists during World War II. After the war both the United States and the Soviet Union stockpiled the chemicals.[59] Government regulations on pesticides did not change. Significant activity on the part of other labor unions throughout the United States following the end of World War II were also occurring. Their efforts raised public awareness of deaths and injuries happening for working employees, but most unions failed to recognize pesticides as part of the same problem.

In 1970, President Nixon passed the Occupational Safety and Health Act (OSHA) making it possible for government to enforce health and safety policies in industries across the country. Farmworkers were excluded from the new law.[60] Though government protections for farmworkers did not emerge during the Senate subcommittee meetings, the hearings served as damaging public relations for growers. They soon chose to negotiate with the UFW.[61]

In 1970, growers in the San Joaquin Valley and the UFW signed union contracts protecting farmworkers for both produce and wine grapes. They were the first such successful national initiatives. Through the contracts, farmworkers received the highest pesticide protections established at the time in the country. Those protections included not only DDT but also organophosphates and other pesticides. The agreements also created a health program, access to social services and medical care, on-site bathrooms and breaks to use them, shaded covers to rest during mandatory breaks, water onsite, and time for eating. None of these provisions existed for farmworkers prior to the union contracts of 1970.

THE LEGACY OF THE UFW

While growers of wine and table grapes in the San Joaquin Valley did not sign UFW contracts until the 1970s, several large wine-specific businesses did so a few years earlier. In 1967, field workers in vineyards for Almaden, the Christian Brothers, Paul Masson, and Gallo, four of the biggest wine producers in California at the time, signed. The contracts demonstrated that vineyard workers deserved representation and protections.

The 1970 UFW contracts with Central Valley grape growers were historic for the United States. The signing marked a moment when consumer rights, environmental demands, and worker protections operated simultaneously, and was also the first time farmworkers were recognized as deserving protections.

In 1973, Gallo did not renew with the UFW and the UFW brought a boycott against their wines. The boycott lasted five years until Gallo again worked with the UFW.[62] The grower contracts also ended in 1973. Growers used legal loopholes to stop negotiating with the UFW and signed with other unions that reduced farmworker protections. Even so, the UFW continued its efforts to improve conditions for farm labor. In 1975, UFW efforts and especially the leadership of Huerta helped secure the California Agricultural Relations Act. It protected the legal rights of farmworkers at the state level for the first time. That meant protections were instituted outside of labor contracts. The contracts of 1970 served as a basis for determining appropriate protections in the new law. Over time, the UFW also succeeded at expanding regulations to other harmful agricultural chemicals.

The UFW declined in power over the decades that followed. It is now rare for vineyard employees and farmworkers to hold union contracts in the state of California. If the UFW success is judged on the existence of contracts alone, their initiative in the long term failed. But the impact of the UFW extends beyond the success of 1970, or the disappointment of 1973. The legal protections of farm and vineyard workers have since been further expanded in state law. Efforts from as recently as 2015 to

2024 included challenges for new minimum wage levels, as well as daily and weekly work limits that allow for overtime pay. Other provisions continue to be debated.

Collaboration in the 1960s and 1970s between Manong, Sikh, Japanese, and Mexican-American farmworkers, with multiple organizations of the Black freedom, civil rights movement, student activists, religious leaders, and numerous white allies represents a unique moment in US history. For racial and ethnic groups, it was a recognition that racism, xenophobia, and poverty create unique forms of suffering. Everyday discrimination, mistreatment, and lack of resources has now been proven to be a factor in disparities of health and the onset of disease.[63] The UFW was among the first organizations to fight for human rights protections on this basis. The UFW also demonstrated that charismatic leadership initiated by people in poverty and without education could create effective change. Their efforts brought together social and civil rights by revealing that poverty and racial inequality both occurred through discrimination and unequal access to resources.

The public engagement with UFW boycotts, and the union's success in collaborating with multiple civil rights organizations, also helped show the public that economic issues like exploitation are always also social, legal, political, and moral issues. Its impact would continue to seep into the national consciousness and change the public's perception of government, food, and wine in America.

The farmworkers' success at educating consumers about the health impact of farming and food laid the groundwork for broader interest in non-chemical agriculture (expanding interest in organic and biodynamic farming). As they put farmworker needs at the center of the conversation, they also helped consumers reconsider the source of what they buy and how it's grown. The notion helped expand a return to smaller, family-run farms and an increase in the planting of vineyards.

In these ways, the farmworker movement helped inform a broader return-to-nature and back-to-the-land ethos in the United States.[64] Altogether, the farmworker effort helped open the way for the growth of fine wine in America.

The rise of California wine

Key dates

1960s	Hippie era
1961	Freedom rides travel the southern United States
1963	Nuclear Test Ban Treaty
1964	Free speech demonstrations begin at UC Berkeley
1965–1973	United States involved in the war in Vietnam
1965	Berkeley protests move against the war
1965	Malcolm X killed
1965	UC Santa Cruz opens
1966	Ronald Reagan elected governor of California
1967	Table wine outsells fortified in the United States for the first time
1967	Alan Chadwick starts farming at UC Santa Cruz
1968	Martin Luther King, Jr. assassinated
1968	Robert F. Kennedy assassinated
1969	1969: Richard Nixon becomes President
1970	United States enters Cambodia
1970–1973	US banks encourage vineyard and winery investment
1971	George Jackson shot
1971	Alice Waters opens Chez Panisse
1972	Alan Chadwick forced to leave UC Santa Cruz
1972	Nixon re-elected
1973	Watergate Scandal breaks
1974	Nixon resigns
1978	Robert Parker launches *The Wine Advocate*
1980	Reagan becomes US President
1991	"The French Paradox" airs on *60 Minutes*

4

VIETNAM, NIXON, AND THE RISE OF CALIFORNIA WINE

The 1960s–2000

Until the mid-1960s, fortified wine dominated the US marketplace, and California was its top supplier. Producers of fine wine steadily lost market share, most going out of business. The number of wineries in the state continued to decline. UC Davis researchers Maynard Amerine and Vernon Singleton predicted in 1965 that the future of California wine would be made by a few high-volume, industrial wineries; smaller family efforts would not have a significant impact.[1] Just over ten years later, their prediction was proven wrong.

In the decade between the mid-1960s to the mid-1970s, vineyard acreage in the state increased. But, more significantly, the volume of wine being made swelled and flipped from fortified wines with higher alcohol levels and plenty of sweetness, to predominantly table wine.[2] It was in 1967 that table wine outsold fortified wines in the American market for the first time since before Prohibition. Without the requirement of a still, production costs went down, allowing smaller volume wines and wineries to return. Larger ventures did not disappear, but the number of family wineries and smaller vineyard plantings proliferated. The 1970s marked a significant expansion of winemaking in California and laid the groundwork for the modern industry.

Public interest in fine wine had grown, it's true, but national and international perspectives were also changing. The social and political upheaval of the 1960s and 1970s called on the educated, white, upper and middle classes to reconsider where they were living, what they ate, and how they invested.

At the start of the 1960s, the state of California was entering a period of unprecedented social change. The decade included the rise of the so-called hippies, free love, and experimentation with both marijuana and LSD, as well as a new era of social activism. Civil rights, the Black freedom struggle, the free speech movement leading into the anti-Vietnam effort, the farmworker-labor rights crusade, and the rise of environmentalism coincided.[3] Activists worked in multiple organizations. The nonviolence principles Martin Luther King, Jr. adopted from both Gandhi and theologian Howard Thurman[4] guided most efforts. The practices of one movement inspired the others. They would soon lead also to the growth of women's liberation and Black feminism, the rise of trans activism that instigated the fight for gay rights, and the formation of the new left.

Fear from the Cold War of the late 1940s and 1950s was partially mitigated by the Limited Nuclear Test Ban treaty signed in 1963 by leaders of the Soviet Union, Britain, and the United States. But new understanding of the harmful impacts of pesticides and chemicals used not only in war but also in agriculture were being realized,[5] only to be closely followed by the reality of the Vietnam War.

Live reporting from Vietnam, photographs of war including images of massacres and war dead, along with daily body counts of US soldiers lost to war were aired on the nightly news. It was the first time in US history the American public was made so persistently and visually aware of the realities of war as they happened. California became the frontline for not only the military draft and infantry training centers, but also production facilities for artillery, gunnery, and military vehicles sent to Vietnam.[6] It also became the epicenter of largely student-led anti-war protests. And so too the testing grounds for police and national guard response to such demonstrations. The magnitude of anti-war activity in California inspired a national migration of young people away from the middle-class lifestyles their parents fought to create. At the same time, it instigated retaliation from then California governor Ronald Reagan, university and other government officials throughout the country, and eventually also President Richard Nixon.

The late 1960s and early 1970s saw some of the largest peace protests in American history, and yet the United States soon sent infantry into armed conflict in Cambodia, and stayed in Vietnam until 1973. Young people who'd devoted themselves to stopping the war became cynical that large-scale social change could ever be achieved through the government, and skeptical of authority, including not only the state or police but also corporations. They began to imagine more independent ways of life. When Nixon resigned due to the Watergate Scandal, their fears were confirmed. The effect was felt broadly. The American public turned from the trust gained by President Franklin D. Roosevelt because of World War II towards greater interest in self-determination and the belief that citizen advocates were required to ensure citizen rights.

In California, the national crisis laid the groundwork for a back-to-the-land movement, farm-to-table restaurants, and the beginnings of organic farming. For those wanting to maintain a foothold in mainstream America, tax incentives, investment projections, and the overall move to return to nature instigated the search for newer, alternative forms of financial investment.[7] The combination spurred a growth in vineyards that established the beginnings of modern wine in America.

And for one man in Maryland, it triggered the idea that independent reporting and assessment of wine was needed. Though the attention of Robert Parker would be on the broader world of wine, it helped build success for California wine specifically.

FREEDOM MOVEMENTS SET THE SCENE

For almost eight months in 1961, an interracial group of young people aged between 18 and 30 joined forces to travel the southern United States on interstate buses. They were known as the Freedom Riders. The group of both Black and white young people (including civil rights organizer and future congressman, John Lewis) challenged still-existing segregation practices in the region with a commitment to non-violent protest. As they traveled, the group was met with significant violence from anti-integration populations. Yet their effort led to the Interstate Commerce Commission banning segregation in waiting rooms and at lunch counters, helping to instigate related legal changes in southern states.[8] The risk the Freedom Riders took also brought national attention to the civil rights movement.

The group included student activists from UC Berkeley. Upon return to campus, they established tables on university grounds, providing information and raising money for civil rights efforts. But the university deemed such activity prohibited on campus. In 1964, students on campus undertook a series of sit-ins and other demonstrations fighting for free speech. It led to the university authorizing police intervention and mass arrests. The result was ongoing student action demanding the university reconsider its policies and negotiate with students. The following year, the protests on campus included actions against the war in Vietnam.

The demonstrations were the beginning of a history of student activism at Berkeley. They also helped instigate an ongoing movement across the country protesting US involvement in Vietnam. The period represented a synchronicity of the civil rights, free speech, and anti-war movements in the United States,[9] as well as a new form of collaboration for social change. Berkeley's activism made national news. The coverage triggered a perception of social unrest at the national level while inspiring young people across the country to reconsider the middle- and upper-class lifestyles in which they were raised. Youth migrations to California instigated increased activism and fed the already established hippie community as well as new left politics.

The fear of unrest triggered a backlash in California, with voters seeking a sense of safety. Ronald Reagan won an overwhelming, and unexpected, victory in his race for state governor by running on a promise to "clean up the mess at Berkeley".[10] After becoming governor, Reagan collaborated with the UC Board of Regents and the FBI to expel UC President Clark Kerr, and sent the national guard to campus to contain, tear gas, and arrest student protestors. When the national guard arrived, students were in the middle of a peaceful sit-in on campus.[11] They were tear gassed anyway.

Police and state retaliation to the students eventually dispersed the most visible protests on campus but the student realization that outside authority was a substantial threat led to a rethinking of how young people participated in society.

FARM-TO-TABLE

Alice Waters of restaurant Chez Panisse joined free speech protests while studying at Berkeley, after returning from a year in France. The political awakening she experienced on campus coincided with her discovery of flavorful country food in France. She fell in love with the simplicity of seasonal, local ingredients. American food at the time was comparatively flavorless. It was dominated by frozen or canned foods and produce distributed nationwide via the same supply networks built for servicing those fighting World War II.[12] Locally grown food, in contrast, supported the local economy, brought focus to the people doing the farming, and improved the freshness and flavor of food. The idea of eating locally became a new approach to revolution.

Chez Panisse opened in the shadow of political strife: the assassination of Robert F. Kennedy, the killing of George Jackson of the Black Panther party[13] (both in California), the assassination of Malcolm X in New York, the increased efforts against social movements by governor Reagan and President Nixon, and the killing of four student protestors by national guardsmen at Kent State University in Ohio. The original vision was to feed Waters' activist friends. But as the specter of national violence increased, elevating food became a way to bring focus to simple joys as a respite from global turmoil.[14] For Waters, food was not mere pleasure, but a way to change the economy, build community relationships, and express one's values. For Waters, what we eat is as political as it is personal.[15]

Chez Panisse was opened walking distance from campus. Waters partnered with local farms for fresh ingredients, naming the providing farms on the menu. The Four Seasons restaurant in New York City (opened with the help of consultant James Beard) was the first to name food sources on their menu, whether from foraging or local farms,[16] but it was Chez Panisse that made farm-to-table a movement. It also helped start a new genre of food named "California cuisine." The new style included a focus on seasonal ingredients, chiefly produce. But importantly it also depended on a lighter touch in the kitchen.[17] The landscape of fine dining in the United States at the time centered around richer aspects of French cuisine, sauces made of butter and cream, slow cooked meats, or ingredients bathed in salt and fat to make confit. California cuisine was inspired by French techniques and its use of quality ingredients, but it relied on shortened cooking times. Vegetables were left uncooked or cooked only enough to lightly soften them. Meats and seafood were served seasonally as well, with little to no sauce and fewer seasonings. The goal was to showcase the fresh flavor of quality ingredients.

In Sonoma, farmer Bob Cannard of Cannard Farms and later Green String Farm became an indispensable partner for Chez Panisse, providing as much as two-thirds of the restaurant's ingredients. He also taught Waters, as well as numerous chefs throughout the state and beyond, about the importance of organic farming.[18] (Cannard would also go on to organically farm hundreds of acres of vineyards in Sonoma County, most notably for Cline Cellars.)

As the success of Chez Panisse increased, its commitment to local, organic, and seasonal ingredients changed the California palate. The new style of food and its support for local farmers helped increase the presence of farmers' markets selling directly to consumers, and that raised interest in where food was grown. The change in perspective helped develop public interest in organic foods. The new cuisine also brought greater attention to California wineries producing wines compatible with these fresh, seasonal ingredients.

For Waters, finding wine she felt well-matched with Chez Panisse cuisine was not a simple feat. Wines available from most stores at the time were sold in higher-volume jugs from California stalwarts such as Almaden and Wente Brothers.[19] On opening night, in August 1971, the Chez Panisse wine list consisted of just two wines: Robert Mondavi Fumé Blanc from Napa Valley, and Ridge Geyserville Zinfandel grown in Sonoma and made in the Santa Cruz Mountains. Her selections came from relationships she already had with people at the wineries. Waters went on to partner with winemaker Walter Schug, to create two house Zinfandels, one a nouveau style served on tap in the upstairs café, and the other a red wine for the downstairs restaurant. Schug was then at Joseph Phelps in Napa, which was founded in 1973 just two years after Chez Panisse.[20] The partnership led to the first Zinfandel festival at Chez Panisse in 1979. The variety continues to be a feature on the wine list. Waters describes it as an ideal accompaniment to their style of food.

Chez Panisse went on to help jumpstart the success of numerous then-niche California wineries. Many such as Qupé, Edmonds St. John, Phelps, Ridge, Joseph Swan, Mount Veeder, Green & Red, Navarro, and more recently Broc Cellars are among those who credit the restaurant for boosting their reputation in wine. According to chef Joyce Goldstein, of Square One in San Francisco, the compatibility of these wines, especially that of Paul Draper at Ridge, who went on to influence the next generations of California winemakers, relied on moderate ripeness levels as well as a more minimalist, pre-industrial approach in the cellar, much like the restaurant used in the kitchen.[21] Bruce Neyers, previously of Phelps, who now has his own winery, Neyers, and worked for a long time for the importer Kermit Lynch, credits Waters with "inventing buying local wine" for restaurants.[22] Former chefs of Chez Panisse, such as Goldstein, followed suit. Her son, Master Sommelier Evan Goldstein visited wineries in Napa and Sonoma directly to woo them to a restaurant wine list for the first time. Soon after, others such as Wolfgang Puck's Spago in Los Angeles took up the practice as well.[23]

What was common between these wine lists was their prioritizing of smaller, family-run California wineries to champion. Though Chez Panisse is closely associated with French importer Kermit Lynch, Neyers argues the restaurant more centrally featured California wines accompanied by French wine options.[24]

The practices of Alice Waters took hold during national social transformation and retaliation against its impact. At the same time, consumers nationally were slowly becoming aware of the potential harm caused by farming practices or food additives, as well as the slowness of government to respond with protections (see chapter 3 for more on this).[25] Her efforts not only inspired the chefs who worked with her, but also spearheaded a larger turn to fresh ingredients and local wines.

Thanks to a series of books as well as reports on the nightly news in the 1950s and 1960s, consumers were slowly becoming aware that farming practices or food additives could be harmful.[26] The massive food distribution systems and industrial farming of food from the post-war effort would not go away. But the farm-to-table movement and focus on organic farming that took hold in the 1960s and 1970s changed public perception of food's value. It emphasized for the American public that *how and what* we eat was not merely a matter of sustenance or even nutrition, but political and moral commitments.

As with the environmental movements of the 1950s and 1960s, Waters' failing was in overlooking the farmworkers while elevating the value of a food's source. Celebrating specific farms through menus helped build the commitment to local food and supported a turn towards organics but the change failed to prioritize the people who make growing our food and wine possible: the laborers. Even so, the farm-to-table movement made how we farm a political and social commitment.

BACK-TO-THE-LAND

Until the end of the Vietnam War, the United States relied on a military draft, a lottery selection based on birthday, of citizens sent to fight as soldiers in war. Young men were required to register for the possibility of selective service at the age of 18. If drafted, they were obligated to join military forces or face the likelihood of federal prison.

Vietnam was essentially the world's first televised war. Not only had the television become a household staple in the United States and elsewhere, but the creation of satellite technology also made daily transmission of news from other countries possible. Nightly coverage included graphic images and frontline reporting not only of soldiers in battle but also of civilians being killed or maimed. It was the first time the American public had witnessed the impacts of war directly, and it turned many against US involvement in Vietnam.

For the generation vulnerable to the military draft, the idea of being conscripted into what they saw as an unjust war called the legitimacy of government into question. Doubts about other aspects of mainstream life increased as well: the police arrested peaceful protestors; corporations were part of the military-industrial complex, making money from the war; higher education prepared people to enter those corporations; the comfort of suburban life depended on that same system.[27] Cynicism regarding the middle- and upper-class lives their parents had created increased.

As protests became more and more violent amid police and military intervention mandated by government officials, many young people were convinced they were not being heard by the government and had few options in mainstream society.[28] This disillusionment among young people instigated a back-to-the-land movement unseen before in US history. In 1967, the *San Francisco Oracle* published a piece announcing, "many of us … are voicing our need to return to the soil, to straighten our heads in a natural environment." The announcement came with an urging to leave the city and find new land in the country.[29]

In every other period, federal census data shows an increase in urban populations over time. The 1970s marked instead a demographic shift from urban to rural areas. Skeptical of government, police, or corporate authority, large portions of the populace sought instead self-sufficiency.[30] Those migrating from the cities to the country were almost entirely young, white, educated, and born of families from the middle and upper classes. Disenchanted with the comforts of what they saw as a corrupt society, they sought to create a new way of life reliant on the land.[31] While they were willing to choose a sort of poverty, non-white Americans were still struggling with more basic freedoms without that ability to choose.

Communes emerged across the country, the largest concentrations in Northern California, Southern Oregon, and New England.[32] As soldiers returned from Vietnam, many joined the back-to-the-land ethos to recover from their experience. Throughout California small vineyards were planted in isolation, some on communes, then by Vietnam veterans, and later by other young people impacted by societal changes. Such sites emerged in the wilds of Sonoma and Mendocino Counties as well as in other less developed parts of the state. Most famously, winegrower Richard Sanford returned from Vietnam and established the first vineyard in what today is the Sta. Rita Hills AVA, as well as one of the first in Santa Ynez Valley. The remoteness and barren landscape of the area meant it was sparsely populated and offered the isolation and imposed self-sufficiency Sanford sought.[33]

As the 1970s continued, homesteaders took up single-family versions of going back to the land.[34] Most took up organic farming as well. Associated expenses were lower. But farming in conjunction with nature, rather than through chemical inputs, also provided a sense of connection to the land for those who couldn't connect to society.[35]

Influenced by the back-to-the-land movement, in the 1970s David Hirsch retreated to the rugged landscape of Sonoma's coastal mountains. There he established his own and neighboring vineyards in what today is the heart of the Fort Ross-Seaview AVA, and the West Sonoma Coast AVA. Over time, he transformed the site into one of the most respected organic and biodynamic vineyards in the state and helped others establish vineyards there as well.

ORGANIC FARMING

In California, the rise of counterculture activities at Berkeley led the University of California to create a new, more isolated campus in the forests of Santa Cruz, in 1965. The hope was to draw some of the more radical students of Berkeley to the overtly progressive, grade-free, cross-disciplinary educational program of Santa Cruz (UCSC) with its surrounded by nature vibe.[36]

After opening UCSC, educators on the new campus sought to further integrate its alternative approach to education with its unique surroundings. English master gardener Alan Chadwick was hired to develop a university farm. Chadwick had studied farming from Rudolph Steiner, considered the father of biodynamics, and believed in life processes beyond mere biological or chemical means. But he also eschewed the most esoteric side of biodynamics in favor of the practical aspects that helped plants grow. Chadwick applied his experience to building the garden and farm on campus. For students, his eschewing of chemical inputs represented a possibility of self-sufficiency with nature akin to the broader desires of the time. The success of the garden attracted state and national media attention. Such articles drew even more interest for the university. But the disdain within the scientific community also grew.

Organic farming was not yet understood as reasonable agriculture. The FDA warned that organic foods were likely full of parasites. A Columbia professor of pathology announced in the *New York Times* in 1970, "there is no convincing scientific evidence that so-called organically grown foods contain any extra nutritional value as compared to the same food grown in a conventional way." The Secretary of Agriculture warned organic farming would lead to mass starvation.[37] Even Rachel Carson, who through *Silent Spring* documented the harmful effects of chemical agriculture and DDT, saw organic farming as too radical.[38] Though rejected by the scientific community, organic farming appealed to those critical of mainstream society. It represented a move away from industrial agriculture that with the use of DDT and other chemicals was too closely associated with the horrors of Vietnam. And as a return to nature, it also meant leaving the larger cultural system to build self-sufficiency.[39]

In 1972, because of a backlash against his use of organics, Chadwick was forced to leave UCSC and the farm project behind. In 1973, he moved to Covelo, California and began the Covelo Village Garden, considered the pinnacle of his efforts in new agriculture.[40] For many, Chadwick is still considered "an early visionary of sustainable food." He was an inspiration too to Waters: "We are in the same movement, which is back-to-nature … There was a lot of … dropping out and growing your own and not buying commercial food." It was the culture of the time, she says, to find connection with nature, whether through farming, food, or in a literal departure from society into remote areas.[41] Waters was not the only one to feel kinship with Chadwick.

Several apprentices as well as former students followed Chadwick to Covelo. Among them were Alan York, Katrina Van Lente, and Jonathan Frey.[42] At the age of 18, York was drafted to Vietnam. As he described, for the first time he was "faced with being killed or killing others," yet could not see "for what purpose." Inspired by anti-war protests, the leadership of Martin Luther King, Jr., and the cost Muhammad Ali faced by refusing the draft,[43] York too refused to go to Vietnam.[44] Once released from federal obligations as a conscientious objector, York joined Chadwick at the Covelo Garden where he met Lente and Frey. York went on to practice biodynamic farming but dismissed the spiritual aspects even more strongly than Chadwick, in favor of a commitment to practical farming. The point was to do what makes plants and soils healthy. Lente and Frey married, then established Frey Vineyards in Mendocino County. Nearby, Barney Fetzer had already established Fetzer Vineyards. As the two vineyards developed, they both hired York to help them deepen their commitment to organic farming. There he converted several hundred vineyard acres and earned them the Demeter certification for biodynamics.

York went on to work with numerous other vineyards throughout California including Mendocino's Bonterra, Quivira in Sonoma's Dry Creek Valley, Robert Sinskey in Napa Valley, and Sonoma's Benzinger Family, among others. He also mentored numerous winemakers and viticulturists who took his practices to vineyards throughout California and beyond. York brought organic and biodynamic farming principles to Chile, and consulted on multiple vineyards in Europe, including one owned by Trudie Styler and her husband, the musician Sting.

While working with Fetzer, York befriended and guided Paul Dolan, then winemaker and eventually president of Fetzer. Dolan went on to help expand production at Fetzer, thereby demonstrating organic farming could succeed at both greater acreage and higher volumes. Dolan's influence on California viticulture was immense. He helped develop the California Code of Sustainable Winegrowing, which became the basis for a statewide certification, as well as Parducci, the first US winery to earn a carbon-neutral certification.[45] Dolan also helped establish the Regenerative Organics Alliance, which went on to create the Regenerative Organic Certification.

The hippie movement of the 1960s was founded by people in their twenties and thirties but it influenced younger people too. As a teenager, Phil Coturri lived in San Francisco. He joined in the activities of the cultural revolution, spending time with musicians, going to concerts, and enjoying marijuana. Worried for their son, his parents ensured he spent weekends and summers at a remote country home in the wilds of Sonoma Mountain.[46] Though his going back-to-the-land was imposed by his parents, it became a way of life for Coturri. While looking for something to do, Coturri befriended neighbors, including retirees, gardeners and war veterans. Asked to take over a neighbor's garden, Coturri borrowed issues of the Rodale Catalog, one of the first organic farming resources in America and applied its guidance to the garden.

When he proved successful at organic gardening, Coturri was pushed by a neighbor to try his hand at farming organic vineyards. In 1979, Coturri took over the Dos Limones vineyard on Sonoma Mountain, applying organic methods to 12 acres of vines.[47] Down the hill, Cannard was already farming, and helped Coturri adapt organic methods to his new crop.[48] In a few years, Cannard would begin providing ingredients for Chez Panisse.

The following year, Coturri started the venture Enterprise Vineyards, providing organic viticulture to sloped vineyards throughout the north coast. His influence has filled the mountains of both Sonoma and Napa Counties and ventured north into Washington state, and beyond. He and Enterprise have managed vineyards for some of California's most celebrated sites, including Sonoma's Rossi Ranch, Laurel Glen, Landmark, Bucklin Old Hill Ranch, and Repris, as well as Napa Valley's Mayacamas, and Oakville Ranch.

During the next decade, Coturri also started planting vineyards for business mavens and celebrities looking for an earthier lifestyle and a new sort of investment. Clients have included director and animator John Lasseter, comedian Tommy Smothers, musician Boz Scaggs, and screenwriter Robert Kamen.[49]

NEW INVESTMENTS

Even the wealthy moved to the country. Most Americans had no desire to go back-to-the-land, by joining communes, but new tax incentives and market predictions alongside the cultural shift towards a relationship with nature spurred interest in agriculture.

The California wine industry gained increased attention in the late 1960s, most notably thanks to marketing maven Robert Mondavi opening his eponymous winery in 1966. He made a splash by committing to his own facility, highly visible on the main highway through Napa Valley. Mondavi's enthusiasm demonstrated a life of sophistication and acclaim could be had through wine.

Then, in the late 1960s, tax exemptions for agriculture were adopted by the state of California. Vineyards would remain untaxed for their first three years, before they became productive. Other incentives and benefits appeared at other stages of growth. Tax incentives alone might not have expanded the vineyard area significantly, but in 1970 a series of investment predictions on the expected return for investing in wine production were released.

In 1970, Bank of America predicted that by the end of the decade, American consumption of wine would increase from 250 million to 400 million gallons annually.[50] Two years later, Wells Fargo responded by asserting it would be 490 million per year.[51] *Time* magazine took the opportunity to describe a California wine boom akin to a new gold rush.[52] The *Wall Street Journal* advised investment in vineyards.[53] Bank of America returned with a new report in 1973, expecting wine consumption rates in the United States to increase to 650 million gallons.[54] Private wine consultants went even higher in their predictions, declaring Americans would drink 800 million gallons annually of California wine alone.[55]

In an interview in the early 1990s, winemaker Warren Winiarski, founder of Stag's Leap Wine Cellars, credits the bank reports as assuring his investors vineyards were a good idea.[56] The predictions from established financial centers gave reassurance to curious wine lovers that they could succeed in a new business. Engineer Dave Stare of Dry Creek Vineyard, oilman Tom Jordan of Jordan winery, and lawyer Jim Barrett of Chateau Montelena were all convinced by the Bank of America reports and various news articles to open their wineries.[57] Similarly, oilman Ray Duncan followed the financial reports to Northern California to open Silver Oak.

The predictions also convinced older generations that wine was a growth market. Many used their financial means to support their children's verve for a new way of life in California wine. Tom Burgess credits the reports for convincing his father to become a partner in his new Burgess Cellars.[58] Father and son Charlie and Chuck Wagner of Caymus Cellars also joined forces.

In 1972 alone, Stag's Leap Wine Cellars and Chateau Montelena (winners of the famed Paris Wine Tasting of 1976; see below) both opened. Jordan launched his eponymous winery, which went on to become the favorite of President Ronald Reagan. Burgess, Caymus, Cakebread, Carneros Creek, Silver Oak, Diamond Creek, Clos du Val,

Franciscan, Mount Veeder, Edmeades, Sullivan, Dry Creek Vineyards Chalk Hill, among others, all bonded the same year. The eruption of new wineries continued throughout the decade.

At the start of 1970, around 230 wineries populated the state of California (versus more than 1,000 prior to Prohibition). Over the ensuing decade, several hundred more wineries opened, and vineyards saw a significant surge in plantings of fine wine grapes.[59] In Napa Valley alone, a little over 1,000 acres were planted to grapes in 1970.[60] By 1980, that number had increased to more than 28,000 acres.[61]

Wagner credits the hippie era for getting him interested in wine, seeing it as "more natural."[62] Many also describe their desire to get back to the land, or the desire to follow an agrarian lifestyle, closer to nature. Though Spottswoode winery wouldn't start until the early 1980s, Mary and Jack Novak left southern California in 1972 to live on what became Spottswoode's vineyard so they could raise their children in a rural environment.[63]

For many, it was also a way to connect with a new community.[64] Caymus used equipment from nearby Raymond Vineyards. Longstanding Beaulieu Vineyards, Inglenook, and Mondavi helped train the new rush of winemakers on how to make wine. Established businesses Shafer's and Duckhorn's helped Mary Novak continue in the wine business when Jack unexpectedly died. Winiarski credited his desire to leave the city for a more farm-based life, as well as his newfound love of wine, for pushing him to California. Once there, the congeniality of the winegrowing community reminded him of the intimacy of the Polish neighborhood in which he was raised.[65]

In 1976, their decision to invest in vineyards was proven to be wise. At a tasting in Paris organized by Stephen Spurrier and Patricia Gallagher, pitting Californian wines against their French counterparts, two California wines came top (see box, p. 164). The result expanded interest in California vineyards (especially in Napa Valley) further. New forms of investment combined with an agrarian lifestyle became a a sign of sophistication and luxury.

The new American ethos looking for greater connection to nature combined with the increase of table wine sales and the urge to invest in wine to instigate the growth of the California wine industry. The same urge motivated a new crop of wine writers and reviewers, including the most influential critic in wine history, Robert Parker.

NIXON, RALPH NADER, AND ROBERT PARKER

Even as significant portions of the population embraced the push for social change in the 1960s and early 1970s, others were disturbed by the perceived chaos it caused. After the liberal leanings of the previous decade, the uncertainty many felt helped push the country in a more conservative direction approaching the 1970s.

The largest protests against US war involvement occurred in the late 1960s and early 1970s. And yet, in April 1970, Nixon announced he was expanding the Vietnam War into Cambodia. Demonstrations continued. In May 1971, civil disobedience activities erupted throughout the country demanding the government bring the country out of Vietnam. Nixon instead kept troops in the war until 1973. The significance of the large-scale protests across the country inspired activists. But Nixon remained in the war, and despondency increased. It appeared the government ignored public appeals.[66] For many, that confirmed there were few options for them in mainstream society. The move towards nature increased.

In 1972, Nixon won re-election. Recognizing the youth vote could impact his electoral run, Nixon campaigned with a rhetorical subtlety. In his famous, "Silent Majority" speech he told citizens that, like them, he wanted peace, but he also did not want US defeat.[67] The strategy worked. It appealed to those who abhorred the social chaos of the previous decade, as well as those confused by the tension between those changes and the appeals against the war.[68]

In 1973, it was revealed that Nixon's re-election campaign had illegally broken into the headquarters of the Democratic National Committee running his opponent's campaign. It was further demonstrated that Nixon supported plans to obscure his administration's involvement in the break-ins and impede investigation into the situation. What became known as the Watergate Scandal led to 69 indictments and 48 convictions, including high-ranking officials working with Nixon. Multiple articles of impeachment were drawn against President Nixon but in August of 1974 he resigned as President to avoid either impeachment or conviction. The collective realization that voters had elected a candidate who went on to break federal law and lie to the populace further shook American confidence in the presidency, the country's government, and the electoral system.

A few years previously, General Motors (GM) had publicly confessed wrongdoing to a US Senate committee in response to a lawsuit brought by consumer advocate Ralph Nader. In mid-1965, Nader's book *Unsafe at Any Speed*, critiquing the US automotive industry, had become a bestseller. In response, the following year, the federal government passed the National Traffic and Motor Vehicles Act. It allowed them to create safety standards for all vehicles sold in the country that led to an 80 percent reduction in auto fatalities.[69] GM went on to hire private investigators to follow Nader to discredit him and also undertook ongoing harassment of Nader. GM was forced to give Nader a financial settlement, which he used to found ongoing consumer advocacy initiatives throughout the country.

Nader's efforts went on to force the government to rebuild the Federal Trade Commission. He also aided the creation of the Freedom of Information Act,[70] Clean Water Act, Whistleblower Protection Act, and Consumer Product Safety Act.[71]

Though Watergate and the GM fiasco were made public two years apart, together they represented the national realization that neither corporations nor the government had the public's interest at heart. In addition, 11 lawyers were disbarred because of the Watergate Scandal, including Nixon himself. His former vice president, Spiro Agnew, was also disbarred prior to Watergate. The legal profession was called into question.

Law students and lawyers throughout the country grappled with uncertainty about the US legal system as well as their own employment. In Maryland, Robert Parker had just finished law school as the Watergate Scandal became public. While he went on to practice law for another decade, the number of lawyers involved in the scandal made him doubt his profession.

Nader continued to gain attention for his consumer advocacy. His efforts on behalf of consumers as much as individuals, and a desire for self-determination fed by the 1960s combined to make him a beacon for those now disillusioned with multiple forms of authority.[72]

As a lawyer effectively changing the relationship between federal law and corporations, Nader became an inspiration to Parker. For Parker, Nader represented an ideal to be brought into all aspects of American life. Parker sought to expose and combat conflicts of interest, and thereby oppose misuse of power on behalf of the general consumer.[73] Having recently fallen in love with wine while visiting France, Parker combined passions and decided to launch a fully independent wine journal giving unbiased recommendations for fine wine to consumers. Over the next few years, Parker deepened his wine knowledge and honed his idea. In 1978, his *The Wine Advocate* launched. By 1984, it was successful enough for him to leave law to become a self-employed wine critic.

Throughout the 1980s, the Reagan administration brought a nationwide fight against drug use in young people. At the same time, the group Mothers Against Drunk Drivers (or MADD) became a national force. Reports questioning the health of wine also emerged. It was another period of neo-prohibitionism. Until the mid-1980s, wine sales were steadily decreasing. At the same time, those who had gone back to the land in the 1970s were starting to merge back to mainstream society searching for ways to stay connected to nature in more urban environments.[74] Then in 1991, *60 Minutes* released a segment on the French Paradox.[75] Baby Boomers across the nation (including my father) turned to red wines for the health of their heart. Those who'd returned from the woods found ways to integrate their values into their new more urban lives while at the same time seeking some conveniences like a cup of coffee and a glass of wine. Wine sales increased. And a new rush of consumers wanted help deciding *which* wines to buy. The power of Parker increased again.

By the 1990s, Parker's palate was influencing sales and styles of wine around the world. He had been strongly critical of California wine in the 1980s, and later called that decade the decade with the best of French wines, but by the early 1990s, Parker's reviews were changing. He announced that anyone denying that California made some of the best wines in the world was a fool. Even so, he said there were fewer than 100 in the state that counted among the best, but many among those became some of Parker's top scored wines in his history as a critic. Parker in this period helped create the idea of a cult wine: smaller production, lavish reds that relied on a direct-to-consumer mailing list for sales. They needed Parker to get the word around about their winery ventures. Parker would go on to call the 1990s and 2000s the golden age of wine in California. His support helped bring the attention of wine collectors as well as outside investors to the state, and thereby build its reputation as well as its industry.

Parker's success for and in California also reflected the unique approach the state took to marketing its wines. While Europe had history and established reputation to provide a consumer base, the California wine industry was too young to have either. California wineries instead turned to building strong relationships with media throughout the United States, including Parker and the *Wine Spectator*. As *Spectator* and the *Advocate* grew in sales, wine columns across the country were bolstered and new ones emerged. The swiftness with which California built success through media relationships became a model Europe would seek to emulate.

5

SOCIAL MEDIA, RECESSIONS, AND WEATHER

Challenges and opportunities of a new era, 2000–2025

It hasn't been an easy millennium, so far. The California wine industry seems to have come up against more challenges than ever before. The first quarter of the century included multiple recessions, a global financial crisis, technological implosions, political instability, and for wine producers, a series of seemingly impossible vintages with profoundly unpredictable weather. More recently, there has been yet another rise of neo-prohibitionism.

At the same time, the volatility of these first 25 years has revealed exciting new opportunities. The rise of social media pushed influence from individual authorities to communities of people across the internet. Cold vintages, and an international thirst for wines with more restraint, expanded the range of styles available.

The natural wine movement has been a community force. It has expanded beyond its earlier focus on winemaking into deeper questions of place, farming, and *what* is being fermented. No longer just about *vinifera*, the natural wine movement has brought attention to grapes native to parts of the world besides Europe, as well as to those species crossed with *vinifera* as hybrids. It's also brought attention to fruit and grape co-ferments, herbal infused wines, honey wine, rice wine, and fermentation in food. As a movement it first caused people to rethink

Challenges and opportunities of a new era
Key dates

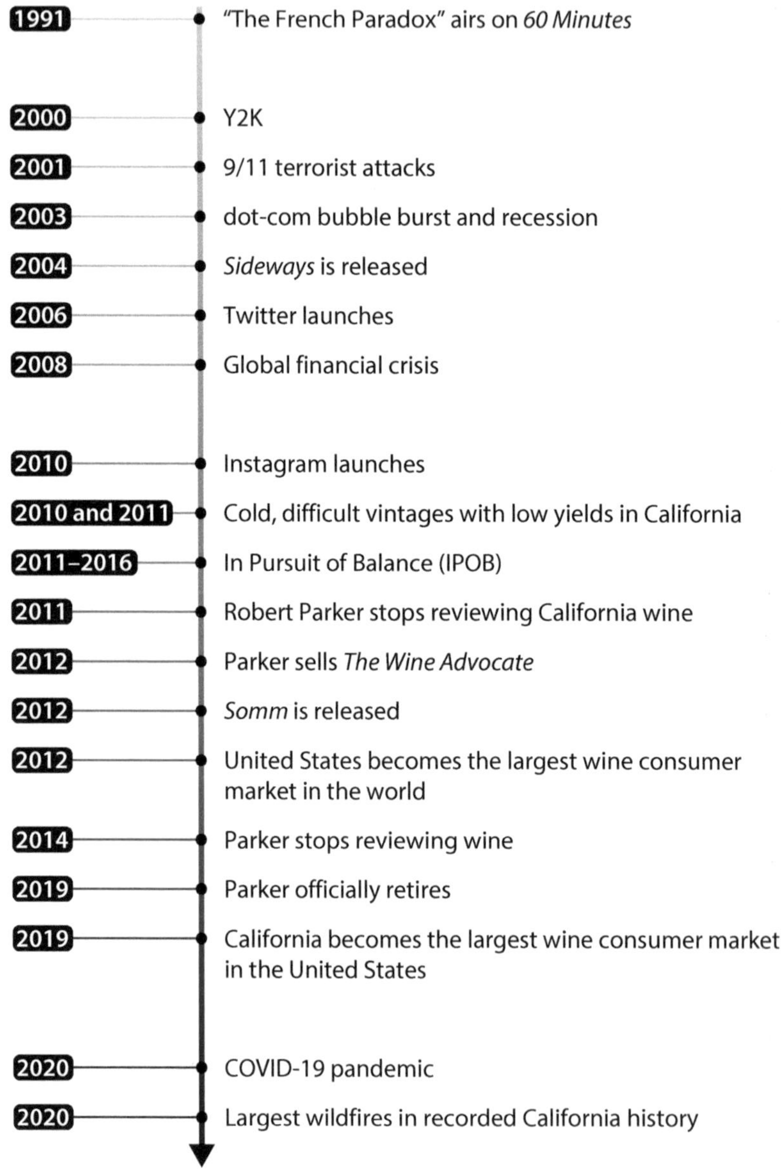

the boundaries of wine, but more recently it's also welcomed in scores of people previously disenfranchised from the industry or even alienated from enjoying the beverage.

Social movements started in the 1960s and 1970s continued to evolve. Deeper interest in egalitarianism has led to women in wine initiatives, recognition for people with disabilities or neurodivergence, racial or ethnic groups, and the LGBTQIA community, and increased understanding of complex notions of diversity like intersectionality. There has also been resistance to such changes. But, even so, these movements haven't just been about political change. They've infused themselves into the wine industry. New organizations have emerged to grow a broader base of support. Companies have invested in diversity initiatives or built new mentoring programs to train people less familiar with wine. These factors have together stretched the notion of wine to include more categories and importantly reach more people in the new millennium. New people mean broader market potential.

The movie *Somm* was released in 2012 and led to the rise of the American sommelier. While California wine relied on print media to grow its base in the 1970s and 1980s, it has more recently had to turn to building relationships with wine buyers and those in direct contact with consumers through the restaurant floor. For wine brands operating outside the persuasive force of Robert Parker[1] restaurants had already become a launching pad as early as the 1980s.[2]

But the California wine industry is often slow to change. COVID-19 collapsed the hospitality industry, decimating the restaurant sales many wineries depended on. Businesses that thought the pandemic would last a few months relied on reserves to get through what became two years of recurring shutdowns and restrictions. By the shift into 2022, many of these reserves were gone, and businesses had to face different questions to survive. The wineries that jumped to new ways of connecting to customers at the start of the pandemic created new sales outlets and sources of income and came into 2022 with greater stability.

Since the pandemic, there has been yet another recession, this is after one in the early 2000s on the heels of Y2K and the dot-com crash, and a global financial crisis that started in 2008 and lasted well into 2009, but that some people in wine feel never fully went away.

Wine descriptors that used to rely on standardized written tasting notes have shifted, with a focus on expanding the lexicon to more visual approaches from illustrations and infographics, along with short form video and even wine movies, as well as direct consumer connections through social media.

Direct-to-consumer initiatives in the last 25 years have similarly shifted. They went from regular newsletters shipped by mail to online updates and website sales. It took until the lack of travel during COVID-19 for many wineries to boost their online presence. Those wanting to keep a feeling of luxury and exclusivity remain offline.

The history of California wine demonstrates that collaboration through trials can advance its success further than previously thought. The volatility of the current millennium presents another opportunity to explore that practice. But unlike past moments of change that have meant consolidating portfolios or advancing the idea of fine wine to the public, today's challenges include the weather, the market, wine's audience, and changes in national and international distribution and regulations.

In the midst of all these challenges, in 2012, the United States became the leading consumer of wine in the world.[3] By 2019, California was not only its leading producer but also its leading sales market.[4] As historians often propose, knowing our past can help keep us from repeating it. These first 25 years and the century that went before might reveal lessons to guide wine forward.

THE MILLENNIUM BEGINS: TECHNOLOGICAL VOLATILITY

In the 1990s, computer experts worldwide warned of the collapse of global systems. The problem was a computer glitch nicknamed Y2K.[5] Banks, weapons arsenals, credit companies, grocery stores, hospitals, and every other system from sewers to public utilities were affected by the way computer chips handled the date on the calendar.[6] Many predicted that the inability of computers to distinguish between the years 1900 and 2000 would lead to chaos on January 1, 2000. The United Nations brought governments together to solve the problem. The United States hired computer programmers in countries around the planet to check code and rewrite it. Wine writers predicted the disappearance of Champagne long before the end of 1999 and promised a sparkle-free holiday.[7] Survivalists prepared for chaos.

We survived. Champagne did not disappear, and California sparkling wine increased its sales.[8] Even so, Gartner Research, a technological research and consulting firm that advises investment companies, technology corporations, and world governments, estimated that altogether the planet spent between $300 and $600 billion to prevent Y2K (figures not adjusted for inflation).[9] The billions spent to repair the problem would otherwise have gone toward production and purchasing around the planet.

Y2K didn't sink the global economy, but it did prepare the way for another sort of crisis. The incredible wealth of the 1990s, and low-interest rates of the late 1990s, bolstered the value of the American stock market. Investments in internet businesses boomed. The ability to share information worldwide, along with the increase in personal home computers made the internet one of the most seductive investments in decades.[10] Search engines, news readers, shopping sites, and communications companies were built through the anticipated financial return of the internet. By 2000, the bulk of new online businesses had burned through their capital, creating and launching new technology but failing to generate a profit. Then in 2001, the September 11 attacks crashed the US stock market. The Dow Jones saw its biggest losses in history at that point.[11] Next, in 2003, a global oil shortage increased the price of gasoline. The combination of a decreased stock market and higher priced gas put the nation in a recession.

The financial wealth of the 1990s swiftly expanded California's wine industry. New wineries bloomed, while existing wineries expanded their portfolios, and most grew swiftly. The success of wine sales made it easier to enter the wine industry. In 2003, when the national recession hit, sales slowed. The average purchase price of a bottle moved from the premium category to below $15. Wine businesses that were poorly managed or had grown too quickly, relying on immediate sales over long-term stability, closed, entering bankruptcy, or were sold.[12] Grape growers also experienced declines. As wineries reduced spending, the price of grapes decreased as much as 75 percent. Some growers sold their vineyards or simply stopped farming them.[13]

California wine altered its approach. Due to the change in wine sales, wineries sought ways to reduce production costs and make their wine more attractive to consumers. Producers worked with restaurants and wine bars to offer affordable by-the-glass options. In a recession, it's easier to invest in a single glass than an entire bottle. The bulk market increased, and wineries began bottling bulk wine or combining it with their own wine to create lower priced bottles. Other segments of the industry let go of their lowest priced wines but also dropped their most expensive. They couldn't compete with bulk prices, and few people were spending on luxury. The approach drew customers who were committed to quality but interested in buying more wines for less rather than one luxury bottle.[14]

By the end of 2003 the economy had improved, and so did wine sales.[15]

THE *SIDEWAYS* EFFECT

In 1991, when Morley Safer aired "The French Paradox" on *60 Minutes*, red wine sales across the United States boomed.[16] The market for California wine escalated and vineyard acreage increased throughout the state. US consumers switched from cocktails to a glass of red wine. The impact that one television show could have so profoundly on the choices of Americans proved prescient.[17]

By the early 2000s, Cabernet Sauvignon, Zinfandel, and Merlot were among the top five varieties being harvested in California. (The other two were Chardonnay and French Colombard.)[18] Merlot sales were steadily increasing. Then in 2004, the movie *Sideways* came out. The main characters set out on a road trip around Santa Barbara wine country drinking Pinot Noir along the way. Long romantic shots of grapes ripening on the vine while love interests spoke poetically of their passion for wine, drew in the audience. Then, during the movie's seduction into the virtues of Pinot Noir, main character Milo turned to his best friend Jack and said, "If anyone orders Merlot, I'm leaving." Sales of Merlot reduced almost immediately (sales of Pinot Noir have increased ever since).[19] In Santa Barbara County, wine tourism surged. Restaurants and wineries featured in the movie grew wine sales without adding promotions.[20] Vineyard acreage in the region also increased.

THE GLOBAL FINANCIAL CRISIS

It was like a new form of the dot-com crash. Under President George W. Bush, in the mid-2000s, the housing market was growing. Lenders offered mortgages at lower rates to interested home buyers with lower credit scores than before. More people across the country owned homes, and home prices soared, peaking in 2006. The banks sold their home loans to outside investors expecting to make enormous returns from the continuation of the home boom. In 2007, when house prices began to fall and interest rates rose, homeowners stopped paying mortgages and investors lost money. Running through the crisis there turned out to be a lack of government oversight.

The collapse of the housing market created a credit crisis that rippled beyond the United States, through the international marketplace, creating a global financial crisis (GFC). America's success in home ownership had increased property values elsewhere as well. When US values fell, so did those in other parts of the world. To stave off further collapse, the US government gave enormous bailouts to banks, insurance companies, and investment firms. The bailouts stopped further collapse but also kept the country in its recession. National funds had helped corporations rather than individual homeowners. By 2009, US unemployment rates had reached 10 percent. Median household income did not return to pre-crisis levels until 2016.[21]

The comprehensive nature of the GFC meant all businesses were affected. In the wine industry, the impact extended beyond wineries and growers to corporate wine businesses, retailers, wholesalers, and more.[22] Sales of the most expensive, luxury bottles decreased.[23] Consumers moved down a tier (or more) in wine spending. They didn't stop drinking wine, but they turned to less expensive bottles. Fewer people went out to eat.[24] Restaurants and the hospitality industry were hit, as were their wine sales. Even the wealthy were drinking what they already had.

THEN THE RAINS CAME

Wineries were already struggling. Grape prices plummeted because of the GFC. Wineries had less money to spend so they tightened their portfolios, moving from curiosity wines to the guaranteed favorites. French international varieties – Cabernet Sauvignon, Chardonnay, Pinot Noir – took centerstage.

Wine styles across the state were still celebrating the bold flavors Parker preferred. His influence was still largely global. Then the 2010 harvest hit.[25] The vintage was cold, culminating in rains during harvest (meaning lesser ripe fruit, lower alcohols, leaner wines) and the 2011 vintage followed suit. But 2010 also included ravaging heat spikes.[26] Thanks to the rains, some fruit rotted on the vine before it was considered ripe enough. Yields went down dramatically.[27]

Wine businesses were forced to adjust again. With wineries and retailers needing cash income to survive the downturn, wines were discounted all over the country. Wineries again tightened their portfolios and sought new ways to sell wine. Behold! The growth of the wine club. More consumers were drinking wine at home to lower spending. They were also gaining curiosity. Wine clubs were a new direct-to-consumer strategy, one step removed from the winery. Clubs brought together usually a mix of wines in a collection available monthly, quarterly, or twice a year. Club memberships were sold over the internet, and through wine retailers. Memberships allowed wine lovers to sample more options.[28] If they found one they liked they often became repeat buyers of that wine. Eventually clubs were also formed directly by wineries. The option gave devoted consumers a steady supply at home for favorites. Internet sales increased, and by 2009 Amazon was looking at venturing into online wine sales. The success of the online marketplace brought further attention to the internet and new regulations emerged.[29] Internet sales did not go away (they would increase significantly within a decade), but they were stalled as businesses responded to new policies.

THE EXPANSION OF STYLE

Economic and weather challenges created hardships for wineries and growers alike. Grape contracts included ripeness agreements. In 2010 and 2011, wineries rejected fruit based on lower sugars. Colder temperatures meant grapes ripened not only more slowly but not as much. Yields were also abbreviated. Wineries saved money by purchasing fewer grapes. Growers lost money from lower yields and canceled grape contracts.[30] Wineries turned from lesser-known varieties to the French stalwarts again, streamlining portfolios for what already worked.

The growing conditions of cold vintages made wineries approach their wines differently. Winemakers simply could not find grapes that reached the ripeness levels they'd relied on. For some, that meant using boosters in the winery to produce the fleshier and fruitier wines consumers were used to. For others, the two vintages offered a glimpse at another style of wine. As winemakers gained familiarity with the restraint of two colder vintages, some began to explore farming to make wines in a leaner style. Throughout California, wineries began stepping back their ripeness levels and considering lower levels of new oak as well – leaner wines often don't carry the oak flavors as well.

In Sonoma, Andy Smith, partner and head winemaker of DuMOL, spoke to me of the evolution in his wine ripeness level. He also described how it took time to understand the farming techniques needed to gain balance in the wines at lower alcohols, and to adjust the tendencies of the vineyard. At White Rock, in Napa Valley, winemaker Christopher Vandendriessche described something similar. He realized he preferred his wines with a lighter touch and lower alcohol.[31] Copain in Mendocino and Sonoma did too. They are just three examples.

From 2011, new brands began to emerge. Growers whose grape contracts were rejected needed a new outlet. So, like Theodora Lee, owner of Theopolis in Mendocino, the best option was to start making wine. Lee partnered with another winemaker to make her first vintages, using Petite Sirah from her own vineyard.[32]

At the same time, trade professionals had already lost interest in the riper styles of California wine. The food-friendly styles that had drawn Chez Panisse to California brands in the 1980s were harder to find in the state thanks to the riper tendencies of the 1990s and 2000s. So two wine professionals decided to make a statement. In 2011, then-head of wine sales for Hirsch Vineyards on the Sonoma Coast, Jasmine Hirsch, and then-sommelier Raj Parr (both of whom now oversee vineyards and make wine in California), joined forces and started what became a movement, through the tasting event In Pursuit of Balance (IPOB). Though it only ran for five years, IPOB stirred controversy, brought wine lovers together, and launched new wine brands. Established wineries that had been successful with the riper styles of California wine were offended by the implication that their wines might be out of balance. Others resented the perception that wines were only balanced below 14 percent alcohol. (IPOB never said this but the statement was associated with Raj Parr's work as a sommelier.) But a new community of wine lovers was also created.

It turned out there were plenty of people who'd grown tired of the state's bolder wines. In parts of the country it had become hard to sell California wine outside of steak houses. What IPOB offered was a way for like-minded individuals to come together in one place. Over its five-year inception, IPOB had global impact. It showed wines annually in California or New York, and made stops in Japan and London, amongst other places. Those who attended were reminded of long-standing wineries such as Au Bon Climat, Hanzell, and Calera. Others found brands that already existed in California gained further attention: Drew, Cobb, Littorai, Matthiasson. Newer wines also found an audience: Chanin, Liquid Farm, LaRue. Most made modest volumes. IPOB brought them attention, but their sales potential was limited by the smaller amounts of wine they produced. Au Bon Climat proved unique in both making wines considered to be in the IPOB style and having the volume to sell. According to founder-winemaker Jim Clendenen, IPOB boosted his wines internationally.[33]

During its five years, IPOB focused only on Pinot Noir and Chardonnay. But it brought the question of balance and ripeness levels to wines more broadly. Winemakers of other varieties benefited from the shifting stylistic interests. In 2011, Jon Bonné, then wine writer for the *San Francisco Chronicle,* named Cathy Corison winemaker of the year.[34] She'd been making Corison wines since the late 1980s that were committed to elegance and restraint. In the years when Parker dominated, she garnered comparatively little media. As people began openly declaring their love for lighter wines, Corison regained attention. Her wines have consistently stayed below 14 percent alcohol. Bonné was also part of the tasting committee that selected wines for IPOB.

Though Chez Panisse in Berkeley had helped launch numerous California wineries in the 1970s and 1980s, the fleshier styles on the 1990s and 2000s never worked with their fresher styles of food. As the IPOB mindset took hold, more food wines emerged from California and found their way onto the restaurant's wine list.

The last year of IPOB was 2016. By that time the wine industry had gained more stability following the GFC, and growing conditions brought a very good vintage. IPOB opened a new conversation in wine. The change in grape contracts in 2010 and 2011, followed by higher yielding vintages of 2012, 2013, and 2014, launched opportunity for newer winemakers to source fruit from previously unavailable sources. Seeking out more affordable grapes, many also turned to lesser-known regions, and less common varieties. New small and micro wineries emerged throughout California. Others simply grew in acclaim. Wineries such as Sandlands from Tegan Passalacqua, Enfield from John Lockwood, Dirty & Rowdy from Hardy Wallace, and Jolie-Laide from Scott Schultz all launched in 2010. Forlorn Hope by Matthew Rorick and A Tribute to Grace by Angela Osborne started earlier but gained attention along with the further changing of the guard.

PARKER FALLS APART

Parker's taste buds dominated the international wine market into the 2000s. But as his influence grew, so did criticism against him. With the rise of the internet, the role of the singular wine authority started coming into question.

Wine discussion boards like Wine Berserkers, Wine Disorder, and others formed online communities with their own views on wine. Parker had started the online version of his magazine in 2000. He interacted directly with readers through a bulletin board and posted both articles and sometimes screeds through the website. In 2006, Twitter created an online community in short bursts of text. News, opinions, and wine recommendations all flowed together, creating a sort of ambient knowledge between its users. A community of wine lovers connected, affectionally called Wine Twitter, including well-known wine writers like Eric Asimov, James Molesworth, and Jancis Robinson interacting directly with wine consumers and other writers of wine.

In 2010, Instagram emerged and created yet another way to communicate wine. The photo-based app brought people together through imagery. As the Instagram community evolved, wine began shifting from text-based descriptions to visual representations. In summer 2011, I launched illustrated wine notes through the online identity @hawk_wakawaka. The concept hadn't been used before: at least, it had never been published. A few months later, Madeline Puckett started a site with far bigger impact, Wine Folly, using infographics to communicate wine concepts. Wine had turned to imagery. Social media became a new way of interacting over wine. It further eroded the authority of the singular wine critic by making it possible for consumers to connect with each other directly, and for other wine writers to connect with them. It also gave wineries the means to communicate one-to-one with consumers.

Then, in 2012, the movie *Somm* was released. Like "The French Paradox" and *Sideways* before it, it created a movement in wine. Not from abstinence to red wine, or Merlot to Pinot, as the two previous films had; instead, a new generation discovered they could pursue jobs in wine. The United States experienced what was essentially a sommelier boom. Study and certifications through the Court of Master Sommeliers, and WSET increased significantly. The appearance of the sommelier shifted from the stalwart sophisticate reminiscent of a British butler with wine expertise, to the cool kid wine ingenue. Sommeliers became increasingly hip and quickly demanded the market share of wine media. They also became the new draw for wineries seeking promotional outlets. The sommelier became yet another source of wine influence outside the wine critic.

In 2011, Parker handed the reins of reviewing California wine to his crown prince, Antonio Galloni. Then, in 2012, Parker sold the *Advocate* and Galloni left, launching his own wine reviewing website, *Vinous*. Parker stayed at the *Advocate* continuing to connect with subscribers through his bulletin board and contribute opinion pieces to the larger site.

As the influence of sommeliers and social media increased, Parker's began to wane. With the emergence of natural wine and the lighter styles represented by IPOB, it waned further. In 2014, Parker took all of them on simultaneously in a very public rant via his website. He criticized what he called crusaders pushing natural wine, challenging the authenticity of more mainstream examples.[35] He described the lower alcohol movement, as heralded by IPOB, as a slam on California wine. Which tells us something about how Parker saw California wine in general, even though California had always had lighter styles. In portions of his rant, he seemed to take the new trends and the people who supported them as personal attacks.

Parker did not hold back. He wrote:

> *... this is an ill-conceived attempt to simply differentiate themselves from me and/or other established critics, because that's the only way they feel they can create attention for themselves. Obviously 35 years of comprehensive writing about the wines of the world doesn't leave too many stones unturned, and so it is difficult to impossible for new wannabes to get attention, and even more unlikely to monetize their internet site. So they do what many people do in many fields when they can't stand on their own merits and credibility – they simply try to discredit the people at the top and use both the producers and their readers alike in their self-serving scheme.*

He insulted the farm-to-table movement describing wines that people like Waters, or Joyce and Evan Goldstein had brought to pair with their foods as "insipid" as well as "hapless, characterless and sterile."[36] He also asserted that the people who love natural wines do so not for their quality, but because of their novelty, and diminished their tasting ability while asserting that his preferences were the right ones. The rant was not unprecedented. In the April 15, 2012 issue of *Sommelier Journal*, Parker had done an interview with David Denton. There, he ranted against the natural wine movement and described those who support or promote it as charlatans,[37] even while contending that the wine he partially owned at the time, Beaux Freres, could likely count as a natural.

Wine writers from around the world responded. From the UK, Jancis Robinson in the *Financial Times*,[38] Jamie Goode on his blog *Wine Anorak*,[39] and Rebecca Gibb for *Wine Searcher*,[40] and in the United States, Alder Yarrow of the wine blog *Vinography*[41] debated Parker's post. Wine Twitter erupted in arguments about the piece and wine bulletin boards held heated discussions. Parker's reputation splintered. His series of rants gave those who'd long thought critically of him the room to do so publicly. Wineries to whom he brought acclaim remained loyal, but more quietly. They'd have to find new avenues to promote their wines. In 2014, Parker handed over his Bordeaux reviewing to Neil Martin. In 2019, he officially retired from the *Advocate*. *The Wine Advocate* continues to publish a host of wine critics. It has a strong following in wealthier parts of Asia, especially Singapore, but its influence in the United States and Europe has declined, if not disappeared.

COVID-19 AND WILDFIRES

The COVID-19 pandemic changed the wine economy. The series of lockdowns caused many to upscale their wine drinking, but by opening wines already in their cellar. Social restrictions in 2020 and 2021 gutted the hospitality industry and decimated restaurant sales of wine.

Swift-moving wineries created new markets for growth. Companies looking for ways to bolster employees as they worked from home hired winemakers to send wine to employees' homes, where they could host online tastings. The Wine Institute launched a 32-week webinar series finterviewing winemakers and wine professionals. The luxury club 67 Pall Mall started wine television. Reels on Instagram and TikTok exploded. Wine was moving to video.

Shortly after COVID-19 created social isolation and new protocols for businesses worldwide, California was hit with what were then the largest wildfires in state history. What was different about the 2020 wildfires was their sheer size, their virulence, and their proximity to agricultural centers, especially wine country.

In 2020, not only California but the entire west coast of North America saw more than three months of wildfires. In California, the fires created a loss of $3.7 billion in wine crop value.[42] According to the California grape crush report for the 2020 vintage, yields were down compared to 2019 by more than 13.8 percent. In addition, Arnulfo Solorio, director of the Napa Valley Farmworker Foundation, estimated vineyard workers lost around $50 million in wages during the fires of 2020. In Sonoma, impacted vineyards left grapes unharvested to reduce costs. Lost crop wasn't removed until pruning the following year.[43]

Wildfires cause immediate economic impact via crop, as well as infrastructure losses. But they also generate longer-term costs that are harder to estimate. Harvested crops might have smoke taint that makes the wine undrinkable. People working outside can suffer lung damage from poor air quality, or worse, loss of life. In California, because of the 2020 fires, insurance providers canceled winery insurance or made it so expensive most can't afford it. But lenders won't provide credit without insurance. Volume loss due to smoke taint in fermented wine won't be known for several years.[44]

The combination of the pandemic losses and the high costs of the wildfires led to a swell of wineries closing and others selling, most to outside investors.[45] Between the stay-at-home protocols of COVID-19 and the anxiety of the fires, wineries that survived were forced to find new solutions. Traditional sales channels were largely ineffective during the pandemic. Wineries that remained with these channels relied on financial reserves to survive. Then, in 2023 signs of a recession caused by COVID-19 became clear and wine sales did not return to pre-pandemic levels. Spending on luxury items in the United States depressed again.

Young people who had been reduced to eating and drinking what they could find at home emerged from lockdowns with a thirst for new flavors. In the meantime, the United States became health-conscious and skeptical of alcohol's impact on health. A new range of beverages took hold: low-alcohol seltzers, non-alcoholic wines and cocktail mixers, and ready-made beverages (pre-mixed cocktails). Millennials and Gen-Z have diversified their drinking, differentiating their habits from those of Boomers and Gen-X. Wine has become only one beverage option among many.[46] As wine sales decreased in 2023 and 2024, the industry suffered new anxiety.

DIVERSIFYING WINE

As traditional wine media and the power of the wine critic have decreased, the means to get wine in front of consumers has changed. The rise of the natural wine movement brought a wealth of natural wine fairs. Such fairs became a powerful way to get new brands and micro-wineries in front of impassioned wine lovers. Internationally, RAW Wine became a standard bearer, with tastings held in London, New York, and California, amongst others. Wine lovers worldwide fly to attend, in order to access smaller volume wines they cannot find otherwise. In California, the Oakland wine shop Ordinaire launched one. San Francisco-based Pamela Busch started WINefare to support women, non-binary, and LGBTQIA people in wine. Other examples started in San Diego, Los Angeles, and throughout the state.

Over time, the expanding influence of the natural wine movement combined with diversity initiatives and organizations to reach wine lovers traditionally overlooked or marginalized in wine.

The Association of African American Vintners, founded in 2002, connects Black vintners with supporting businesses and allies. Membership has steadily grown, to mentor a new generation of Black, Indigenous, and people of color (BIPOC). Black Vines, started in 2010 by Fern Stroud, brings together artists, musicians, and Black-owned wineries for an arts and culture event that has inspired numerous others. Christopher Renfro, an organizer and urban farmer in San Francisco, has successfully partnered with viticulturist and winemaker Steve Matthiasson to create a training program for BIPOC stepping into wine. The program exposes fellows to long-standing vintners, growers, and professionals throughout Napa Valley to learn across all aspects of vineyard farming. Renfro also brings attention to previously unknown vine libraries and historic field blends and makes wine from them in partnership with North American Press in Sonoma.

Owner–winemaker Matt Niess of North American Press left his job of more than a decade as a winemaker for one of Sonoma's most celebrated Pinot wineries, to instead make wine from hybrids, apples, and grapes infused with wild plants from the same site. Niess and Renfro have both found audiences for their wines through wine fairs, most notably ABV, which celebrates fermented beverages from anything but *vinifera*.

Outside the confines of the wine industry, other festivals are also putting wine in front of new audiences. Two of the biggest music events in the state of California, Coachella and Outside Lands, partner with wineries and sommeliers to pour wine for attendees. La Crema has been quietly sponsoring Pride events around the United States. It's brought an increase in sales and new customers. Wineries have also begun gaining new followers through professional sports. NBA players have made a splash with their love of fine wines, especially California Cabernets.[47] Their enthusiasm has expanded into winery ownership. Most famously, Dwayne Wade started Wade Cellars in Napa Valley in 2014. As the public has become used to the idea that the best of basketball drinks the best of wine, sports lovers have become more interested. In 2024, Kendall-Jackson partnered with the NBA as an official sponsor. Now that basketball fans know their idols drink fine wine, K-J is offering them an affordable option. At the same time, La Crema partnered with the WNBA: it's a way of showing they support women's sports, and gaining new audience there. Both businesses have reported increased sales and interest from new consumers.

Within wine companies, diversity initiatives have also emerged. Carlton McCoy of the Lawrence Wine Group in California helped start The Roots Fund, bringing BIPOC into internships at wineries and providing funding for wine courses and certifications. Gallo quietly started a diversity, equity, and inclusion (DEI) program within the company. Jackson Family made their commitment to diversity part of their overall sustainability commitments. Business longevity is supported by the people who work in it. Silver Oak partnered with the Veraison Project and created a year-long, paid internship bringing BIPOC professionals into the wine industry. The California chapter of The Hue Society, the Batonnage Mentorship Program, and The Roots Fund became connecting forces to help BIPOC professionals find wine jobs as well as professional mentorship and connection to a diverse community. The Diversity in Wine Leadership Forum was launched in 2020 to build stability for the plethora of new diversity organizations in the wine industry.

Significant pushback against DEI efforts have also emerged. Even so, DEI programs have opened wine to new audiences.

THE RETURN OF TRUMP

The US presidential election of 2024 led to the re-election of Donald Trump. His first presidency included increased tariffs on international wines, greater fear among immigrant communities, crackdowns on migrant labor communities, and an increase in the value of the American dollar.

As the value of the dollar increased, wines from California became more difficult to export. While it became more affordable for Americans to travel, the cost of goods reduced international sales. The rise of tariffs on international commodities in turn increased the cost to American consumers. International goods entering the United States from tariffed countries received a duty tax, paid by American consumers and sent to the US government. While tariffs are often seen to punish a foreign economy, the impact on the local economy can be severe. The cost of European wine increased, with few countries treated as exceptions.

In 2025, Trump has promised to impose tariffs more broadly. He is working on undoing established trade policies with Mexico, Canada, and China by imposing tariffs, and is suggesting tariffs on other countries as well. Economists have demonstrated that tariffs can have some impact on the foreign country's economy (the smaller their gross domestic product the greater the effect). Even more, tariffs operate to tax the American middle class and increase revenues for the national government. The impact on the wine industry could be far-reaching.

While pro-tariff advocates argue duty on European wine can benefit the California industry, previous periods with such taxes make the suggestion questionable. In the fine wine community, it is rare for a wine lover to simply replace one beloved wine brand or cuvée with a new one they hardly know. In other words, it's not an obvious or easy switch for wine drinkers to simply move from European to California wine. Perhaps even harder, the increased price of international wines slows their sales in the US market, creating a surplus of wine in the distribution system. As the distribution system becomes clogged, sales slow, prices often reduce, and the entire international industry is impacted. The slowing of sales for European wine can also slow the sale of US wine.

Trump's distaste for immigration offers another challenge. President Barack Obama increased deportation significantly during his presidency, as did President Biden. Trump promises the greatest number and most visible deportations in US history. The approach increased fear in the farm labor community even among workers who were already citizens of the United States. Some farmworkers simply left the country or the agricultural industry, even if US citizens, to find other, less stressful opportunities. Surprise raids again took hold in early 2025 to deport laborers in construction, hospitality, and agriculture. More are planned. As US history has repeatedly shown, immigration crackdowns consistently lead to labor shortages, which then negatively impact the nation's economy.

At the same time Trump is working on a way to retain the legal right to stay for people brought into the United States when young, who have gone on to become college-educated contributors to important national industries. If he succeeds, it could help stabilize portions of the economy.

As Trump exercises his new policies, the pressure on the wine industry to adapt increases. The result will be new challenges requiring new problem solving. But if the history of California wine demonstrates anything, it is the power of the state's people to innovate.

NEW OPPORTUNITIES

As California entered the current millennium, the wine industry faced numerous challenges. And yet, thanks to collaboration and innovative thinking, a new wealth of wineries emerged ushering in new styles, celebrating lesser-known varieties, and grabbing the attention of a new generation of wine drinkers.

The 2020s represent a period of experimentation and curiosity amid uncertainty that arguably exceeds what wine has seen before. So much volatility, after the incredible growth of the 1990s, has made the changes especially challenging. Yet California wine has consistently faced insuperable difficulties and found solutions through collaboration. At the same time, younger generations show more excitement, enthusiasm, and curiosity than seen in previous consumer generations. Their interest in new flavors and other beverages can be seen as an expansion of their palates and curiosity. Just as Alice Waters argued in the 1970s, simple pleasures and taste can be a way to claim joy during global turmoil. Millennials and Gen-Z might be leading experts in the practice.

Traditional sales channels, media marketing, and established categories of wine continue to decline. It would seem they no longer work, or at least not at the same scale. But from another perspective, the diversifying of the market, the room to experiment, and the expansion of wine's audience present more opportunity than ever before.

PART 2

WHERE WE GROW

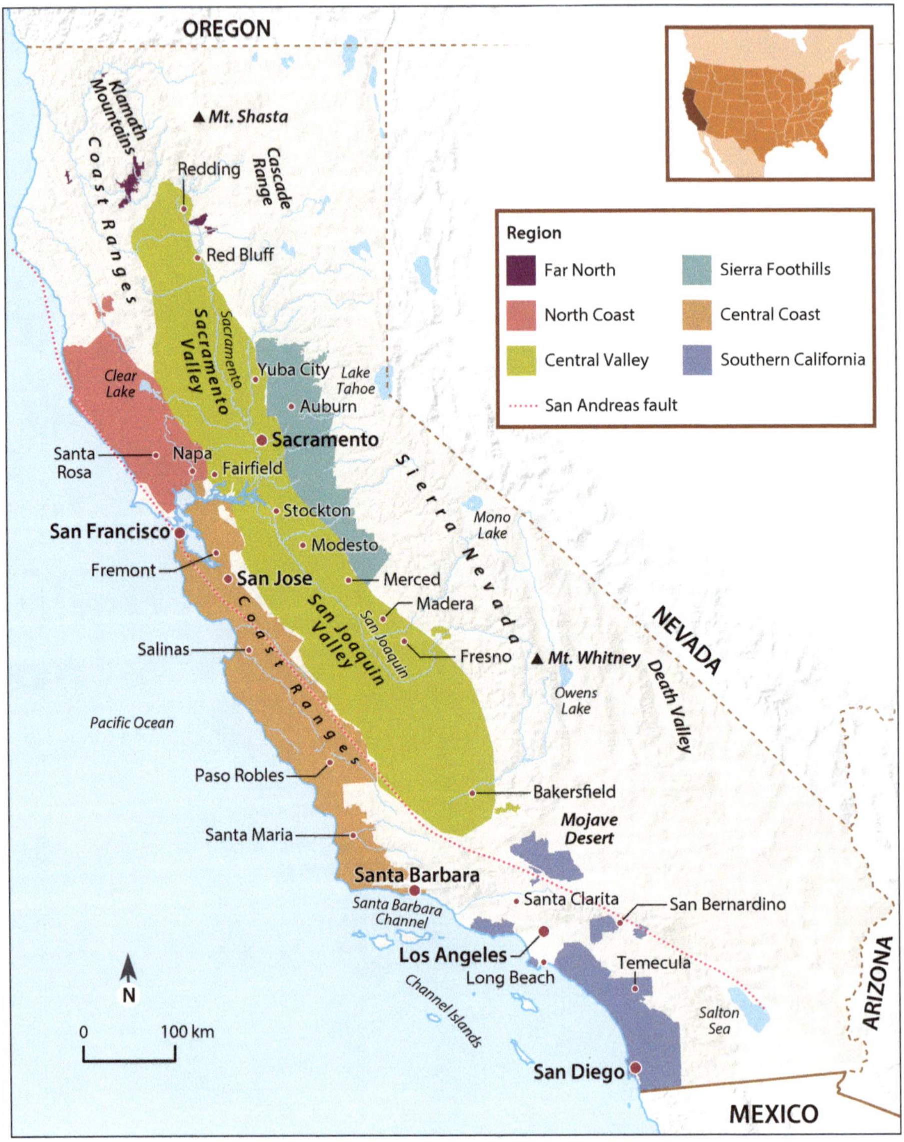

Map 1: California wine regions and topography

6

CALIFORNIA GROWING CONDITIONS

California, if it was a country, would be the world's fifth largest economy. It is also the fifth largest supplier of food and agricultural products in the world, with around 400 distinct crops, including wine. The state is the fourth largest producer of wine in the world. When it comes to wine, it includes 154 legally recognized appellations, or American Viticultural Areas (AVAs); there are 276 in the entire country. The area planted to vineyards as of 2023 stands at 610,000 acres (246,858 hectares).[1]

California's geology is complex. The intersection of multiple tectonic plates uplifted mountains along its coast as well as its eastern border. Volcanic activity caused by this same motion formed more mountains. The Vaca Range in Napa Valley, the Cascade Range in the state's north, the Mammoth Mountains halfway between Los Angeles and San Francisco, and an ancient range, mostly now eroded, in the Mojave Desert were all formed by volcanic activity.

The state's Mediterranean climate creates two seasons instead of four. Dry summers lower disease pressure for grapevines, while risking vine stress from too little water. Wet winters divide the state into two parts: the west relies on collecting winter rain and the east gets its water from snow melt in spring. California's topography, formed through various geologic processes, creates wind tunnels and temperature variation that give the state the full five climate regions described by the Winkler Index (see box, page 104). The coldest, region 1, can be found in coastal parts of Monterey County and small portions of Sonoma.

The hottest areas, region 5, can be found in the southern portions of the Central Valley.

Soils in California have incredible diversity. The combination of their geological origin with erosion and weathering means the state hosts seven of the world's 12 soil orders, and 1,500 soil series.

Vitis vinifera arrived a decade after Spanish explorers colonized the central and southern coasts in the late 1700s and planted what became known as the Mission grape. The number of cultivars expanded in the 1820s, as Mexico invited new settlers from around the world to develop land in exchange for citizenship. Franciscan monks remained and brought new selections from their homeland, Spain. Soon after, settlers from central Europe brought vine collections from the Austro-Hungarian empire. French cultivars arrived at the end of the 1800s and took hold in the 1960s as primary grapes in the state.

Together, the climate, geology, topography, and state history guided the planting of California wine. But the state includes another unique growing condition and is not found in the same way anywhere else in the world. It's one of California's human elements: its acumen for research and innovation. Near to the wine regions of California stands University of California Davis (UC Davis), one of the top wine and viticultural education and research programs in the world. It is home to one of the most advanced and sustainable wineries in the world, used as a teaching facility for wine students at the university. Nearby grows Silicon Valley.

California wine in numbers

Total planted area: 610,000 acres/246,858 hectares
Commercially grown varieties: around 130
Number of AVAs: 154 (as of 2024)
Number of counties growing wine: 49 of 58

RESEARCH IMPACT

Silicon Valley is one of the top technological innovation centers of the world, housing some of the largest and most successful global companies. Its stature makes it a hub for startup companies, as well as tech investment money. As of 2022, its gross domestic product value, or GDP, was $840 billion, making the economy of Silicon Valley alone larger than that of Switzerland or Saudi Arabia, and approaching that of Turkey or Indonesia.[2]

Successful tech entrepreneurs lead some of the state's most important wineries. Ridge and Clos de la Tech, both in the Santa Cruz Mountains, were founded by technology innovators of the region. In Sonoma and Napa Counties, the wineries Peter Michael, OVID, Vineyard 29, BRAND, and Momento Mori were all started by owners who first found successes in Silicon Valley technology. Elsewhere in California, there are more.

The technological innovations of Silicon Valley have influenced wine in other ways. The industry's companies developed many of the solutions at the core of the UC Davis teaching winery, and numerous other advances for the wine industry at large. Refractometers to determine the sugar levels for harvest, mass spectrometers for analyzing tannin levels as well as assessing a wine's progress in fermentation and aging, and liquid and gas chromatographs to measure aspects of wine quality were all advanced and made affordable in Silicon Valley.

Vineyard equipment to improve water management and vine health also came from Silicon Valley. Satellite imagery and thermal imaging to guide watering, treatments, or replanting; sap flow sensors

A vineyard crew harvesting Cabernet Sauvignon from Napa Valley, seen from hot air balloon

and evapotranspiration technology to improve the timing of irrigation and track the effectiveness of climate change solutions; wireless weather stations, wireless fermentation tanks, software to manage direct-to-consumer sales, and compliance regulations for the three-tier system – each effected through technology collaborating with wine in California.[3]

UC Davis researchers led by plant geneticist Carole Meredith (aka. Scientist Legend) solved how to determine the parentage of grape varieties genetically, providing the first such relationship, for Cabernet Sauvignon. Its parents, Sauvignon Blanc and Cabernet Franc, got together and birthed one of the world's most prolific wine grapes. Davis researchers' techniques for determining grape parentage helped advance other steps in studying grapes genetically as well. To mention just a few of the other achievements of UC Davis researchers, they have developed rootstock for grapevines and cultivars resistant to Pierce's disease, created water capture and reuse systems, and are working on improving and honing fining techniques in winemaking.

The university also houses one of the largest wine archives in the world, with a selection of books spanning the history of wine, some of which do not exist anywhere else. Within the archive is the world's largest special collection of wine writers' notes, materials, and other writing. Hugh Johnson, Jancis Robinson, Karen MacNeil, Dorothy Gaiter, and John Brecher, plus podcasts like *The Color of Wine* and *Love and Vines*, all donated their libraries. They're a resource for studying the history of wine and learning solutions from our past.

And surrounding the acumen of UC Davis and the ventures of Silicon Valley, California boasts 610,000 acres of wine grapes.[4] Forty-nine of the state's 58 counties grow wine. California grows well over 100 different *vinifera* varieties. Lodi alone boasts 136.[5] Included among California's wine regions is one of the world's most famous, Napa Valley.

It's a unique combination. Nowhere else in the world has academia, the tech industry, and agriculture in such proximity. Within a mere 90 minutes of each other are a top wine education and research institution (UC Davis), a top technology innovation hub (Silicon Valley), and one of the most famous wine regions in the world (Napa Valley), each supported by the state's enormous capital investment.

The rise of California wine through its history came, partially, thanks to the partnership of these three industries. Post-Prohibition, when winemakers struggled to understand wine chemistry – its stability, controlled malolactic fermentation, ripening, the balance of innate acidity to tannin to sugar levels – UC Davis researchers helped find solutions. When the solution was technology, Silicon Valley helped create the instruments and machinery required. The future of California wine amid economic fluctuations and climate change depends on ongoing collaboration. UC Davis and tech researchers have developed means to do carbon capture from fermentation tanks to reduce the amount released into the winery or atmosphere. They're also working to convert the captured carbon into usable goods and carbon sequestering materials. Electric tractors are already in use, built in partnership by a roboticist, a systems manager, and a designer, each from Silicon Valley, and a North Coast wine grower.

The human factor in wine regions is often overlooked as one of its growing conditions, but it might be its most important. Only through our choices do we venture to grow wine. The culture and availability of resources in an area determine what choices we can make with the topography we've been given.

CALIFORNIA GEOLOGY[6]

The John Muir trail that reaches 213 miles (343 kilometers) through the Sierra Nevada range from Yosemite National Park south to Mount Whitney is considered one of the most beautiful hiking trails in the United States. It passes through multiple forests, and several national parks. Mount Whitney, at its southern end, is the highest peak in the continental United States (14,495 feet/4,418 meters). A mere 85 miles (137 kilometers) east, the landscape dips to the lowest point in North or South America: at 283 feet (86 meters) below sea level sits the Badwater Basin of Death Valley.

California's unique placement at the point where the Pacific plate and the North American plate meet created the state's coastal mountains as well as the Sierra Nevada and a run of other smaller mountains in between. Some were formed due to volcanic activity caused by the movement. Others resulted from the accretion of fragments of the earth crust dragged by a tectonic push and later attached to part of the continent by that same movement. The encounter line between the two plates forms the San Andreas fault, which runs northwest like a zipper up the coast to the boundary of Sonoma and Mendocino Counties. Its activity created the Gulf of California along the Baja Peninsula of Mexico.

The San Andreas fault

The San Andreas fault runs for 800 miles (1,287 kilometers) along the lower two-thirds of California, north to south at a slightly westward tilting angle. Its northern terminus is in Cape Mendocino. Its bottom point within California is the Salton Sea that sits above the border of Mexico, but it continues south and created the Gulf of California on the eastern side of the Baja Peninsula. Though the name implies the San Andreas is a single fault line, it is a parallel series that follows a significant portion of California coastline. The faults help absorb the movement of the Pacific and North American plates sliding past each other.

A fault line system of this length does not move all at once. Instead, portions of it will move more frequently leading to tremors or smaller earthquakes, while other portions might have only occasional stronger movement. The 1906 earthquake in San Francisco is an example of a sudden, significant rupture. The fault moved a 270 mile (435 kilometer) portion of the state from San Juan Batista, in San Benito County of the Central Coast, to Humboldt County above Mendocino. The impact of the 1906 quake was significant in San Francisco, which had been an epicenter for urban winemaking. The earthquake and resulting fire forced those facilities to the eastern side of the Bay.

In 1986, a portion of the fault in Santa Cruz County moved suddenly, causing damage north in San Francisco. Wineries as far south as Paso Robles experienced the tumbling of wine barrels and loss of wine as a result.

At the northern end of the San Andreas fault, the Pacific and North American plates encounter a third, the Gorda plate. Cape Mendocino experiences regular, comparatively smaller, quakes as a result.

Famed vineyards grow atop and along the fault line. Hirsch Vineyards in the Fort Ross-Seaview portion of the West Sonoma Coast sits atop the San Andreas, lending its name to one of Hirsch's best-known Pinot Noirs. In San Benito County, Randall Grahm's Popelouchum is bordered by the fault and Eden Rift straddles the San Andreas. Ridge and Rhys in the Santa Cruz Mountains overlook it from either side of one of the fault's ravines. Where the San Andreas fault moves off the continent and into the ocean rumbles the Mendocino Triple Junction, the point where the Pacific and North American plates conjoin a third, the Gorda plate.

A concentration of earthquakes follows the San Andreas fault and gains frequency along Cape Mendocino thanks to the intersection of

California includes a plethora of geological parent materials that form rocks and eventually soils of the state. Here, compressed volcanic ash in Sonoma County

the three plates. Their movements create unique complexity in the rocks of California. Uplifted sandstone and siltstone, mudstone, and shale appear from the Pacific plate. Mylonite and serpentine are found along the fault lines. Metamorphosis under the pressure of tectonic activity has made serpentine one of the most abundant rocks of the fault area, so abundant, in fact, it is California's state rock. Serpentine appears abundantly throughout the Sierra Nevada range, and along fault lines in the Sonoma coastal mountains as well as in other areas of fault activity. The heavy metals and lack of nutrients in soils derived from serpentine make it especially difficult for grapevines.

The uplift of the Pacific plate has created a history of volcanic activity. The result? Basalt, andesite, compressed volcanic ash, and rhyolitic tuff swirled through sedimentary uplift of the coast. The Vaca Range in Napa Valley, Sonoma Mountain, and large portions of the Mayacamas between Napa and Sonoma, and running into Mendocino, are dominated by volcanic bedrock and stones.

The tectonic activity also formed inland valleys and mountain gaps. The enormous inland valley, also known as California's central valleys, formed through tectonic activity as well. Millennia ago, a large ocean

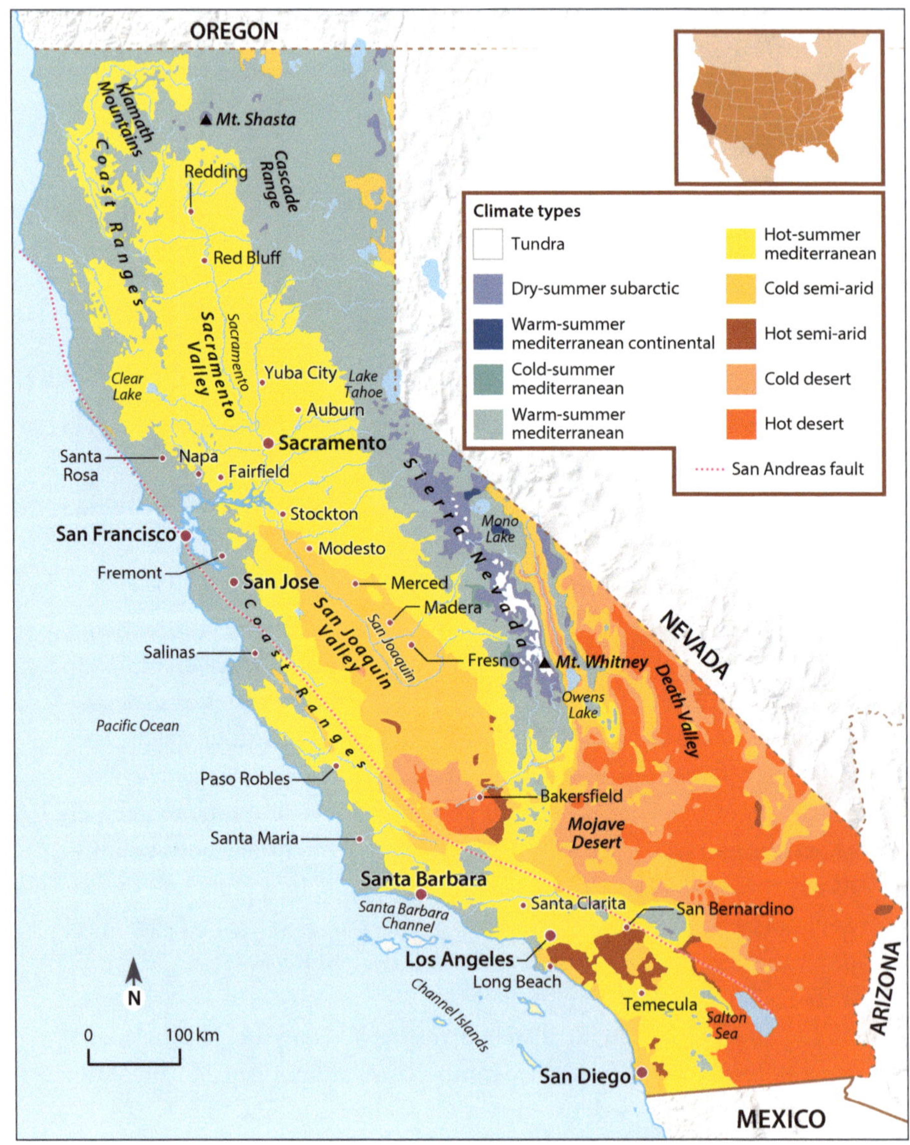

Map 2: California climate types

plate, the Farallon, existed between the Pacific and North American plates. As it was subducted (almost entirely) beneath the North American plate a low inland basin was formed. Its opening on the western side through a break in the coastal mountains let in the sea. The sea created limestone that can be found in much of southern California, parts of the Sierra Foothills, the Santa Cruz Mountains, the Gavilan Range (where Josh Jensen of Calera found limestone atop one of its peaks), the western side of Paso Robles, and Ballard Canyon in Santa Barbara County. The chalky tannin character of Ridge Monte Bello reveals the effect of growing Cabernet Sauvignon in calcareous soils.

California's soils are the result of erosion through landslides, earthquakes, and colluvial deposits, as well as weathering from winter rains, a cycle of freezing and thawing, sun exposure, and extreme heat. Weathering, in other words, created by California's climate.

THE CALIFORNIA CLIMATE

California sits on the western edge of the North American continent. It stands as the first area to greet weather formed over the Pacific in the earth's spin from west to east. Its proximity to the world's largest ocean creates the state's climate through the year.

The series of mountain ridges lifting across the state, and the valleys between, filter and move the influence of the ocean through California. The state's most impactful landforms are the Sierra Nevada range in the east, the coastal mountains of the west, and the central valley between. The undulating topography throughout the rest of California generates microclimates and temperature variations that help build the diversity of wine. The Sierra Nevada mountain range stands as a barrier protecting California from the more extreme continental climate to the east. It also creates a more contained area for ocean influence. Along the coastal side, the cold California Current pushes down from Alaska along the northern two-thirds of the state, turning into the ocean again off Point Conception at the southwestern corner of Santa Barbara County. A deep upwelling of cold water from under the surface of the ocean lifts into the current along the coast, cooling its impact further. Summers stay cool, and winters remain moderate.

Between the islands of Hawaii and the coast of California sits a high-pressure system known as the North Pacific High. In summer, it pushes the atmospheric jet stream south, below California, ensuring the region remains dry during that half of the year. In winter, the North Pacific High moves south, exposing California to greater storm conditions. Moisture gathers in the atmosphere off the ocean as the earth spins east. When it hits the California coast, precipitation dominates the region. Rainfall fills the coastal side of the state, snowfall hits the east. The result is a Mediterranean climate.

What is a Mediterranean climate?

Mediterranean climates occur on the western side of some continents but generally only between 30° and 44° latitude. Such regions are found near the Mediterranean itself and in California, south-western Australia, Chile, South Africa, and northwestern Baja, Mexico, as well as a small portion of southern Oregon and western Nevada. They share dry summers and rain events only in part of the year, either autumn, winter, or both.

Temperatures can vary in Mediterranean climates, creating a range of microclimates within them. Plants native to these environments have adapted to rely on little water most of the year and appreciate it during rain cycles. Such plants are generally hardy and have developed ways either to store water, or to use and lose less of it. For example, smaller leaves, thicker stems, and often denser bark. Olive trees and eucalyptus are two trees that have adapted well to such environments. In California, various native oaks, madrone, and cypress are other examples.

Dry summers generally reduce disease pressure on vines. However, natural drought cycles carry the risk of dehydration. Additionally, soil exposure can increase aridity through warmer summer months and lead to soil loss, erosion, and increased regional temperatures. Ground cover can help alleviate the problem.

Winds

The climate also brings wind. The mountains create natural barriers, allowing temperatures to rise compared to the colder Pacific. As temperatures increase inland, air is pulled from cooler areas to replace it. Along the western side of the state, the Petaluma Gap in Sonoma County, the Santa Lucia Highlands in Monterey (including the Santa Lucia Highlands and Arroyo Seco), and Edna Valley in San Luis Obispo, are all inundated with daily wind for this reason. In the Santa Lucia Highlands, winds are strong enough that houses are built to include courtyards facing south to provide a windbreak for families in summer.

The most powerful example can be found through the California Delta and over Lodi. As the Central Valley heats during the day, temperatures lift, raising an enormous mass of air off the valley floor. The vacuum it creates pulls air from the cold Pacific through the Carquinez Strait near Martinez and over the California Delta into Lodi. The Delta breezes, as they're called, arrive daily in the afternoon through the warmer half of the year. They are persistent and powerful enough to provide significant energy needs for the state through wind turbines placed along the air flow.

Further south, Santa Maria and Santa Ynez Valley feel a similar effect. As the inland side of Santa Barbara County heats up, cold air is pulled from the ocean to the mountains in the east. It is said that temperatures increase a degree Fahrenheit for every mile (0.6°C every 1.6 kilometers) traveled away from the ocean. This is partially thanks to wind.

Sonoma and Marin Counties share the only appellation in the country defined by wind. The Petaluma Gap AVA is outlined by the outer limits of where the wind from the coast travels east at speeds of at least 8 miles per hour (13 kilometers per hour). Scientific studies have shown that at that speed vine respiration slows. As vine respiration slows, natural acidity stays up. This occurs, to some degree, in other windy appellations as well. The Santa Lucia Highlands naturally creates Pinots of power thanks partially to the influence of wind. In vineyards directly in the wind's path, berries are smaller, creating a greater skin-to-juice ratio, and sometimes thicker skins. The effect is higher natural acidity levels and greater structural presence in the wines. Wine experts often mistakenly think the aroma of the region's Pinots indicates that they include whole clusters when in fact the spiced-resinous components are common to destemmed reds of the area.

The wind can also lower disease pressure. As it blows, humidity is pulled from the area removing the conditions that allow mold or mildew to grow.

Drought versus flooding

California has the most variable climate year to year in North America.[7] Scientists able to research climate history through tree rings, coral reefs, glaciers, and sedimentary deposits from ancient lakes have determined California has always had drought, as much as it has always had floods. Annual rains go through wetter and drier periods.

The Winkler Index

UC Davis researchers A. J. Winkler and M. A. Amerine developed their Winkler–Amerine Temperature Summation Index, also known as the Winkler Index, Winkler Scale, or Winkler Range. The index may be used as a way to correlate potential wine quality and grape types to climate regions. While considered overly simplistic today, prior to the Winkler Index planting grapevines in any region was a complete guessing game. The index helped narrow the field to a range of potential grape types, and pointed producers to areas of the state that might offer a style they wanted.

The scale was groundbreaking in its time. Released in 1944, it focused on temperature summation alone of a region. Winkler's later work considered fog, overall humidity, sunshine duration, and precipitation. But in charting the climate scale, Winkler and Amerine believed that, of these, temperature plays the greatest role in the ripeness and development of wine grapes. In very broad strokes, at a time when a climate map of California had never been done and considering that California has no summer rain (and therefore precipitation is not a worry during the growing season), the Winkler Range makes historical sense.

More recent research recognizes the importance of multiple other factors in grape development. Wind plays a role in disease pressure and retention of natural acidity as well as influencing yields. Fog reduces sunlight duration in the morning and provides increased humidity that may cool the vines. Different soils and rocks have different temperature retention potential; clay is colder than sand, dark cobbles warmer than light stone.

In its original design, the scale depends on daily accumulation of growing degree days (GDD). To calculate GDD the average daily temperatures over 50°F (10°C) for every day from April 1 to October 31 are added together. April to October is considered the viticultural growing season.

The total for every day in that period is considered the GDD. For example, if the average temperature of a particular day is 80°F (since the index was developed in Fahrenheit), then the accumulated heat for that day is 80 – 50, which equals 30, meaning 30° above the baseline of 50°, and so 30 would be added to the overall heat accumulation for the growing period. A day that is 50° would add 0 to the overall heat accumulation. The total at the end of the growing period determines the climate range of that region. Winkler and Amerine then matched best grape varieties for each temperature range.

Winkler and Amerine determined 5 climate ranges:

- Region I: 2,500 GDD and under, marginal climate, hybrids and early ripening *vinifera*;
- Region II: 2,501–3,000 GDD, early and mid-season *vinifera*, good quality wine;
- Region III: 3,001–3,500 GDD, favorable for standard or high quality wine;
- Region IV: 3,501–4,000 GDD, favorable for high production, acceptable quality at best;
- Region V: 4,001 GDD and over, favorable for extremely high production, or table grapes.

Over time difficulties over the best method to determine average temperature have emerged, creating issues with the application of the system. It is easy to see that the recommendations for wine quality from the index are too limited, but purely in terms of temperature ranges the index provides a useful way to start identifying regions.

During droughts, high pressure over the Pacific stays north later into the year, preventing storms from reaching the coastline. California's location between the Pacific and the continent east of the Sierra Nevada range gives it a natural drought cycle. Flooding cycles have also been natural occurrences. Development for fine dining started around Yountville in the Napa Valley rather than the town of Napa itself because until 2016 the town of Napa remained a flood zone. That year, the Army Corps of Engineers completed the Napa River Flood Project creating an overflow system that serves as a public park in the dry months, and an enormous river catchment when it rains.

Yet, as climate change gathers pace, the movement of the North Pacific High has become less predictable. Weather events are more erratic. Drought versus flood years are harder to forecast. And weather events have become more intense. The frequency of atmospheric rivers flooding the coastline has increased. Vineyards have been impacted by landslides created by excess rain, flooding, hailstorms, and spring freeze all occurring unexpectedly later in spring when vines are no longer dormant.

Paso Robles, notoriously dry, flooded in 2023. Along the coast of the Santa Cruz Mountains and parts of Santa Barbara County storms caused significant erosion. Napa Valley got snow that same winter over the mountains and into the valley floor. No one could remember seeing it in the region before. The vines were dormant, so it caused them no harm, but in 2022, frost appeared in May, more than a month later than expected, and damaged vines throughout the North Coast, up into the Pacific Northwest and British Columbia.

Mesoclimates

As noted, California's Mediterranean climate enjoys an overall pattern of wet winters and dry summers. Its western side is hit by winter rain, while the mountains in the east gain snow. The state's topography creates wind tunnels that lower yields, lessen disease pressure, and increase natural acidity. Yet the varied topography of smaller mountain ranges, rolling hills, canyons, and valleys creates more variation. Cold air settles to lower elevations. The fog line generally only reaches heights between 400 and 800 feet (122–244 meters). Higher parts of the mountains experience an inversion layer where temperatures between day and night are more even. Lower elevations experience large temperature swings from day to night.

The combination creates a range of mesoclimates (climates of a large area) across the state. Though it is understood today as an imperfect system, the Winkler– Amerine Heat Summation Index (see box, p. 104),[8] more commonly known as the Winkler Scale or Winkler Index, provides a simplified way to identify ripening potential in various parts of the state. More precision is needed to understand a region thoroughly.[9] The scale clarifies mesoclimates through cooler and hotter parts of the state. The coolest are found along coastal Monterey and parts of Sonoma, the hottest in the southern parts of the San Joaquin Valley. Either can ripen grapes, but differing types for wines in differing styles depending on overall temperature accumulation. Although limited, the index offers a useful way to start. It maps California's full reach of temperature ranges: five different zones spread through varying parts of the state.

CALIFORNIA VARIETIES

The wide variety of microclimates and soil types in the state of California means the area can grow just as great a mix of grape varieties. More than 100 *vinifera* cultivars are found in California. The region of Lodi has counted around 130 that are commercially farmed and has the greatest varietal diversity in the state.

The state of California includes a wealth of old vine vineyards planted in the late 1800s and early 1900s, including these, in Lodi

Vinifera vines planted in California originate from across Europe. The most prolific of the state's varieties are the expected 'international' grapes of France: Cabernet Sauvignon and Chardonnay consistently top the list as the most planted varieties. Sauvignon Blanc, Pinot Noir, Pinot Gris, Merlot, and Syrah appear next, along with Zinfandel. And one surprise, French Colombard, can be found among them as well. As public prejudice against varieties like Merlot or Zinfandel has grown, clever producers have begun using them in red blends that need not name every grape. Studies show consumers love drinking these wines but as the grapes have been kicked from the cool-kids club, wines that bear their moniker see lower sales. Even so, some of the most sold red blends of California feature them prominently.

Historic plantings of Zinfandel often include intermixed vines of Petite Sirah, Alicante Bouschet, Carignan, Grenache, and sometimes Mission, Mataro, and Peloursin. Genuinely obscure varieties such as Green Hungarian, Grand Noir, Muscadelle, Muscat of Alexandria, Burger, and others are found among them as well. Studies of older field blend vineyards have helped uncover parts of the history of California, and some producers, including Sandlands, Bedrock, and others, have made a point to proliferate them again. The range of eighteen grapes approved for

Châteauneuf-du-Pape now grows in California. For some of the whites, the world's planting area doubled when planted in Paso Robles by Tablas Creek. Among the eighteen, Cinsault has gained a passionate following. It offers another option for anyone seeking a fresh, midweight, flavorful red. The selections imported by Tablas have spread throughout the west coast of the country. Chenin Blanc and Riesling have been in the state for decades. Chappellet brought attention to the former with their wine, still an international favorite. The oldest Riesling in California grows in the vineyard of Stony Hill, but it can be found from Potter Valley in Mendocino to the coldest parts of Santa Ynez.

Besides these, a feast of cultivars can be found, most in modest quantities. An expected California Port boom in the 1990s led to the planting of Touriga Nacional and Souzao, along with very small plantings of other Portuguese cultivars. Arnot Roberts helped bring attention to such varieties by making rosé from Touriga Nacional that has been growing in Lake County since 2010. Others use them for red table wines, as well as fortified wine, in small quantities. Since 2010, there has been a mini boom of planting varieties from the Jura and French Alps. Trousseau has taken the main slot in popularity among these, Arnot Roberts earns part of the credit there as well, making it from the same Lake County site with a portion from fruit in the Russian River. Their success sparked Sandlands to get it planted in the West Sonoma Coast and Rootdown to establish it in Cole Ranch. Iruai sources it from the far north near Mount Shasta. Jolie-Laide uses it in blend with the equally obscure Cabernet Pfeffer, Gamay, and Poulsard. California enthusiasm helped instigate plantings of the variety in Oregon as well. Also bitten by the obscure grape bug is Phelan Farms along the Cambria area of the San Luis Obispo (SLO) Coast, which grows Trousseau, Poulsard, Mondeuse, Savagnin, and Mencia, amongst others. Pax has worked with some of these in Sonoma County. Cole Ranch is establishing plantings of them in Mendocino, and Iruai has helped establish them near Mount Shasta. Each of these is at relatively small quantity.

In Lodi, a wealth of Iberian and Austro-Hungarian (aka. German) varieties can be found: Verdelho, Albariño, Garnacha (white, red, and gris), Dornfelder, Tempranillo, Kerner, Lemberger, Zweigelt, Bacchus, Regent, and Optima. Aromatic varieties like Symphony, Sauvignon Gris, Malvasia, and various Muscats are also found here. In the 1980s, California went through what was called a Cal-Ital boom: Italian varieties made by California producers. The results initially were mixed and producers turned away from them but in the years since, some producers have mastered making successful wines from Italian grapes grown in the right place. Montepulciano, Teroldego, Nero d'Avola, Ribolla Gialla, Friulano, Nebbiolo, Greco, and Aglianico have all taken hold in small quantities. Sangiovese and Barbera dominate part of the state.

Interspecific hybrids are being grown for wine to good effect in some areas. In Sonoma County, Matt Niess founded North American Press for exactly that purpose. He farms a Baco Noir vineyard in the prime Pinot country of western Sonoma County. He has shared cuttings to other parts of the state as well. What makes Baco Noir worth considering is its resistance to powdery mildew, a real concern in more humid coastal areas. Niess also experiments with co-ferments of grapes with other fruits and plants all grown in the same site.

Interspecific hybrids

When a vine species successfully breeds with another vine species, the offspring produced is called an interspecific hybrid. That simply means a hybrid between two different species (rather than two different cultivars of the same species, for example). Hybrids can occur naturally through proximity of vines. They are also created intentionally. Breeders can harvest pollen from one species and introduce it to the flowers of another species. Hybrids can also be developed in a vine nursery through more technical means.

In the late 1800s, a significant effort to create interspecific hybrids developed in the international wine industry as one possible solution to the grape louse phylloxera that was decimating vineyards at the time. Some species native to North America were found to be resistant to phylloxera. Not all North American species are. Phylloxera originates from the continent east of the Rocky Mountains. Species that developed west of the Rockies are generally *not* resistant to phylloxera.

As breeders gained experience, hybrids were made for the sake of developing disease resistance. Different hybrid grapes have different types of resistance. Many hybrids are resistant to rot and mildew, others to winter cold. More recently, research centers such as the University of Minnesota, Cornell University, and UC Davis have developed hybrids resilient against specific diseases. A group of cultivars released in Europe known as PIWI grapes carry disease and climate resistance. In 2022, UC Davis researcher Andy Walker released five new hybrids resistant to Pierce's disease.

UC Davis researcher Andy Walker has relied on interspecific crosses to create five new cultivars resistant to Pierce's disease, which is an ongoing threat to wine industries worldwide. It led to the demise of the original southern California wine industry in the late 1800s and continues to damage vines at comparatively smaller scale throughout the state. *Vitis arizonica*, a species native to the southwest United States and Mexico, has a genetic resistance to Pierce's disease. Walker bred it with *vinifera* varieties over several generations to ensure the new plants have the protective gene yet are more like *vinifera* in their wine characteristics. After decades of effort, the result is five new varieties, three red and two white, available to the wine industry. They have been planted in both California and Texas. In Ventura, Adam Tolmach of Ojai has planted a test block of 1.2 acres (0.5 hectares) with four of the varieties. The grapes grow in the same site that was destroyed by Pierce's disease decades ago. In Napa Valley, Whitehall Lane blends them with *vinifera* to good results. They have also been established in Sonoma and Temecula.

California also grows its own heritage hybrid. Planted in the late 1700s, the grapevine known as Viña Madre still grows on the grounds of San Gabriel where it was planted during the mission era. It's likely the oldest cultivated vine in California, and certainly one of the oldest in North America. Incredibly, the vine survived multiple fires, earthquakes, phylloxera, Pierce's disease, and a series of floods. Originally believed to simply be the Mission grape, its survival was a mystery. The advent of genetic mapping identified Viña Madre as a self-made cross of the Mission grape with southern California's native variety, *Vitis girdiana.* Since being identified, Viña Madre has been discovered through portions of downtown Los Angeles, and cuttings have been sent north to Matt Niess in Sonoma. A partnership of three Los Angeles producers – Byron Blatt, Angeleno, and Caveletti – now maintains Viña Madre and they now make a mission-era Angelica sweet wine from its grapes.

As climate change alters growing conditions, producers are becoming ever more curious about alternative varieties. Experimental vineyards to determine what varieties grow best in various locations as temperatures and weather events change can be found in multiple areas: Napa Valley, the Central Valley, and San Benito Valley all include them. Historic field blends are offering other opportunities as well. Pre-existing archival vineyards such as the one beside the St. Helena public library, farmed by Turley, and the Filoli vineyard south of San Francisco, maintained by

Christopher Renfro of the Two Eighty Project, both revealed varieties that California producers didn't know the state had. Other producers continue to bring in still other grape types to trial as the next intriguing grape of California.

California's most planted wine grapes

(As of 2023, CFDA.)

Total planted area in California: 610,000 acres (246,858 hectares)

Cabernet Sauvignon: 95,638 acres (38,703 hectares)
Largest total plantings by county: Napa, San Luis Obispo

Chardonnay: 88,063 acres (35,638 hectares)
Largest total plantings by county: Monterey, Stanislaus, San Joaquin

Pinot Noir: 46,608 acres (18,862 hectares)
Largest total plantings by county: Sonoma, Monterey

Zinfandel: 37,947 acres (15,357 hectares)
Largest total plantings by county: San Joaquin, Sonoma, Fresno, Madera

Merlot: 33,443 acres (13,534 hectares)
Largest total plantings by county: San Joaquin, Napa, Sonoma

Sauvignon Blanc: 17,261 acres (6,985 hectares)
Largest total plantings by county: Sonoma, Napa, Lake

Pinot Gris: 16,061 acres (6,500 hectares)
Largest total plantings by county: San Joaquin, Fresno, Sacramento, Yolo

French Colombard: 15,960 acres (6,459 hectares)
Largest total plantings by county: Fresno, Madera

Syrah: 14,293 acres (5,784 hectares)
Largest total plantings by county: San Luis Obispo, Madera, San Joaquin, Sonoma, Santa Barbara

Petite Sirah: 11,653 acres (4,716 hectares)
Largest total plantings by county: San Joaquin, San Luis Obispo

Counties with greatest planted area of red grapes: San Joaquin, Sonoma, Napa
Counties with greatest planted area of white grapes: San Joaquin, Sonoma, Monterey, Fresno

7

THE NORTH COAST

The North Coast grows California's most famous wine. Napa Valley and Sonoma County lead the charge. Sonoma is one of two birthplaces of fine wine in the state, while Napa Valley is one of the world's top regions. Together, they produce just 10 percent of the wine of California yet gain most of its media coverage.

The regions of the North Coast encircle the central pair of Napa and Sonoma. Sharing their borders to the north are two counties: Mendocino and Lake County. Mendocino is most famous for Anderson Valley, Pinot Noir, and, inland, old vine red wine. Lake County, surrounding Clear Lake, is volcanic, and full of Cabernet. Lake County is also the birthplace of Kendall-Jackson's Vintners Reserve, the most sold Chardonnay in the United States, and arguably part of the variety's boom worldwide. To the south are two more counties, each only lightly planted for wine. Below Sonoma is Marin County. Wine geeks drive through it to regions further north, but sci-fi lovers rejoice: here George Lucas makes wine from his Skywalker Ranch, perched along ridgelines in the mountains of Marin. Below Napa, Solano County is home to the second oldest AVA in California, after Napa Valley. A landscape for wine tasting as much as farm visits, and olive oil production, it is also home to the first Six Flags Amusement Park in California.

North Coast

Established: 1983

Includes 6 counties: Mendocino, Lake, Sonoma, Napa, Marin, Solano

Main varieties: incredibly varied

The North Coast was recognized as an official appellation in 1983, two years after Napa Valley. It encompasses more than 3 million acres (1.2 million hectares) of land, around 65,000 acres (26,305 hectares) of which are planted to vines. Vines first arrived in the North Coast with kayakers following the coastline from the far western reaches of Alaska. Unangan, Sugpiaq, and Tlingit people from the islands along Alaska's south, guided Russian colonists seeking to occupy a warmer locale to grow resources. In the early 1800s, they planted *vinifera* vines on the coast of Sonoma near today's Fort Ross (aka Russ, as in Russian). Soon after, in the town of Sonoma, Mexico established its only independent mission. All others were instead built by Spain. The outpost was settled to fight the Russian encroachment towards Mexican territories. Within another fifteen years, Mexico was providing vine cuttings to residents of nearby Napa Valley, and eventually proliferated them through the reaches of northern California. In the mid-1800s, the town of Sonoma also became one of two regions founding fine wine in California. The other was Santa Clara.

Sonoma's first vines

The first vines in California's North Coast were planted by Russians in the early 1800s. Threatened by the advance of its western counterparts, Mexico built a lone mission near the town of Sonoma, where they also planted vines.

Until 1844, General Mariano Guadalupe Vallejo led a garrison of troops under Mexico in the town of Sonoma. Just two years later, American settlers in Mexican territory launched an uprising. They arrested General Vallejo and took down Mexico's flag, replacing it with the Bear Flag of the California Republic. California was independent of outside countries for 25 days. Then American troops came in and claimed all of California as American land. The US flag, then with 18 stars, was lifted in the Bear Flag's place.

General Vallejo remained in Sonoma, the town he founded. He became a member of California's first Constitutional government, serving as part of the territorial and then state legislature. The series of changes in nation-state did not dispel Vallejo's love of wine. Vallejo was known for using vines planted for the Sonoma mission to make his own wine. He also gave cuttings to those willing to grow them in other areas to increase the availability of wine. Most significantly, Vallejo's vineyard provided the first *vinifera* vines for the next door Napa Valley.

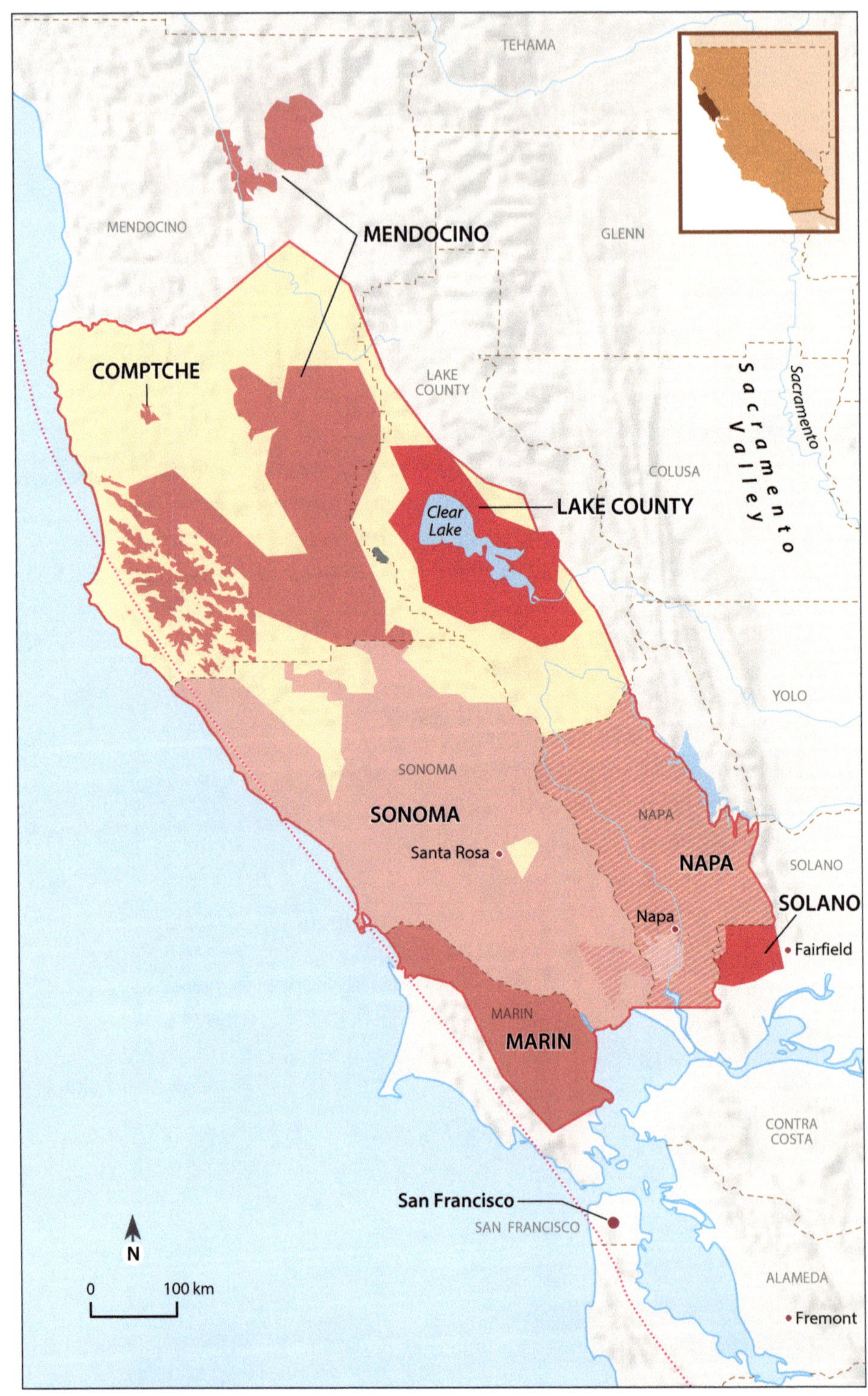

Map 3: Major wine regions of the North Coast

Three of California's legendary winemakers who have ushered important change into the state. From left: Cathy Corison was one of the first women winemakers in Napa Valley and today is known for her Corison Cabernet; Theodora Lee was one of the first Black women to own her own vineyard in the United States and found a winery, Theopolis Vineyards; (me); Carole Meredith, aka. Scientist Legend, through her research lab at UC Davis helped solve how to determine genetic parentage of grapevines

The state's first viticultural movement centered around the Los Angeles area. But with the arrival of new European immigrants from central Europe, and the encouragement of resettling spurred by the gold rush, vines moved to the north. Quickly, it was Zinfandel (then of unknown heritage) that dominated planting instead of Spain's original Mission grape. Then vines spread east to Napa and north to Mendocino. At the same time, they surrounded the gold fields of the Sierra Foothills to the east and filled in the lowlands between the gold fields and San Francisco. The economic and viticultural center had moved to the north.

The North Coast is riddled with mountain ranges, river cut canyons, wind-blown valleys, and redwoods, with a rainforest on one side and arid conditions on the other. Its western boundary greets one of the cold ocean currents pushing south from Alaska. In the summer, parts of the coast have maritime fog so thick walking through it leaves a person wet, as if it had rained. The coastal mountains, the Mayacamas, the Vaca Range, and several configurations of volcanic hills and mountains shield the inland side from ocean influence. The volcanic plateau of Lake County is warm and dry, moderated by the influence of Clear Lake. The eastern side of Napa County, over the Vaca Range, is all chaparral and oak woodlands, plants adapted to survive with less water. To the south, the San Francisco and San Pablo Bays provide a cooling influence.

The North Coast can thank activity along the San Andreas fault for its rugged terrain. The slippage of the Pacific plate along the edge of the North American created the fault, its mountains, volcanoes, and earthquake activity. It also generated a wealthy mix of geological parent material (read: bedrock not brimstone), eroded rocks, and the weathered soils that come from it. Uplifted seabed offers sandstone and siltstone. Volcanic stretches reveal basalt and compressed ash. Areas under the most pressure include serpentine, gneiss, and blue schist. Serpentine presents difficulties for vine growth.

The western rim of the North Coast, along the ocean and into its coastal mountains, is the coolest part of the region. The northern half of Sonoma up into Mendocino even has rainforest that continues all the way north into Canada. In the cool of mornings, the needles of redwoods gather moisture. It drips to the ground over the course of the day, feeding surrounding soils with extra water. Growers near their boughs swear the redwood drippings help their vines. The further east

Morning fog fills a Cabernet Sauvignon vineyard in the North Coast of California

one ventures in the North Coast, the more water becomes an issue. Wineries and growers across the region are learning to capture winter rains in outdoor reservoirs and indoor tanks. The practice helps ensure vineyards have what's needed in summer heat spells.

California's Mediterranean climate dominates the North Coast. No rain in summer means golden grasses on rolling hills, desert shrubs, and enormous, slow-growing trees. The landscape seemed familiar to the first waves of Italian immigrants in the late 1800s, who settled into the area and created a community in northern Sonoma now considered a historic landmark. The success of the venture drew more Italian Americans to the area, many planting vineyards in Napa, Sonoma, and Mendocino. As California became part of the United States, vines were also ushered into the area by German, Austrian, and Hungarian immigrants. When Prohibition ended in 1933, Italian families were among the first to restart wine in America.

Even given all its acclaim and success, much of the North Coast can also seem rugged and remote. The feeling is strongest in the forests of Mendocino – though getting lost in the winding roads of Sonoma mountains is not recommended. Many of the wine growers of these areas moved there precisely because of the remoteness and are there to be left alone. When it's hard to see your neighbors through the trees, it's also hard to be influenced by their decisions. That sense of independence is strongest in Mendocino.

MENDOCINO

Known for: Anderson Valley, Pinot Noir, aromatic whites, sparkling wines, redwoods
Lesser-known strengths: old vines, organic and biodynamic farming, hearty reds
Leader in: organic, biodynamic, sustainable, and fish-friendly farming
First peoples: Pomo, Micewal, Ukomno'm and Ukohtontilka
First vines: 1850
First AVA: 1983, Anderson Valley
Number of AVAs: 13
Planted vineyard area (2024): 17,470 acres (7,070 hectares)

Every direction into Mendocino includes a winding road. The coastal mountains on its Pacific Ocean boundary form a series of ecological terraces complete with steps of unique vegetation, soils, and microclimates. It's also one of the most rugged coastlines. A long portion of it is known as the Lost Coast, named for its isolation and because it is difficult to follow. On Mendocino's eastern boundary, the Mayacamas carve the border. From there, they continue south, reaching more famously into the boundary between Napa and Sonoma.

Until the 1960s, Mendocino was mostly regarded as too cold for grapes. Its distance from state population centers stunted its growth as a region. It remains today a three-hour drive from San Francisco. For a century, the county instead grew timber. Attention for the area grew when, in 1960, UC Davis researcher A. J. Winkler (see box, p. 104), recommended Anderson Valley as a smart choice for vines. It would be, he reported, an ideal place for cooler climate wines: Pinot Noir, Chardonnay, maybe Riesling. Slowly, vineyard plantings expanded in the area, supported by a 1990s boom, and then again by the post-*Sideways* rise of Pinot Noir in America. The inland sides of Mendocino were already planted. Warmer, less humid than the west, the higher elevation steps of the eastern side welcomed head-trained vineyards as early as the mid-1800s.

Micewal, Ukomno'm and Ukohtontilka (named by outsiders the Wappo and Yuki) resisted outside settlement until the influx of people from the gold rush overwhelmed the area. The Pomo on the county's ocean side peaceably kept settlers out until the 1850s. The county of Mendocino was circumscribed with statehood. California was rushed past territorial status as soon as gold was discovered. County seats for representative power were named as soon as the state was.

Those who failed at the gold rush moved into Mendocino and launched a new rush for timber instead. The old growths of the region commanded top dollar for the American logging industry. In the 1990s, timber harvesting changed as objections against logging gained media attention. Activists trying to save the redwoods began protesting. Their actions were ineffective until one day in December 1997 a woman named Julia Butterfly Hill climbed a tree she called Luna and didn't come down again for 738 days. Supporters sent up food and supplies with rope pulleys while the logging company sent helicopters and fire hoses to harass her. Eventually, Hill's more than two-year commitment brought national media attention to the value of old growth trees and redwoods in general. The logging company agreed to preserve parts of the redwoods.

California's ecological farming epicenter

One-third of all organic grapes grown in the state of California come from vineyards in Mendocino County. One quarter of the county's vineyards are certified organic. It also celebrates ten times the amount of biodynamically certified acreage than any other region in the state. In 1980, its Frey vineyard became the first certified organic winery in the United States. In 1996, it became its first biodynamically certified.

The ecological commitments of growers in Mendocino extend beyond organic and biodynamic farming. Parducci was the first winery in the United States to be certified carbon neutral. Bonterra is the largest winery to achieve Regenerative Organic Certification (ROC), and is also TRUE Waste certified, successfully diverting more than 98 percent of its waste from incineration or landfill. Philo Ridge was the first 100 percent off-grid, solar-powered winery in the country, and has one of the largest solar arrays in the wine industry. The county also has the greatest number of acres dedicated to Fish Friendly Farming.

Mendocino has long been considered the northern boundary of wine growing for California. In the new millennium, stalwart ventures brought vines to higher elevations near Mount Shasta, further north. French alpine grapes have done well there, and growers continue to experiment with other varieties. Even so, the total plantings in California's far north are sparse enough that Mendocino can still be called the northernmost established region.

Parducci has been the longest operating winery of Mendocino. Started in 1932 as the announcement was made that Prohibition would be repealed, it is the only winery from that time to have survived. In 2024, it was sold to a Central Coast company, WarRoom Cellars, which continues to operate it today. Fetzer and Frey have historically been the region's most influential producers. The two families ushered in a new era of organic and regenerative farming that made Mendocino the epicenter of such efforts in the state of California. Then Fetzer expanded; renamed Bonterra it became the largest grower of organically farmed wines in the country.

But sparkling wine also drew attention to Mendocino. In the early 1980s, both Scharffenberger and Champagne's Roederer established themselves in Anderson Valley. Reliant on Pinot Noir and Chardonnay, the two houses demonstrated how chiseled and delicious California sparkling wine could be.

Fetzer and Frey

One of the greatest influences in Mendocino wine has come from the Fetzer family, as well as the Freys. Lumberman Bernard Fetzer moved with his wife and family of 11 children from Oregon to the forests of Mendocino in the late 1950s. As the timber industry shifted through the next decade, he and his family began to plant vines.

By 1980, organic influences had taken hold in Mendocino. In the 1970s, master gardener, and one of California's first practitioners of biodynamics Alan Chadwick started a garden school in the Mendocino town of Covelo. Some of his apprentices followed. Katrina and Jonathan Frey as well as Alan York were among them. In 1980, the Freys founded their eponymous vineyards and winery in Redwood Valley, not far from Fetzer. Frey Vineyards would become the nation's first organic and biodynamic winery.

Within a few years, York and the Freys also persuaded Fetzer to change its approach to farming. In the 1980s, Fetzer converted fully to organic farming. In 1999, the winery became the first in the country to rely entirely on renewable power, building the industry's largest solar array. It began measuring its greenhouse gas emissions in the late 1990s and reporting them to The Climate Registry in 2005. Fetzer became the first winery to ever achieve the TRUE Zero Waste certification in 2014 and has since earned the Regenerative Organic Certification, which former Fetzer winemaker and president Paul Dolan helped found.

One of California's first biodynamic farming consultants, Alan York, learned from master gardener Alan Chadwick, then helped convert both Fetzer and Frey to achieve Demeter certifications. While working with Fetzer, York befriended Dolan, mentoring him in organic and biodynamic viticulture. Dolan went on to help build Fetzer, and then Bonterra, into the largest organic winery in the state, traveling the country to promote the health benefits of organic farming, food, and wine.

Dolan and his brother-in-law Jim Fetzer created the Fetzer Food and Wine Center in 1984, in partnership with organic gardener Michael Malthus. It became a destination for foodies and chefs worldwide. Julia Child and James Beard both taught classes at the center while using Fetzer wines.

Before them, it was Donald Edmeades who first followed Winkler's advice and planted in Anderson Valley along the banks of the Navarro River. His success demonstrated the growing potential for the region and led to the growth of the first family wineries in the area; important among them were Husch, Handley, Lazy Creek, Greenwood Ridge, and Navarro. Wineries such as Littorai, Radio-Coteau, and William Selyem brought further attention to the quality of Anderson Valley's local wineries. The success of their single vineyard Pinots from the area brought acclaim. While best known for Pinot Noir, the Anderson Valley has also begun showcasing Chardonnay, and, uniquely, it has always done well with Gewurztraminer and Riesling in smaller quantities.

Outside interest has steadily brought attention to the old growth vineyards on Mendocino's inland side too. Jolie-Laide, Martha Stouman, Dirty & Rowdy previously and now Extradimensional Wine Co. Yeah!, Calder, and Pax have all made exciting wines from old vines in this part of the region. In the west, Turley makes old vine Zinfandel from Mendocino Ridge.

Small, local wineries continue to reveal themselves. Drew from Mendocino Ridge, and Baxter from Anderson Valley have both demonstrated their resolve and quality over the years. Near Ukiah, Rootdown has made a home in the second smallest AVA in America, Cole Ranch.

DuPratt vineyard on Mendocino Ridge was planted in 1919 where vines planted on their own roots persist without the influence of phylloxera. The humidity of the area has allowed the farm to form its own miniature eco-system. The vineyard is surrounded by redwoods

Rootdown started with a focus on Trousseau Noir, and now Cole Ranch is steadily being replanted from Cabernet to grapes of the Jura and French Alps.

Near the south of Mendocino, stands the Mendocino AVA. It appears in the shape of a letter V, tipped slightly to the west. Its western arm includes most of the coastal AVAs of Mendocino: Anderson Valley, Cole Ranch, Yorkville Highlands and small portions of Mendocino Ridge. North of and outside the Mendocino V but still within the county, sits Comptche. Unusually, the Comptche AVA was approved in 2024 and considered so distinctive from its surroundings that it falls within the geographical boundaries of the North Coast but is excluded from its legal boundaries. Wines made from within Comptche cannot be labeled a North Coast wine. Along the eastern arm of Mendocino's V grow the McDowell Valley, Potter Valley, Redwood Valley, Dos Rios, Covelo, and Eagle Peak of Mendocino County AVAs. At the very bottom of the county, Pine Mountain-Cloverdale Peak AVA also continues into northern Sonoma. These areas are all higher elevation plateaus or mountain tops, as well as warmer and drier than the west. A plethora of old vine plantings, especially of red varieties, can be found in the east, but a mid-1900s era vineyard of Chenin Blanc can also be found on Mendocino's eastern side.

On Mendocino's western side stands the county's highest peak. Anthony Peak reaches 6,959 feet (2,121 meters), a testament to the ruggedness of Mendocino's coastal terrain. Further south, the Mendocino Ridge AVA only takes in vines grown at 1,200 feet (366 meters) and above. It's the only non-contiguous AVA in the United States, circumscribing not a complete area but instead mountain islands within it. Potter Valley is the highest viticultural region in the east, starting at 948 feet (289 meters), 200 feet (61 meters) higher than surrounding areas.

Mendocino County AVAs

Mendocino (1984): reaches both the coastal and inland side of Mendocino
Main varieties: varied

Anderson Valley AVA (1983): the best-known and planted AVA in Mendocino
Main varieties: Pinot Noir

Cole Ranch (1983): was the smallest AVA in the US until 2021
Main varieties: Riesling, Cabernet Sauvignon

Mendocino Ridge (1997): the only non-contiguous AVA in the United States
Main varieties: Pinot Noir, Zinfandel, Syrah

Yorkville Highlands (1998): connected to but warmer than Anderson Valley
Main varieties: Cabernet Sauvignon, Syrah

Comptche (2024): excluded from the North Coast appellation
Main varieties: Pinot Noir

McDowell Valley (1981): inland Mendocino
Main varieties: Zinfandel, Grenache, Syrah

Potter Valley (1983): unique cooler, elevated section of inland Mendocino
Main varieties: Sauvignon Blanc, Riesling, Chardonnay, Pinot Noir

Redwood Valley (1996): the area where Mendocino's first vineyards were planted
Main varieties: Zinfandel, Petite Sirah

Dos Rios (2005): all terraced slopes for vineyards
Main varieties: Cabernet Sauvignon

Covelo (2006): has four distinct seasons, but no rain in summer
Main varieties: Tempranillo, Grenache, Mencia

Eagle Peak of Mendocino County (2014): much cooler than nearby Redwood Valley
Main varieties: Pinot Noir, Chardonnay

Pine Mountain-Cloverdale Peak (2011): in both Mendocino and Sonoma Counties
Main varieties: Cabernet Sauvignon, Malbec, Sauvignon Blanc

Wineries

Baxter

Philo

www.baxterwinery.com

Phillip and Claire Baxter live with their two children on a rural ridgeline in this idyllic example of an Anderson Valley family-owned winery. Here they harvest and make long-aging Pinot Noir, sometimes including whole cluster fermentation for an additional layer of savor and aromatics. Phillip was inspired by his winemaker father and learned from both studies at UC Davis and internships in Burgundy. He and Claire met in France, though she comes from England, and run Baxter together.

Drew

Elk

www.drewwines.com

The epitome of reclusive Mendocino winemakers, Jason and Molly Drew live tucked amongst the redwoods of Mendocino Ridge, growing heirloom apples and making wine. Beyond their own vineyard, the Drews also source fruit from sites they admire along Mendocino Ridge and in Anderson Valley. They're known for their Pinot Noirs and Chardonnays, which are among the best in California, but if you can find their Syrah, bring home a case.

Roederer

Philo

roedererestate.com/

The California outpost of Champagne's Louis Roederer, Mendocino's Roederer is considered one of the top sparkling wines in both California and the country. The company's choice of Mendocino in the early 1980s was a huge hat tip to the growing potential of the region. The Brut multi-vintage cuvée is easily one of the best value and most delicious sparkling wines in the state.

Theopolis Vineyards

Yorkville

www.theopolisvineyards.com

In the early 2000s, when Theodora Lee planted her Mendocino vineyards, she just wanted to own land and grow grapes. It was a way to build a long-term investment and honor the advice of her father to have property. A cold vintage less than 10 years later led to the launch of her winery, Theopolis. When her fruit was considered too lean for the winery that usually bought the grapes, Lee took it and made what is one of the most enticing and aromatic Petite Sirah wines in the state. Lee continues to work as an employment lawyer advising other wineries in California. She also stands as a mentor and inspiration to numerous women winemakers. She is one of the first Black women in the state to own a vineyard and winery (Iris Rideau was the first).

Back-to-the-Land and Amigo Bob Cantisano

The counterculture of California in the 1960s rolled into the back-to-the-land movement of the 1970s. The isolation of Mendocino meant more affordable land and reduced the watchful eye of governments or corporations that younger generations wanted to escape. Communes populated the forests of Mendocino. Such communities depended on shared resources and combined skills. Food had to be grown on site, housing built by hand. The combination inspired some of California's most influential organic farmers.

Amigo Bob Cantisano grew food for a commune where he lived in the late 1960s. He'd learned gardening with his grandmother, updating some methods through the gardening magazine of Rodale Press.

The first Earth Day occurred nationwide in 1970, inspired partially by the anti-pesticide efforts of the United Farm Workers Movement. The event in San Francisco featured a series of workshops and talks as well as rallies centered around environmental efforts, pesticide-free practices, and the like. Cantisano attended. It was there he had an epiphany. He'd learned organic gardening from his grandmother, and he'd devote his life to it. Cantisano became one of the pioneers of organic farming in California and expanded its presence in California wine.

In 1973, he co-founded the California Certified Organic Farmers organization, helping to establish standards in the state. In 1976, he launched Peaceful Valley Farm Supply to make sourcing organic inputs and supplies easier. It continues today. He founded the Ecological Farm Conference (Eco-Farm) in 1980, an annual event that brings together ranchers, farmers, environmental activists, and farmworkers for several days of workshops and idea sharing. It celebrated its fortieth anniversary in 2020. He became one of the first organic agricultural advisors in the country in 1987.

Before his passing in 2020, Cantisano advised orchard and nut growers, dairy and livestock farmers, viticulturists and wineries throughout California. The wineries he advised on organics include Shafer, Tres Sabores, the former Araujo vineyards, Frog's Leap, and Long Meadow Ranch in Napa Valley, Hudson in Carneros, Preston, Cornell, Benzinger, and Slipstone in Sonoma County, larger wineries like Sutter Home and E&J Gallo, and Fetzer of Mendocino.

LAKE COUNTY

Grapes were slow to arrive in Lake County. Its first vines were probably planted in the 1870s, but wider development didn't start for another

100 years. Much of the history of California wine depends on the advent of the railroad within the state. But the train never steamed its way to the area surrounding Clear Lake, and so vineyards didn't expand until after trains were no longer necessary. Pears and orchard fruits formed the region's economy.

With the arrival in Lake County of influential trial attorney Jess Jackson things started to change. Seeking a countryside haven from the exploits of the city, Jackson and his then-wife Jane Kendall bought a pear and walnut orchard near the western shores of Clear Lake, at the center of the region. There, they planted Chardonnay. Soon after, Jackson brought in winemaker Jed Steele, and the two began making wine. It was the start of Kendall-Jackson Vintner's Reserve, still the most sold Chardonnay in the United States. Jackson and Steele's innovation was not in making Chardonnay as much in making a new category of wine. American wine existed in a two-tier model at the time. Highly affordable wines of underwhelming quality sat on one side. Pricier, quality wines on the other. Jackson targeted the middle ground. Vintner's Reserve was one of the first wines of the period to sell for slightly more than the entry level wines, while offering quality closer to those in the top tier. His wine opened an entirely new market and launched a period of California wine known since the 1980s as the fighting varietals.

Fighting varietal wines featured a single grape type, were affordable but not cheap, and taught the nation's consumers what Chardonnay and Cabernet Sauvignon were. They arguably shaped the direction of premium and fine wine in America since. Some argue Jackson's new style of slightly sweet, flavorful, but not huge Chardonnay even influenced some of the comparatively higher volume whites of Burgundy. The fighting varietals phenomenon that took America by storm did at least open the US market to similar wines from other countries, especially Australia.

Known for: hearty reds, Clear Lake, volcanic soils, mineral springs
Lesser-known strengths: Sauvignon Blanc, ultraviolet light intensity, air purity
Leader in: one of the fastest growing wine regions in California
First peoples: Pomo, Micewal, Miwok
First vines: 1854 or 1871
First AVA: Guenoc Valley, 1981
Number of AVAs (2024): 10
Planted vineyard area (2024): 11,307acres (4,576 hectares)

As Jackson succeeded, he began purchasing property in other cooler climate, coastal areas of California. The combination of tropical flavors from the warmer Lake County, steely tones from cold Monterey, and the addition of other sites in between gave distinctive complexity to Kendall-Jackson's cuvées. It became a model he followed with other varieties as well.

In the 1980s, other significant North Coast producers also invested in Lake County vines. Parducci from next door Mendocino, Beringer from Napa Valley, and Louis M. Martini from Napa and Sonoma all made an impact on the region. In the 1990s, famed Napa Valley grape grower Andy Beckstoffer also invested north. Beckstoffer made a name for himself buying the best of Beaulieu Vineyards (BV) along the Mayacamas bench of west Napa Valley. He partnered with winemakers willing to use the vineyard name on their label and created an empire of sites considered among the best of Napa Valley. His move into Lake County was a sign of the region's expansion. Providing grapes to other winemakers made it possible for brands in other regions to explore the unique expression of Lake County fruit. In 2004, Beckstoffer, Beringer, and Kendall-Jackson were among the growers that successfully created the Red Hills AVA, one of the most celebrated in the region.

In the 1990s, Kendall-Jackson (KJ) grew too large to produce in Lake County. They bought a historic property off Highway-101 in northern Sonoma and turned it into more than a production facility. Today, it serves as a tourist destination complete with its own food program, gardens, gift shops, and tasting rooms. Jackson kept the Lake County vineyard. When KJ relocated the winery to Sonoma, Steele remained in Lake County and launched his own winery Steele. Prior to his work with Jackson, Steele had made wine in the historically influential winery Stony Hill of Napa Valley, the first to launch solely with white wines, and then Edmeades in Mendocino. With KJ, his familiarity with white wines helped build Vintner's Reserve to a million-case (12 750-milliliter bottles each) wine. Steele used his experience sourcing from multiple regions for KJ to establish his own brand, anchoring himself in Lake County fruit while combining it with vineyards the entire length of the US west coast, from Washington state to Santa Barbara County. Steele's reputation spurred further interest in Lake County wine and inspired a winemaker ethos of the staunch, independent, persistent, craftsman, a hard worker with vision. In 2021, Shannon Family, a more recently established Lake County winery, purchased Steele wines. They've since made sure all their vineyards are organic.

In addition to local wineries, outside producers have also brought attention to Lake County. In 2010, Arnot Roberts began making a distinctive rosé from Touriga Nacional grown in the region. The wine was so distinctive, the grape variety such a surprise it was like honey for the wine-geek hive. It triggered other niche winemakers to look for lesser-known varieties, arguably helping instigate the Gen X winemaker drive for oddball grapes, as well as California's thirst for rosé.

Lake County appellations mostly encircle its central feature, Clear Lake. It's the largest naturally formed, freshwater lake in California with a surface area of 43,520 acres (17,612 hectares). A plentiful home for fish, reptiles, and bird life, the Lake was well populated by Pomo, Micewal, and Miwok peoples until the arrival of outside settlers in the 1840s. The expansion of the immigrant population in the area led to the creation of the county itself in 1861, taking parts of Napa Valley to the south, and Mendocino to the west to form it.

Two AVAs of Lake County diverge from the circle of Clear Lake. Adjacent to the border with Napa but inside Lake County, Guenoc Valley sits at moderate elevation between 700 and 1,500 feet (213–457 metres) in gravelly, well-draining soils, and is dominated by hearty reds, mainly Cabernet Sauvignon. To the west, Benmore Valley proves the coolest part of Lake County. Historically it grew Chardonnay, but today no vineyards are farmed there.

Clear Lake provides a moderating influence for the daytime heat of the inland Mediterranean climate. The higher elevation of Lake County makes the region drier and warmer than its southern neighbor, Napa Valley. Even with warmer daytime temperatures and abundant sun exposure, it enjoys cool nights, providing unique growing conditions for structural reds.

The region's more northern latitude creates a shorter growing season compared to the North Coast's counties to the south. The effect is more compressed ripening, but often the harvest period avoids other problems. The drier conditions and shorter growing season reduce vine vigor but also the growth of surrounding vegetation. The effect is lower pest pressure on vines, while the drier conditions also lessen disease. The high elevation also means colder temperatures, killing off pests that could otherwise survive the winter. Lake County continues to feel remote. Its volcanic hills and mountains offer distinctive character most especially for red wines. Its Sauvignon Blanc is also worth seeking out. The region served a distinctive purpose. Some of the early trials to find phylloxera-resistant rootstock occurred there. And it's attracted unexpected celebrities.

Actors Jackie Gleason and Jack Haley (Tin Man from the original *The Wizard of Oz*) formed a property investment company there and singer Jimmy Durante lived there briefly in the 1950s. Even earlier, in the late 1800s, British actress Lillie Langtry became an American citizen and started a vineyard and winery in Lake County. She was one of the first women to own land and make wine in the region and used her property to gain independence from her husband and assert her American citizenship. Langtry, the winery she founded, still operates today.

Lake County AVAs

Guenoc Valley (1981): just north of Napa Valley and Mount St. Helena
Main varieties: Cabernet Sauvignon, Cabernet Franc, Zinfandel, Tempranillo

Clear Lake (1984): completely encircles Clear Lake
Main varieties: Cabernet Sauvignon, Sauvignon Blanc

Benmore Valley (1991): outside the Clear Lake AVA, very close to Mendocino
Main varieties: not currently planted; previously Chardonnay

Red Hills (2004): best-known AVA in the region
Main varieties: Cabernet Sauvignon, Cabernet Franc

High Valley (2005): surrounded by mountains and just east of Clear Lake
Main varieties: Cabernet Sauvignon, Syrah, Zinfandel

Big Valley District – Lake County (2013): directly across from High Valley on the west side of Clear Lake
Main varieties: Sauvignon Blanc, Chardonnay, Viognier

Kelsey Bench (2013): directly south of Big Valley with similar climate
Main varieties: Sauvignon Blanc, Chardonnay, Viognier

Upper Lake Valley (2022): on the north side of Clear Lake
Main varieties: Sauvignon Blanc, Cabernet Sauvignon

Long Valley – Lake County (2023): expanded the reach of the North Coast AVA
Main varieties: Cabernet Sauvignon, Cabernet Franc, Petite Sirah

Wineries

Hawk & Horse

Lower Lake

hawkandhorsevineyards.com

One of the wineries of Red Hills, Hawk & Horse offers ageable, finessed Cabernet Sauvignon, and robust, flavorful Petite Sirah. Their reds showcase the distinctive sapidity and terroir of the region's volcanic origins, and savory thrust. The property was established close to the advent of the state of California and houses the largest valley oak tree in the state.

Obsidian Ridge

Kelseyville

obsidianridge.com

Established in both the southern edge of Napa Carneros and the volcanic fields of Lake County, Obsidian Ridge relies on its Lake County location to produce red wines of intensity that retain lift. The high elevation property provides structure combined with fresh acidity, and the savory, palate-stimulating core of the wines brings intrigue. Obsidian Ridge has successfully brought national attention to the region's wines.

Shannon Family

Lakeport

shannonfamilyofwines.com

Farmed organically, Shannon Family vineyards were early adopters of the practices of integrating sheep into the vineyards and exploring regenerative techniques for their soils. Raised on a farm in northern Sonoma, pursuing a life of farming but in a less developed region than Sonoma made sense for Clay Shannon, founder of the venture. Shannon Family makes a range of wines including structured, flavorful reds and juicy, robust whites. They also provide classes and workshops on viticulture and wine.

SONOMA COUNTY

The expansiveness of Sonoma County enables its diverse growing conditions. Butted by the cold current of the Pacific and the uplift of coastal mountains, paralleled by hills, Sonoma Mountain, and the Mayacamas, Sonoma enjoys a range of microclimates. Fog laden and chilled in the west, warmer with lush wines in the north, structured and poised along the east, it has a full range of stylistic potential between. The warmer zones of the county grow flavorful Cabernet and Zinfandel. The cool pockets are mostly Pinot, though some of the warmer areas are as well. Head-trained, old vine vineyards speak to the history of the region and instigated the Historic Vineyard Society effort to increase public awareness of their value. The San Andreas fault cuts through the western stretch of the county, leading to the uplift of old seabed in the coastal mountains, as well as volcanic formations to its east. The result is a cascade of soils brought together in one location: eroded sandstone, siltstone, and some mudstone too, basalt, volcanic tuff, quartz, clay loams, and serpentine, sometimes mixed through one vineyard planting.

Outside California, Sonoma is mainly known for Pinot Noir. The variety didn't show itself and the region's potential until the 1970s, but it expanded in the 1990s, and then boomed in the new millennium after *Sideways* changed American wine and inspired its love of Pinot. Arguably, Sonoma is prime Syrah country. Much of its landscape begs for the variety, and the examples that are grown there are exciting and delicious. But the market wants Pinot, and so Pinot Sonoma has. The challenge for Sonoma at large is its success with varietal and stylistic diversity. The American market, the story goes, wants clearer messaging. It's easier to pass tests for wine certifications or appear intelligent while ordering at a restaurant when a region carries a signature grape and style.

The launch of Sonoma Pinot took place through the western parts of the Russian River Valley on the benches beside its winding waterway. William Selyem released one of the first examples to gain international attention. But before that Joseph Swan had proven its worth, starting in 1968. He also delivered impeccable Zinfandel. His plantings of Pinot Noir established a winemaker favorite heirloom selection of Pinot, the Swan clone. Its small clusters are straggly, and inconsistently set, with berries of all sizes. They bring little color to the wine. But the aromatics are enthralling, the flavor naturally spiced, and the structure hard to forget. A sense of rusticity in fineness makes the best of Swan clone Pinot Noir, and the best from Joseph Swan is something close – to be dramatic – to life changing. Davis Bynum and Dehlinger further opened the way for excellent Sonoma Pinot throughout the 1970s. Then at the end of the decade, William Selyem launched, making Pinot from Rochioli Vineyard on westside road. To use modern parlance, if there had been Instagram, the William Selyem wines would have gone viral. Even Burgundy was impressed.

Known for: Pinot Noir, Buena Vista (California's oldest premium winery), quality of produce, redwoods
Lesser-known strengths: Zinfandel, farmers' markets, natural beauty, quality of life
Leader in: diversity of growing conditions
First peoples: Coast Miwok, Micewal, Pomo
First vines: 1818
First AVA: Sonoma Valley, 1981
Number of AVAs (2024): 19
Planted vineyard area (2024): 59,000 acres (23,876 hectares)

In the late 1970s, Sonoma wine expanded in other ways too. The Bohan family planted the first vines, head-trained Zinfandel and Riesling, on the peaks along Bohan-Dillon Rd in what is now the Fort Ross-Seaview section of the West Sonoma Coast. A couple of years later, in a desire to get back to the land, David Hirsch started planting what has become one of California's most celebrated sites. Hirsch grows Pinot Noir not far from Bohan, surrounded by redwood forest, in view of the Pacific, biodynamically farmed, structured in what many consider one of California's *grand cru* vineyards. Next door, he also helped his neighbors plant the vineyard today known as Hellenthal. The Bohan and Hirsch families helped draw attention to an entirely new area, one better known for sheep farming, isolation, and cannabis at the time. The Martinelli family helped open the road as well. Their vineyard of Chardonnay and Pinot Noir was among the first in the mountains. A little to the east, just beyond the coastal mountains, the Martinelli vineyard also includes some of the oldest vines in the North Coast.

The 1990s brought another influx of wine growers. Flowers was established, even closer to the ocean than Hirsch. Failla was planted with not only Pinot Noir but also Syrah. In 1999, an obscure but exceptional planting of marginal climate Cabernet was even established: Waterhorse Vineyard, planted and farmed by Jesus and Patricia Velasquez, has become a landmark secret for the limits of California Cabernet. Contemporary wisdom tells us Sonoma Mountain is as far west as you can reliably ripen Cabernet. Waterhorse, made brilliantly by John Lockwood of Enfield Wine, demonstrates otherwise. It's a decidedly different style of Cabernet from wines grown further east.

Tegan Passalacqua serves as vineyard manager and head winemaker for Turley, as well as winemaker for his own Sandlands. Passalacqua has traveled throughout California locating old vine, organically farmed vineyards, occasionally finding rare historic varieties as a result. Here, he tours some of these old vine sites with Spanish winemaker José Luis Mateo, who does similar work in Galicia

But Sonoma County was founded on head-trained, field-blend vineyards. While these are most often assumed to be Zinfandel based, many take a different turn. The complexity or core variety of Sonoma's historic vineyards proves instructive about its founder's personality. Some are orderly, clean rows of head-trained vines, set in geometric precision, and the first owner turns out to have behaved accordingly. Others, more haphazard, nevertheless reveal exceptional fruit and resilient vines. And others include still a heart of Zinfandel, counterbalanced with a good dose of Carignan and Alicante Bouschet, but then dotted through with more other varieties than easily imagined. There, it turns out, the owner liked to meet everybody who moved to the region. Every new variety, a new friend made.

The legacy of the Italian-Swiss Colony

In 1881, in northern Sonoma, and eventually also southern Mendocino, a group of Italian and Italian-Swiss immigrants established themselves. Together, they created a community meant to provide livelihood for its residents.

Italian immigrant Andrea Sbarboro sought to build a money-making enterprise to bring together other new Italian-Americans. Drawing on the farming experience shared by many of the new residents, the Italian-Swiss Colony was founded to grow wine grapes. Within a few years, the volatility of grape prices led to the building of a winery for the colony to also make its own wine to sell throughout the country.

In its early years, the colony operated in a cooperative model, with individual growers selling fruit to the larger winery. As the winery became more successful, it was eventually sold to outside corporations. Prior to Prohibition, the Italian-Swiss Colony was the largest winery in California. After Prohibition, the winery became a mass-market brand and one of the first in the nation to expand advertising campaigns beyond print into television. By the 1980s, the company had new ownership and a new brand name.

The colony holds a powerful legacy in the history of the North Coast. It created a unique destination for Italian immigrants to return to farming, build business, and share heritage through community. Its presence drew Italian-Americans into Northern California, even when not becoming part of the venture. And following Prohibition, members of the Italian winegrowing community became leaders in the North Coast wine industry. Many continue to be today.

In its day, the Italian-Swiss Colony was also a tourist destination. Visitors came from the San Francisco Bay area and beyond to see the production facility and visit what was the largest wine tank in the nation in 1897: 500,000 gallons (1.9 megaliters). It was made of concrete. The facility also included beautiful gardens, and picnic areas providing rest spots for travelers. It brought about the advent of wine tourism in the North Coast.

The original site of the Italian-Swiss Colony in Asti, California has since been recognized as a historic landmark in Sonoma County. Their San Francisco warehouse, built in the early 1900s, has also been made a historic landmark.

These mixed vineyards were the founding of Sonoma County. They appear throughout the Russian River area, Sonoma Valley, the slopes of the Mayacamas, and especially along the benchlands of Dry Creek in northern Sonoma. There, Zinfandel develops a distinctive spice note with dark fruit, restraint, and acidity that's become the hallmark of the region. But as you get to know the locals, you learn Dry Creek Valley has far more nuance than that. Four distinct areas grow four variations on the theme. Here, in Dry Creek, just outside Healdsburg off the Geyserville/Lytton Springs exit from Highway-101, sits the best taco truck in Sonoma County. Los Plebes is worth a stop for a few tacos of your favorite meat and the grilled onions. It's especially worth the drive since only a minute away, Ridge Lytton Springs tells another part of California history. At the northernmost part of Dry Creek Valley, along the edges of the reservoir known as Lake Sonoma, rises the Rockpile AVA. No wineries exist here, but a plethora of vineyards showcase the distinctive microclimate of this nook of Sonoma. At higher elevation than the valley floor of Dry Creek, Rockpile temperatures are more moderate and even, cooler during the day, slightly warmer at night. The classic Zinfandel grows here, but so does some Cabernet and a host of Portuguese varieties better known for Port.

The Historic Vineyard Society

Witnessing the historic head-trained vineyards of Sonoma get regularly replaced by more lucrative plantings of Pinot Noir, several North Coast producers got together and founded the Historic Vineyard Society (HVS). The non-profit organization works to document old vineyards, including as much as possible, their original planting date and founders as well as their varietal collection. The registry of these vineyards can be found on the HVS website. The HVS also strives to bring broader attention to the value of older vineyards for quality as well as for historical cultural reasons.

The HVS is led by Bob Biale of Robert Biale in Napa, David Gates of Ridge in the Santa Cruz Mountains and Sonoma, Mike Officer of Carlisle, Morgan Twain-Peterson of Bedrock, Joel Peterson formerly of Ravenswood and now of Once & Future, each in Sonoma, Tegan Passalacqua of Turley and Sandlands, as well as Emily Rasmussen, Ryan Brennan, and Larry Piggins. As part of their efforts, the HVS provides education to the wine industry, trade, and consumers on California's historic plantings, and hosts dinners in some of these vineyards featuring wines made from the site, including older vintages.

In 2024, the General Assembly for the International Organization of Vine & Wine (OIV) officially recognized and defined both old vines and old vineyards. The EU is required to factor OIV recommendations into future policies or rule changes. The US is not a member of the OIV so the recognition does not directly influence US regulations on agriculture for wine.

Joel Peterson, founder of Ravenswood winery and now of the winery, Once & Future, helped bring acclaim to Zinfandel of California, and co-founded the Historic Vineyard Society

To the east of Dry Creek Valley, Alexander Valley starts. Arguably the hottest part of Sonoma County, it is known for bold, brooding Cabernet Sauvignon. But on its cooler, western side, a little northeast of Los Plebes, one of California's most important historic vineyards grows. Geyserville was made famous by the bottling from Ridge, one of the best aged wines in California. Even forty-year-old vintages can be beautiful. They have occasionally been my best wine of the year. Planted in 1882, the 'Old Patch' of Geyserville still grows. A field blend of Zinfandel, Mataro, Petite Sirah, Carignan, and Alicante Bouschet, the vines are lower to the ground and more straggled looking than some head-trained vineyards of the area, another glimpse of its founder. In 1891, near the old patch, stands a block named "old Carignan." It became the basis for new insight. Carignan does well in heat, and it keeps its acidity. Older

The Alexander Valley includes vineyards planted up the slopes of Alexander Mountain as seen here

vineyards in warmer places consistently include it as a complement to Zinfandel to help keep a wine's backbone as the Zinfandel ripens its flavor. As Ridge planted around Old Patch and old Carignan, the bulk of their vines have been Carignan.

Sonoma and Santa Clara Counties together represent the dual birthplaces of fine wine in California. Old Almaden, the first winery in the state, established in 1852, launched in the Almaden Valley of Santa Clara County. New *vinifera* cultivars were established in its vineyard, bringing a new possibility for fine wine to California. A few years later, in 1857, Agoston Haraszthy opened Buena Vista Vineyards in the town of Sonoma, north of San Pablo Bay, far south of Dry Creek Valley or Geyserville. Though Haraszthy is credited with a wealth of first things for fine wine in America, his exploits are more limited than the claims. He did bring new grapevines from Europe. He did not bring Zinfandel. While he launched Buena Vista, he did not continue it; he was fired from the venture a mere three years later. But he did travel the world promoting California wine. And Buena Vista continues today in the foothills outside the town of Sonoma. The cooling influence of San Pablo Bay, and the protective shield of surrounding forest and hillsides, made Buena Vista an ideal place for growing fine wine. Enough warmth to get the flavor, enough protection to freshen the fruit. Today, Jean-Charles Boisset, of the French vigneron Boisset family, leads Buena Vista. He's invested not only in restoring the historic vineyards and caves, but also in capturing its history. In partnership with Boisset, writer–historian Charles Sullivan wrote one of Sonoma Valley's best wine histories, including the trajectory of Buena Vista. Boisset also rebooted DeLoach as a biodynamic vineyard and garden in Sonoma County. In Napa he also owns Raymond, JCB, and Elizabeth Spencer. More recently he also partnered with musician John Legend to make Legend's wines, LVE.

One of Sonoma County's hallmark developments has been recognition of the Petaluma Gap AVA. Named for a low spot in the coastal mountains in the west, the Petaluma Gap is the first AVA in the country to be delineated by wind. Full of fog, and with winds surpassing 8 miles per hour (13 kilometers per hour), the area gives rise to wines that are decidedly structural, textured, and savory. Dominated mainly by wineries elsewhere in Sonoma that buy fruit from the region, Petaluma Gap has gained acclaim partially due to the sterling efforts of Keller Estate. One of the few wineries based in the Petaluma Gap and growing its own fruit, Keller also sells fruit to other wineries bringing outside experience and recognition to the region.

Gallo has also played an important role in Sonoma County. Known historically for its affordable wines, Hearty Burgundy being one of its great successes, Gallo began building a fine wine portfolio as early as the 1970s, when it invested in Frei Ranch in Dry Creek Valley. Over time Gallo created vineyards in cooler parts of the Russian River Valley. Today the brand also owns the vineyard purchased and further developed in the 1930s by Louis M. Martini, Monte Rosso, at the top of Moon Mountain. It's one of the region's iconic sites, first planted in the 1880s. Perched atop the ridgeline, looking south to the Bay, the vineyard includes some of California's older Zinfandel vines, and the oldest Semillon vines in North America. A host of producers have showcased the character of Monte Rosso over the years: Bedrock, Carlisle, Louis M. Martini, Ravenswood, LaRue, MacRostie, Robert Biale, Lasseter, Repris, Ridge, Rockwall Wine Co., Turley, Mount Peak, and Orin Swift have all made distinctive and compelling wines from Monte Rosso. Those made by Louis M. Martini were among the first after Prohibition to demonstrate the profound ageworthiness of California's mountain sites, and arguably some of the first long-aging Cabernet fine wine in the state.

Alan York and the rise of biodynamic farming

California has been a leader in organic, biodynamic, and regenerative farming for decades. The origins of such practices can be traced back to the founding influence of Englishman Alan Chadwick, who immigrated to the United States having already studied organic and biodynamic principles. Chadwick first created the experimental farm and garden at UC Santa Cruz, then established Green Gulch Farm in Marin, Covelo Village in Mendocino, and others in Napa, Sonoma, Monterey, and Saratoga.

Chadwick served as a mentor to Alan York, who went on to become a leading biodynamic practitioner in both North and South America. He adapted what he learned from Chadwick into practical techniques applicable to vineyards. In this way, he helped develop the tangible side of biodynamic farming regardless of any spiritual beliefs a practitioner may or may not have. York's influence in California wine growing is hard to overestimate. He helped convert both Fetzer and Frey in Mendocino to organic and biodynamic certifications. While doing so, he befriended and mentored Paul Dolan. Dolan helped build Fetzer, now named Bonterra, into the largest organically farmed wine brand in the United States. Dolan also helped create the foundation for the California Sustainable Winegrowing Alliance, which delivers guidelines and practices to help growers enact their sustainability commitments. More recently, Dolan founded the Regenerative Organic Alliance, which developed the Regenerative Organic Certification used not just by wine businesses but across a range of industries. York helped producers like Fetzer and Dolan recognize that regenerative practices like those found in biodynamics could be done at larger scale. They then went on to demonstrate it.

In California, York helped convert some of the first biodynamic vineyards in the state, going on to do so in Oregon, Chile, Argentina, and Israel as well. While doing so, York mentored other vignerons who have in their turn also mentored others. He also helped convert vineyards in Europe, including one owned by Trudie Styler and her husband Sting in Tuscany. Besides Frey and Fetzer, York also worked with Benzinger in Sonoma, Quivera of Dry Creek Valley, Robert Sinskey of Napa, Dark Horse Ranch and Ceago in Mendocino, amongst others. He also mentored Rodrigo Soto, who has worked with and converted vineyards to biodynamic farming in Chile, Sonoma, and Napa Valley. And he worked with Pedro Parra earlier in Parra's career as an internationally known Terroir Consultant. York was also the president of the Biodynamic Farm & Gardening Association of North America in the 1990s.

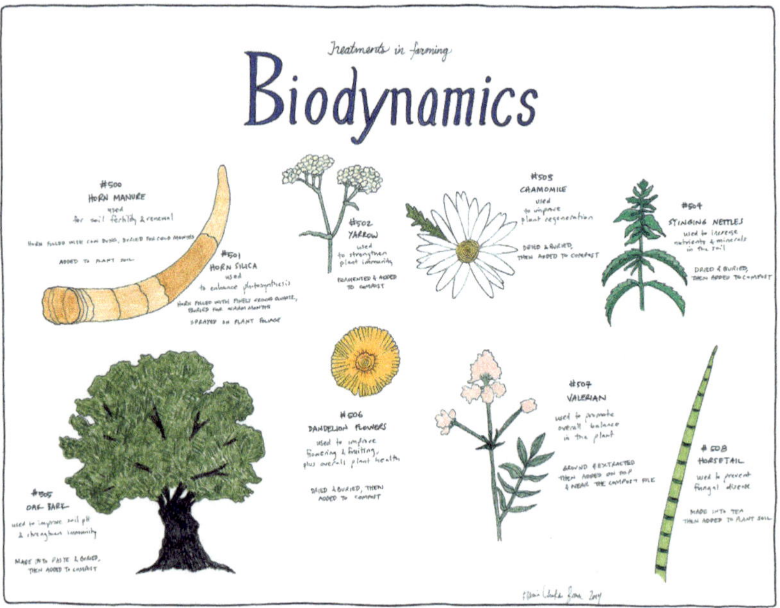

Biodynamic farming includes reliance on multiple treatments prepared over the year and used to strengthen the health of the soils and vines while farming. Here, treatment 500: horn manure; 501: horn silica; 502: yarrow; 503: chamomile; 504: stinging nettles; 505: oak bark; 506: dandelion; 507: valerian; 508: horsetail

Further downslope, Hanzell launched in the 1950s. Founding winemaker Brad Webb brought multiple innovations to winemaking in California. He was among the first to create temperature-controlled tanks, get to know winemaking with Burgundy barrels, and use controlled malolactic fermentation in red wines. Ten years later, each of these practices was adopted and made famous by Robert Mondavi. The vineyard of Hanzell faces the cooling influence of the Bay and continues to grow the oldest Pinot Noir and Chardonnay in the state.

The Carneros region reaches across the southern parts of both Napa and Sonoma Counties, just north of San Pablo Bay. The cooler influence of the bay brings natural acidity, while the clay-ridden soils give power. Pinot Noir and Chardonnay were first planted in Carneros; Louis M. Martini established them there in the 1940s. Plantings of Chardonnay slowly grew in the 1970s. By 1973, there were still only small quantities planted throughout the state. When Mike Grgich decided to make some for Chateau Montelena in the early 1970s, he had to source grapes from three different vineyards in order to get enough fruit. The Bacigalupi family in Sonoma provided around 40 percent for the 1973 Chardonnay that scored highest in the 1976 Paris Tasting. The reputation of California white wine was built through a combination of Napa and Sonoma County vineyards.

In 1981, Sonoma Valley became the county's first AVA. More quickly followed. Knight's Valley, Chalk Hill, Russian River Valley, Green Valley of Russian River Valley, Dry Creek Valley, and Los Carneros (including the Napa Valley side) were all established in 1983. By 1989, wine grapes were Sonoma County's top agricultural crop, outstripping dairy and apples. Though its fame started on its western side, the complexity of the Russian River Valley (RRV) is well established. Vintners have worked together to identify unique growing areas of RRV, identifying what they call six neighborhoods offering distinctive character with Russian River charm. Green Valley of Russian River Valley on its far western side is its coolest. Chalk Hill on its east, the warmest.

Mount St. Helena on Sonoma County's eastern side is its highest peak, reaching 4,342 feet (1,323 meters). Pine Mountain-Cloverdale Peak is its highest AVA. Atop Pine Mountain, the appellation reaches 1,600–2,600 feet (488–792 meters). It reaches north into Mendocino County as well. A little further south, Rockpile reaches 1,900 feet (579 meters), higher than Dry Creek Valley, and thus also more moderate and even in temperature. The far western side of Sonoma in the Coastal Mountains, Fort Ross-Seaview features vines growing from 900 to 1,200 feet (274–366 meters), while Moon Mountain in the southeast stretches to 2,200 feet (671 meters).

Sonoma County helped foster the rise of organic and biodynamic viticulture. Phil Coturri got started on Sonoma Mountain and in the 1970s his became one of the first vineyard management companies to offer organic, mountain-specific viticulture. Down the road on the western side of Sonoma Mountain, Bob Cannard farmed organic gardens as well as Cline Vineyards. In Sonoma Valley, Benzinger worked with Alan York to farm biodynamically. Phil Coturri has farmed vineyards throughout Napa and Sonoma Counties. Laurel Glen on Sonoma Mountain, Kamen on Moon Mountain, Old Hill Ranch in Sonoma Valley, Mayacamas and Oakville Ranch in Napa Valley. He has also overseen vineyards for actor Danny Glover, Pixar maven John Lasseter, musician Boz Scaggs, and comedian Tommy Smothers. When he's not farming he's usually stage side, the lone audience member watching the Grateful Dead rehearse. Coturri and York have both influenced other winemakers and viticulturists too. York worked with Rodrigo Soto at Benzinger, then at Matetic in Chile. Soto then returned to California and led Quintessa, another biodynamic vineyard in Napa Valley, and brought in terroir specialist Pedro Parra, and geologist Brenna Quigley. Both work with vineyards, especially biodynamic ones, worldwide. Both have worked with Hamel on Moon Mountain, and Flowers on the West Sonoma Coast. Soto now has his own Far Mountain wines, and partners with Coturri for fruit sourcing.

Recently, Sonoma has begun to expand its reputation through the rise of Millennial winemakers working with lesser-known varieties and experimental methods. Most unusually, Matt Niess of North American Press has revitalized an early 1900s-era French–American hybrid with characteristics akin to Pinot Noir, Baco Noir. The effort is unique in that his vineyard is found in prime Pinot country but surrounded by the threat of powdery mildew – the Baco requires no mildew treatments. Niess's example is pointing to new considerations amid climate change. Other vintners demonstrate other possibilities. Enfield Wine Co. makes classic style wines with a lighter touch, mouth-watering, refined, and beautiful, by working with vineyards exceptionally farmed in marginal areas. Hardy Wallace, who made a name for himself through the brand Dirty & Rowdy, has moved on to a new project partnering with his wife, Kate Graham, through the brand Extradimensional Wine Co. Yeah! What's different about the project is that instead of making varietally specific wines, they try to make the best wine they can from whatever vineyard they work with, then create blends from varieties less typically found together, with a focus on verve and flavor. It's a unique approach that offers greater flexibility during climate change and economic crisis. Vignerons lose vineyards to billionaire investors, wildfire, low yields, amongst a range of factors. Creating blends less rigidly dependent on specific vineyards or varieties gives winemakers room to pivot through change.

Sonoma County AVAs

Sonoma Valley (1981): the birthplace of fine wine in California
Main varieties: Cabernet Sauvignon, Chardonnay, Zinfandel

Dry Creek Valley (1983): a historic region for old-vine Zinfandel
Main varieties: Zinfandel, Cabernet Sauvignon, Sauvignon Blanc

Knights Valley (1983): hidden in the northeastern corner of Sonoma, west of Napa Valley
Main varieties: Cabernet Sauvignon, Syrah

Chalk Hill (1983): one of the historic homes of Chardonnay
Main varieties: Chardonnay, Cabernet Sauvignon

Russian River Valley (1983): the first area to demonstrate Sonoma's potential for Pinot
Main varieties: Chardonnay, Pinot Noir, Zinfandel

Green Valley of Russian River Valley (1983): the coldest section of the Russian River Valley
Main varieties: Chardonnay, Pinot Noir

Alexander Valley (1984): one of the warmest parts of Sonoma County
Main varieties: Cabernet Sauvignon, Chardonnay, Zinfandel

Sonoma Mountain (1985): the birthplace of organic mountain viticulture, and where the rise of Cabernet began
Main varieties: Cabernet Sauvignon, Chardonnay, Pinot Noir

Sonoma Coast (1987): an enormous AVA that covers the expanse of ocean fog influence
Main varieties: Chardonnay, Pinot Noir

Northern Sonoma (1990): does not include Carneros, Petaluma Gap, or Sonoma Valley
Main varieties: Cabernet Sauvignon

Rockpile (2002): at the north side of Dry Creek Valley and more temperate
Main varieties: Zinfandel, Cabernet Sauvignon

Bennett Valley (2003): hidden at the north side of Sonoma Mountain
Main varieties: Chardonnay, Syrah, Merlot

Fort Ross-Seaview (2012): in the highest part of the West Sonoma Coast
Main varieties: Pinot Noir, Chardonnay

Moon Mountain District of Sonoma County (2013): includes the oldest Pinot Noir and Chardonnay vines in the state
Main varieties: Cabernet Sauvignon, Zinfandel, Pinot Noir, Chardonnay

Fountaingrove District (2015): on the Sonoma side of the Mayacamas mountains
Main varieties: Cabernet Sauvignon, Merlot, Syrah

West Sonoma Coast (2022): encompasses the Coastal Mountain Range of Sonoma
Main varieties: Pinot Noir, Chardonnay, Syrah

Petaluma Gap (2017): defined by wind, overlaps Sonoma and Marin Counties
Main varieties: Pinot Noir, Chardonnay, Syrah

Pine Mountain-Cloverdale Peak (2011): in both Mendocino and Sonoma Counties
Main varieties: Cabernet Sauvignon, Malbec, Sauvignon Blanc

Wineries

Arnot Roberts

Healdsburg

www.arnotroberts.com

Nathan Roberts and Duncan Meyers were best friends, growing up in the earlier days of North Coast wine country. As adults they started a winery partnership, launching Arnot Roberts. The venture revitalized interest in cool-climate California Syrah, inspiring a generation of wine lovers and other winemakers to return to the variety. The attention paid to them in the *New York Times* was among the first signs that a lighter style of California wine was emerging after the heydays of the 1990s and early 2000s. The duo also inspired interest in forgotten varieties. Their Trousseau Noir instigated a wave of interest from winemakers in both California and Oregon, and helped launch a continuing desire for other mid-weight reds. They also released a rosé of Touriga Nacional grown in Lake County at a time when consumers weren't drinking California rosé, no one knew the grape Touriga Nacional, and few brought outside attention to Lake County. Their wines brought a new generation of wine drinkers to California wine.

Bedrock

Sonoma

bedrockwineco.com

Morgan Twain-Peterson, as the story goes, made his first barrel of wine aged 5. It was Pinot Noir sourced from the vineyard of a family friend, a variety chosen precisely because his famous father Joel Peterson did not make it. The elder Peterson worked with Zinfandel and coined the motto "No wimpy wines." It sounds more grandiose now than it did in the 1970s when he started. At the time, Joel Peterson was comparing his wines to the underripe, imbalanced ones made by inexperienced producers of the time. Now that we have survived the ultra-ripe wines of the 1990s and 2000s, the saying today may sound far too big for its britches. Peterson never made that kind of wine. As time went on, Twain-Peterson aligned more closely in interest and style with his father. The two now farm one of Sonoma Valley's iconic old vine sites, their own Bedrock Vineyard. It includes a mix of surprising varieties along with the classic old vine field blends. Through Bedrock, Twain-Peterson has worked to save some of California's other iconic vineyards, such as Evangelho in Contra Costa, and Katushas in Lodi. He and his father also co-founded the Historic Vineyard Society (see box, p. 136). Additionally, Twain-Peterson and his best friend Chris Cottrell also make single-vineyard sparkling wine under the label Under the Wire.

Darling

Sonoma

www.darlingwines.com

Transparent, pure, thoughtful, clear, lifted: all words to describe Darling wines. Making Pinot Noir, Chardonnay, and Syrah, Tom and Ashley Darling bring a light touch to their wines, minimizing extraction while still capturing flavor, making wines almost lacy in structure yet impactful. They source fruit from the coastal areas of Sonoma and Mendocino Counties, relying on vineyards that are all sustainably farmed. Their efforts are also helping some growers place reliance in organic, biodynamic, and regenerative techniques.

Dry Creek Vineyard

Healdsburg

www.drycreekvineyard.com

After Prohibition, most wine from Dry Creek Valley was made by cooperatives, or through the efforts of brothers Ernest and Julio Gallo. It wasn't until 1973 that the first new winery took hold in the appellation. Dry Creek Vineyard was founded by Dave Stare to make a range of flavorful, crisp white wines and a mouth-watering version of Zinfandel. The wines also greatly over-deliver on quality for the price. Today, led by the second generation, Kim Stare-Wallace, Dry Creek Vineyard has also helped build wine tourism in the region by creating a destination winery and tasting room.

DuMOL

Windsor

dumol.com

DuMOL grows wine in the coldest part of the Russian River Valley. The estate brings closer vine density to Chardonnay and Pinot Noir in the Green Valley side of the appellation. They also source fruit from vineyards they admire to make single-vineyard examples of Pinot Noir, Chardonnay, Syrah, Cabernet, and, more recently, Mencia. Their wines tend to be smaller lot bottlings. DuMOL deftly balances the advantages of oak with a commitment to natural acidity, poise, and texture.

Enfield

Santa Rosa

enfieldwine.com

John Lockwood makes some of the best wines in Northern California, yet they are still under the radar. As natural wines took over the market, Enfield went classic. John Lockwood's winemaking is still lower intervention but the wines are also beautiful, energizing, clean, and stable. The resonance of the Waterhorse Cabernet has few parallels. It's comparatively lighter bodied, yet flavorful; restrained and mouth-watering, while seductive. The Enfield 2011 Haynes Vineyard Syrah is still one of the best wines I've had from California. His single vineyard Chardonnay is also a standout.

Extradimensional Wine Co. Yeah!

Sonoma

www.winecoyeah.com

Hardy Wallace first made a name for himself writing about wine on a blog from Atlanta, before winning a social media job for a major winery in Sonoma County, and then learning the ropes for natural wine projects in the North Coast of California. He launched Dirty & Rowdy wines in 2010 and the first wines were a tactile, attractive, skin-contact Semillon, and a series of single-vineyard Mourvèdres. In 2021, Wallace stepped away from Dirty & Rowdy and with his wife, Kate Graham, launched an entirely new project, Extradimensional Wine Co. Yeah! (the exclamation point is legally part of the name). The wines are worth trying. They force

From left: Ryan Glaab of Ryme Cellars, Hardy Wallace of Extradimensional Wine Co. Yeah!, Pax Mahle of Pax

a rethinking and reckoning with what we take for granted in wine. Single vineyard? Probably not. Single variety? Only occasionally. Single appellation? In rare cases. Tasting the multi-regional, multi-varietal blends is a fascinating experience in re-experiencing wine before you knew anything about it. There are no benchmarks to compare. The tasting room offers a range of unique tastings. There is a scent experience where the wines are paired with aromas that favorably transform the wine, and musical pairings where wines are tasted with music meant to elevate the experience. It does.

Flowers

Healdsburg

www.flowerswinery.com

Flowers wasn't the first to plant at the peaks of the Sonoma coastal mountains, but it did get closer to the Pacific Ocean, and helped bring acclaim to the region at a level unseen before. The producer has celebrated an impressive history of winemakers including Steve Kistler, Greg LaFollette, Ross Cobb, and Jason Jardine; Chantal Forthun fills the role today. They have worked with consultants Brenna Quigley and Pedro Parra to deepen their understanding of the vineyard sites and home in on the best techniques in the winery to express the rugged coastal growing conditions. Their Pinot Noir and Chardonnay have been hallmarks of California's coastal mountain character.

Hanzell

Sonoma

www.hanzell.com

Still hosting the oldest plantings of Pinot Noir and Chardonnay in the state, Hanzell carries both historical significance and contemporary quality. Founding winemaker Brad Webb brought important innovation to fine wine in California. He was among the first to focus on aging in French oak barrels, and to mitigate its effects through a variety of techniques including blending with wine aged in stainless steel. He created the state's first temperature-controlled tanks, an innovation then taken and made famous by Robert Mondavi in Napa Valley. He also worked with researchers at UC Davis to perform the first controlled malolactic fermentation of Pinot Noir in the state. Later, Bob Sessions became the winemaker to establish the contemporary Hanzell style. He also recognized the need to get to know clonal materials and rootstocks early, planting an experimental vineyard on site with as many combinations as possible. The current winemaker, Michael McNeill, continues to be guided by techniques instigated by Sessions while bringing a touch more freshness and a little less oak influence to the wines. He also created the earlier drinking Sebella Chardonnay and, most recently, he's brought back Hanzell Cabernet Sauvignon, one of the most exciting, classic examples of the variety in the county.

Hirsch

Healdsburg

www.hirschvineyards.com

Established by David Hirsch in the early 1980s, Hirsch Vineyard is surrounded by redwoods and sits directly atop the San Andreas fault, overlooking the Pacific Ocean. The second generation is now involved, as Jasmine Hirsch now co-farms the biodynamic vineyards with her father, and longtime vineyard manager Everardo Robledo. She also leads the winemaking. Under her direction, the wines have developed some of the greatest clarity and sophistication in their history. The quality of farming has also attracted other winemakers, who have made single-vineyard bottlings from Hirsch Vineyard: they include Williams Selyem, Littorai, Failla, Peay, Twomey, Under-the-Wire, Kistler, Siduri, Flowers, and Rochioli.

J Winery

Healdsburg
www.jwine.com

The daughter of the founders of Jordan Vineyards, Judy Jordan, launched her own venture in 1986, making traditional method sparkling wine from Sonoma. It became one of the top wines of Sonoma. Part of the J secret rested in making enticing wine at multiple price points. Cuvée 20 is designed to be a more affordable house wine, full of flavor and lengthy acidity. And then there are the vintage wines, and most profound among them, the late disgorged cuvées like the 2015 late disgorged brut. When Jordan launched her J project it was one of the few wineries in the state owned and led by a woman. Today, it is still made and led by a woman, head winemaker Nicole Hitchcock.

Jolie-Laide

Sonoma
jolielaidewines.com

Winemakers Scott Schultz and Jenny Schultz coincidentally have the same last name. They also co-own and make the wines for Jolie-Laide. The translation, Pretty-Ugly, captures the founding spirit of the winery. They work with lesser-known fruit from lesser-known places and make many people's favorite wines. My mom used to tell me I tended to make friends with ugly ducklings. It took me years to realize she was saying I knew how to find swans. Scott and Jenny Schultz do too. Their wines exhibit it.

Joseph Swan

Santa Rosa
www.swanwinery.com

In the 1960s, the Russian River Valley did not yet have an international reputation for Pinot Noir. Yet, soon after discovering old vine plantings of Zinfandel, Joseph Swan decided to establish a vineyard of Pinot and Chardonnay as well. Incorporating techniques learned from France, Swan became one of California's first fine wine producers of Pinot Noir, also creating some of the state's most delicious examples of Zinfandel, made in a Pinot-like style. Swan's first plantings are also the origin of the winemaker's favorite, the heritage Swan selection of Pinot Noir. Characterized by small clusters of varying berry sizes, the Swan clone carries a naturally spiced flavor, concentration, and savory length. It's been spread throughout the west coast of the United States. Admittedly, its yields are small, which has limited its attraction for growers. Swan remains in the family led by son-in-law Rod Berglund, who has been the winemaker for several decades.

Keller

Petaluma

www.kellerestate.com

Planted facing the wind on the rolling slopes of the Petaluma Gap, Keller is one of the few estate wineries growing and making their wine fully inside the unique appellation. Because they also sell fruit to other wineries, Keller's vineyards have helped extend interest in and understanding of the unique character of Petaluma Gap wines. Both the Chardonnay and the Pinot Noir are savory, structural, deeply toned, and mouth-watering. Ana Keller, who leads the family estate, has helped hone the wines over time and made the winery a destination in a still developing region.

International Wineries for Climate Action (IWCA)

The romance of family wineries and small-scale vineyards often combines with a disavowal of larger scale projects. But in the fight against climate change, to improve resource management, and shift to renewable resources, larger scale makes the biggest impact. Finding the ability to use greater capital for the sake of broader benefit proves valuable.

In 2019, California's Jackson family partnered with the Torres family of Spain, two sizeable wineries for their regions, to form International Wineries for Climate Action (IWCA). The initiative includes member wineries from countries around the world who have committed to reducing carbon emissions and adopting other climate action practices to reduce their overall impact. Together, IWCA has created tools for carbon tracking and calculating previously unavailable for smaller producers simply because of expense. By expanding the availability of such applications and sharing research findings, IWCA is expanding the effectiveness of its program.

IWCA partners with leading experts on sustainability, climate action, regenerative farming, and wine while relying on science-based guidelines to problem solve and create practical tools for resource conservation, renewal, and climate action.

Kendall-Jackson

Santa Rosa

www.kj.com/

Known for the most sold Chardonnay in the country, Kendall-Jackson Vintner's Reserve, Kendall-Jackson makes comparable wines from other varieties as well. The producer has also been a leader in creating solutions to improve sustainability and reduce climate impact. Kendall-Jackson Chardonnay is fermented entirely in oak barrels. They've developed water capture and reuse techniques, as well as waterless techniques to clean barrels and thus reduce water usage. They've also created a rain-water capture system to use for cleaning the winery. They are integrating sheep and regenerative practices into the vineyard. And in 2019 the parent company, Jackson Family, launched International Wineries for Climate Action with the Torres family of Spain in a commitment to offer solutions for climate action to the wine industry.

Littorai

Sebastopol

www.littorai.com

Before starting Littorai in 1993, Ted Lemon served as the first American winemaker in Burgundy. After returning from France, he made wine in multiple parts of California's North Coast before settling in the Sebastopol Hills of Sonoma County. In search of quality, organic farming, Lemon was one of the first in the state to form contracts with growers based on area rather than tonnage, and then pay more for organics. The practice inspired other winemakers to do the same. He helped encourage more growers to convert to organic farming as a result, and in the case of Hirsch Vineyard, the result was so successful that the next year David Hirsch began converting his entire site to biodynamics. Littorai has made single-vineyard wine from vineyards throughout the cooler, western side of Sonoma County as well as from landmark vineyards in Mendocino, thus bringing attention to the quality of Pinot from Anderson Valley as well.

North American Press

Sebastopol

northamericanpress.wine

Revealing a new era of wine, Matt Niess launched North American Press because after making some of the finest Pinot Noir in California for ten years, he fell in love with French–American hybrids. Passionate interest in fine wines made from hybrids is growing across the United States and parts of Europe, and the wine world would do well to take note. Hybrids represent an alternative response to climate change to altering farming or creating new blends. Many hybrids are already disease resistant, and the Baco Noir Niess works with requires no spraying to prevent powdery mildew. Besides Baco Noir, Niess also makes distinctive site-specific co-ferments combining things like apples, grapes, and herbal plants all grown in the same area. He also partners with Christopher Renfro of the Two Eighty Project to make a rosé from the field-blend historic vineyard at Filoli Farm south of San Francisco. The site includes more different varieties than they've been able to count, all harvested and fermented together.

Occidental

Bodega
www.occidentalwines.com

Steve Kistler became famous for making market-changing Chardonnay from the vineyards of Sonoma County. His Kistler wines became legendary among California wine enthusiasts and older vintages are still brought out to share for events featuring coveted wines of the state. After selling his namesake winery, Kistler began planting his own Pinot Noir vineyards in the far southwestern stretch of the west Sonoma coast, around the villages of Freestone and Occidental. His Pinot has become some of the most enticing in California. The winery doesn't even own a wine press and so the wines are made only with free-run juice, bringing a more open weave with less extraction to the wines. Kistler's vineyards face the ocean, receiving its cooling winds directly. They remain among the closest to the Pacific in the state. Today, Occidental is a family venture since Steve's daughter Catherine Kistler has begun making the wines with him.

Pax

Sebastopol
www.paxwine.com

Getting his start as a sommelier in the eastern United States, Pax Mahle made a name for himself with bold, fully committed Syrah using grapes grown throughout Sonoma County. Intense and structural, loaded with whole cluster influence, the original Pax wines were wines of their era,

Steve Kistler became a well-known name especially for Chardonnay from Sonoma County thanks to his first winery, Kistler. Today, he makes wine alongside his daughter Catherine for the winery Occidental on Sonoma's western coast

the early 2000s, when California wine resembled American cars – the bigger the better. By the mid-2000s, Mahle was taking a new direction. He shifted gears into a new brand, Wind Gap, still making Syrah but from some of the cooler, windier parts of Sonoma County. The effect was transformative. He made wines of finesse, intricacy, and mouth-watering length. More recently, Mahle has returned to using his own name for his wines, Pax. He continues to make some of the best Syrah in California, while also experimenting with no-sulfur cuvées, unusual varieties like those from the Jura, racy, tight Chenin Blanc and more. Mahle has mentored and inspired a host of winemakers now with their own brands.

Radio-Coteau

Sebastopol

enj.b6a.myftpupload.com

Winemaker and owner Eric Sussman brought a wealth of experience to his efforts in west Sonoma. He studied at Cornell, then made wine in Washington state, Burgundy, and Bordeaux, before returning to California to work with Randall Grahm at Bonny Doon. He then moved up to Sonoma and made Pinot Noir for Dehlinger; the experience helped show him the potential of the cooler side of Sonoma County. In 2002, Sussman launched Radio-Coteau making single vineyard wines from the coastal sides of Sonoma and Mendocino Counties. He also farms his own biodynamic estate vineyard, complete with old-vine Zinfandel. Radio-Coteau Syrah is excellent. It's one of those wines I carry around in my head and get thirsty for. Sussman is one of those quiet, more reclusive winemakers, but when you get to know him you realize his quiet demeanor carries a depth of intelligence and feeling, as much as opinion and insight. His brand County Line offers comparatively younger drinking wines at more affordable prices that still manage to offer interest and sophistication.

Ramey

Healdsburg

www.rameywine.com

David Ramey's winemaking career has centered around Chardonnay, although he also does well with Cabernet Sauvignon and Syrah. He was one of the first to rely on whole cluster pressing and honed his own approach to aging in French oak barrels. Ramey has influenced international Chardonnay growth and production, partnering with producers in Australia and France to deepen understanding of the variety and its clonal material. His wines are considered among the best in California, offering a balance of rich flavor with poised structure.

Read Holland

Santa Rosa

www.readhollandwines.com

Ashley Holland made wine for four years in New Zealand, and has worked vintages in Uruguay, Australia, and Oregon. She now works for some of the top wineries in California but also has her own Read Holland label. I first heard about Holland's wines from a friend who wouldn't stop talking about them. Every time I saw him he mentioned the Riesling until I finally agreed to taste it, with another Riesling winemaker, no less. Holland's favorite wine is the Marcel Deiss Schoenenbourg Grand Cru field blend, biodynamically farmed in Alsace, which might give you some indication of her wine style. If you're not familiar, it is ageworthy and rich yet showcases purity; it is textural but floats across the palate; it is mineral, concentrated, profound, and somehow something close to racy. At its best, her Riesling reveals such inspiration.

Rochioli

Healdsburg
www.rochioliwinery.com
Joe Rochioli, Jr. grew up on the family farm (established by his parents) at a time when he had to ride a horse to the one-room schoolhouse because there were not yet cars or roads in the area. The multi-generational family property became one of the founding vineyards of Pinot Noir in the Russian River Valley when he planted a risky, lesser-known grape, Pinot Noir, there in the 1960s. He had already established another then lesser-known variety, Sauvignon Blanc, in the 1950s. Soon after, he planted Chardonnay. Rochioli vineyard got a name for itself when Davis Bynum bottled the first Rochioli Pinot Noir. Later, Williams Selyem bottled one too. Joe's son, Tom Rochioli, started the family winery in the early 1980s, and today the fourth generation, his daughter, Rachael, and son, Ryan, work alongside their father. Together, they make standout Pinot Noir from the heart of historic Russian River Valley.

Ryme

Forestville
www.rymecellars.com
The wines of Ryan and Megan Glaab helped usher in what Jon Bonné called New California. They not only eschew higher alcohol but work with less common varieties and focus not only on freshness and length but also texture, with just enough weight on the palate without being heavy. The His and Hers Vermentinos, two different wines each made in the style preferred by either Ryan or Megan, is the only example, they tell us, of them disagreeing in wine. The His includes skin contact and a little more aging. The Hers is straight to press and sells a lot more

wine. They make a classic expression of Cabernet Franc, aromatic, with moderate ripeness, mouth-watering, savory. They make wines from a range of more obscure Italian varieties and made one of the first orange wines in California. It included extended skin contact of Ribolla Gialla grown in Vare Vineyard, the first to plant the variety in California, and aged for two years before bottling. In late 2024 I had a bottle of its first vintage, 2010, and it was still aging beautifully.

Silver Oak

Healdsburg
silveroak.com

Founded in 1972, Silver Oak was considered among the first great steak house Cabernets of California, along with Caymus and Jordan. Which you preferred served as a sort of personality test for wine lovers. Silver Oak makes two wines, each a Cabernet Sauvignon blend from different parts of the North Coast. The Napa Valley cuvée is a blend of multiple Bordeaux varieties. It's plush, richly flavored, and verging on decadent. The Alexander cuvée tends towards mostly Cabernet Sauvignon, is more structural, less brooding, and fuller of refreshing acidity. It's my preferred of the two, but both have become flagship Cabernets, among the few California wines recognized by consumers in any country I've traveled in the world. Silver Oak has also made a sizeable commitment to sustainability efforts. Their Alexander Valley winery has earned the most difficult sustainability recognition for structures in the world, the Living Building Challenge. It relies on water capture and reduction, minimizing waste, maximizing recycling and reuse, energy efficiency, carbon reduction, recovered materials, and comfortable space for employees.

Sixteen600

Sonoma
www.winerysixteen600.com

Founded by viticultural consultant Phil Coturri, the winery Sixteen600 is run along with his two sons, Sam and Max. Sam leads the winery effort while Max works in the vineyards. Sixteen600 wines are made solely with fruit grown by Coturri, from organically farmed mountain vineyards. The Sixteen600 style is ripe, giving, lip-smacking flavor. Although the wines are large, they manage to stay poised and light on their feet. It's a style that takes control of the palate but then refreshes it. In 2018, Coturri partnered with legendary winemaking consultant

Philippe Cambie to make Grenache rosé and red wines for the project Á Deux Têtes. Isabel Gassier, of Famille Gassier in the southern Rhône, worked closely with both Cambie and Coturri and now serves as consultant and guide for the project.

Ultramarine

Petaluma

ultramarinewines.com

Michael Cruse was a laboratory scientist before starting his winery. After attending a lecture on wine, he took the bold step of changing industries and got a job working as an enologist in labs instead. He began experimenting with the traditional method, used for making sparkling wines like Champagne. Soon after, he launched the sparkling only label, Ultramarine. The *blanc de blancs*, made with Chardonnay from Charles Heitz vineyard in the Occidental region of the West Sonoma Coast, is the standout. In addition to his own wines, Cruse also consults for other wineries for both sparkling and still wines. He also makes a broader range of both sparkling and still wines under his label Cruse.

Vérité

Healdsburg

www.veritewines.com

Founded in 1997, Vérité was the result of the then-unusual collaboration between American entrepreneur Jess Jackson of the Jackson Family, and French-born winemaker Pierre Seillan. The duo had the idea that they could grow some of the finest Cabernet in the world in Sonoma County. Vérité makes three distinct wines, each centered on a different Bordeaux variety while incorporating the others into the blend. La Muse is built around Merlot, La Joie Cabernet Sauvignon, and Le Désir Cabernet Franc. They work with around 50 different small selections of premium vineyards in Knights Valley, Chalk Hill, Bennett Valley, and Alexander Valley and vinify them separately to then blend the three wines.

Williams Selyem

Healdsburg

www.williamsselyem.com

Burt Williams and Ed Selyem joined forces to launch Williams Selyem in 1979. Working with wineries along the western bench of the Russian

River to make Pinot Noir, the success of Williams Selyem brought international interest in the idea that the Russian River Valley could grow wine at least as good as Burgundy. They helped make the Westside Rd section of RRV renowned, which led the Rochioli family to move beyond grape growing into having their own winery, just up the road from Williams Selyem. Known for Pinot Noir, Williams Selyem also makes ageworthy Zinfandel, and Chardonnay.

NAPA VALLEY AND COUNTY

Known for: Cabernet Sauvignon, wine tourism
Lesser-known strengths: Sauvignon Blanc, Zinfandel, farmworker health and education
Leader in: founded the first agricultural preserve in the country
First peoples: Micewal, Coast Miwok, Patwin
First vines: 1854
First AVA: Napa Valley, 1981
Number of AVAs (2024): 18
Planted vineyard area (2024): 46,128 acres (18,667 hectares)

It took time for Napa Valley to become one of the most famous wine regions in the world. Its successes in the late 1800s were quashed, first by phylloxera and then by Prohibition. Regrowth was initially slow going. In 1970, only 40 wineries were open in Napa Valley. By 1980, it was over 100. What is remarkable about the region is perhaps less its fame than its size. It grows a mere 4 percent of the state's wine (0.4 in the world) while contributing around a quarter of its revenue. In 2023, the Napa Valley wine industry contributed more than $2.5 billion to the local economy. Despite the gloom of COVID-19, wildfires, and the fear of recession, 2023 revenue saw around a 13 percent increase from 2018. The great ability for the region to sell wine has fostered its growth and made it partner to other industries. It is located within a mere 90-minute drive from UC Davis, one of the leading wine and viticultural research institutes in the world, and Silicon Valley, one of its technology hubs. The combination has made California a global leader in sustainability solutions and the fight for climate friendly farming. It's also honed California's acumen in winemaking.

It is easy for many to be cynical about the success of Napa Valley. According to the Direct-to-Consumer Wine Survey from Silicon Valley Bank, in 2018, the average price for a bottle of Napa wine was $108. Napa Valley has suffered in some ways from its own success in both sales and winemaking. The investment draws plenty of outsiders who made money elsewhere. The problem with outside investors is they haven't made wine before, an acceptable history but one that means they usually rely on consultants for viticulture and winemaking. There are, of course, talented consultants well worth partnering with for wine. As a few have become consultants for numerous wineries, their style of wine has become more recognizable and more common. Like any region, a portion of the region's wines taste largely the same, made to succeed with a style affirmed by the American consumer and wine scores from critics. But the growth of homogeneity feeds boredom. Couple that with high prices and people lose interest in wine.

The partnership with research and technology has also made Napa Valley a leader in sustainability and climate action. The implementation of Fish Friendly Farming helped restore the Napa River. The rise of weather stations and vineyard sensors in Napa decades before other regions had them is helping better understand climate change. Growers throughout Napa are now partnering with UC Davis researchers to share farming and weather data alongside wine quality studies to better track how winegrowing interacts with climate change. Napa has also been an epicenter for improving waste management. Wineries are working on greater water conservation, and viticulturists are planting experimental vineyards to track what varieties and rootstocks best handle climate change.

The best of Napa Valley wines really are world changing. Producers like Louis M. Martini, Inglenook, Beaulieu Vineyards, Charles Krug, Souverain, Heitz, Stony Hill, Joseph Phelps, and even sparkling wine from Schramsberg demonstrated the profound ageability of wine from the region. Their legacies continue. Since the 1980s, projects such as Continuum, Corison, Spottswoode, Shafer, Stag's Leap Wine Cellars, MacDonald, Dominus, Quintessa, Hyde de Villaine, Chappellet, Frog's Leap, and Domaine Carneros's sparkling wines have demonstrated aging potential of the region as well. But what makes Napa Valley truly interesting is how every ten years or so a new guard cycles in. Zelma Long and Cathy Corison were among the first women to make wine in the valley, at the end of the 1970s. At the same time, Smith-Madrone and Dunn were showing off quality grown in the mountains. Their

Floating over the Napa Valley in a hot-air balloon in the early morning

successes, as well as the earlier examples from Louis M. Martini and Souverain, helped create a new category of wine, mountain Cabernet.

In the 1980s, Rosemary Cakebread now of Gallica, Julie Johnson of Tres Sabores, Celia Welch of Corra, and Françoise Peschon from Araujo, Accendo, and Vine Hill Ranch, among others, arrived. What they have in common is elegantly expressed wines with power, finesse, length, and respect for structure, yet each distinctly different by vineyard location.

In the 1990s, Andy Erickson arrived and became a consulting winemaker, over time working with Bond, Screaming Eagle, Staglin, Dalla Valle, and Mayacamas. He then started Favia and Leviathan with his viticulturist wife Annie Favia. Up on Spring Mountain, Cain appeared. The 1990s also brought the rise of Robert Parker, full throttle reds, and another expansion of Napa Valley. But the success of that style opened the way for a swing in the other direction.

In the mid-2000s, Steve and Jill Klein Matthiasson, Dan Petroski of Massican, and Abe Schoener (then making Scholium Project) all challenged the regional norm. Matthiasson and Massican made wines with flavor but lower alcohol. Massican is entirely focused on white wine. Schoener experimented so thoroughly that his wines re-educated newcomers about what could happen with wine. In 2010, a new crop appeared with natural leanings. Hardy Wallace launched Dirty & Rowdy with Semillon and Mourvèdre. John Lockwood made shockingly good Chardonnay, Syrah, and Cabernet through his Enfield Wine Co. Brothers Graeme and Alex MacDonald took a risk and started MacDonald Wines from one of the oldest parts of To Kalon vineyard on their family's property.

Now there's another new wave. DiCostanzo delivers detailed and complex, fine-boned, single-vineyard expressions of Cabernet. La Pelle, made by a partnership between winemaker Maayan Koshitzky and viticulturists Miguel Luna and Pete Richmond, reminds wine lovers of the savory, restrained side of Napa. Cathiard, between St. Helena and Rutherford, relies on regenerative farming to make incredible red blends. Will Harlan and Cory Empting have partnered to make Promontory, a Bordeaux blend of energy and depth unlike any other in the region. They each represent a return towards Cabernet worth taking. Beyond Cabernet, Robert Biale, Frog's Leap, Storybook Mountain, and Turley for Zinfandel have a fresh take on historic vines.

Sauvignon Blanc might be the region's secret weapon. It's far more expressive of where it is grown than is often believed. In Napa Valley it comes in a full range of styles. Steely wines, like drinking lightning, come from Coombsville. Citrus fruits appear in Oak Knoll. Tropical flavors come from St. Helena and further north. In the middle of the valley, Sauvignon Blanc offers citrus, mineral, and floral notes. Spottswoode, Accendo, Eisele, Illumination, St. Supery, and Mondavi are stand-out examples.

At the same time, full-throttle wines still appear. Napa Valley today includes the broadest range of styles it has ever had. Yes, there is plenty of Cabernet, and yes, in the middle much of it tastes the same. But the best wines of the region are a reminder of how Napa Valley got its fame.

The first vines

Napa Valley's first vines came from Sonoma. Until 1844 General Vallejo led the Mexican garrison in the town of Sonoma. Vallejo was known for his love of wine, he made it himself and gave cuttings to others nearby. If everybody he gave a cutting to grew vines, he figured, there would be more California wine. In the 1830s, George Yount was given a land grant by the country of Mexico that covered 11,887 acres (4,810 hectares) of Napa Valley. The grant included the stretch from what is now Yountville to just below St. Helena and across the full width of the valley. At that point the Micewal still lived in Napa Valley; George Yount arrived as the first outsider. Vallejo welcomed Yount to the region and provided resources to start his own settlement along the Napa River. There, he established the village of Yountville and planted the region's first vines.

Yount's and Vallejo's efforts were never commercial. The wine they made was traded and enjoyed locally. Napa Valley's first commercial vineyard arrived in the 1850s, after California celebrated its official statehood, and was established by the entrepreneur John Patchett. In 1856, in what is now the town of Napa, Patchett built the region's first winery from its first commercial vineyard. At its peak, Patchett's Grove produced 6,000 gallons (22,712 liters), which was sold to restaurants in San Francisco. Patchett stopped making wine in 1870.

In 1861, Charles Krug launched his eponymous winery near what is today St. Helena. Krug got his start assisting winemaking over the hill at Sonoma's Buena Vista. When launching Charles Krug, his problem-solving ability laid the groundwork for quality wine in the region. When his Bordeaux vines came of age (Krug was one of the first to plant them in Napa), Krug used a cider press to make wine. The North Coast already had a history with apples and pears. The cider press was essentially what came to be known as a basket press, the gold standard in winemaking until the arrival of pneumatic presses almost a century later. Krug was one of the first to age Napa Valley wine in French oak barrels, and a pioneer in bottling wines dated by vintage.

As Charles Krug winery's success expanded, so did outside interest in growing wine in Napa Valley. Founded in 1879, Inglenook, near the village of Rutherford, gained acclaim when it won gold medals for its Napa wines at the 1889 Paris Exposition. And Eschol vineyard (today's Trefethen) from the town of Napa also won awards at the same world's fair. Other historic vineyards soon took root: Schramsberg, Mount Veeder, and Beringer all started. H.W. Crabb planted To Kalon in Oakville and began making wine as Hermosa Vineyards, which was soon renamed To Kalon Vineyard Company. And then Alfred Tubbs planted Chateau Montelena near Calistoga.

In the 1870s, phylloxera hit Napa Valley. More than 80 percent of its vineyards were lost. Tubbs had the good fortune of purchasing land to establish vineyards after the phylloxera crisis was resolved. In 1882 he became one of the first in the region to establish his vineyards on rootstock. Beaulieu Vineyards soon followed, in 1900. The phylloxera crisis hit Napa Valley later than neighboring Sonoma. By that point Napa Valley's reputation had already grown, its plantings expanded.

Phylloxera hitting later meant the cost to rebuild the industry was greater. By 1920, Prohibition had begun. It would take until the 1970s for the region to truly rebuild to pre-phylloxera levels.

After the repeal of Prohibition in 1933, several wineries joined the charge to make wine stable again in Napa Valley. Beaulieu Vineyards hired European André Tchelistcheff who had winemaker training and understood chemistry. The Mondavi brothers, Robert and Peter, convinced their parents, Rosa and César, to invest in fine wine by buying Charles Krug. Louis M. Martini moved north and founded his winery in St. Helena. John Daniel, Jr. resurrected Inglenook. Together, they became new pioneers in a post-Prohibition world. Their efforts, often in collaboration, helped establish a slow recovery for Napa Valley. In the uplands of Mount Veeder, Jack and Mary Taylor started Mayacamas in 1941. In 1944, Lee Stewart opened Souverain, near the peak of Howell Mountain. In 1952, a new small winery devoted to white wine opened, Stony Hill, on the slopes of Spring Mountain. Then in 1966, the world changer, Robert Mondavi opened his eponymous winery on the benchlands of To Kalon.

The Robert Mondavi winery became a winemaking and cultural incubator. Mondavi traveled the world to promote the idea of California fine wine. He took winemakers and viticulturists to Europe to improve their techniques. The winery team drank the best of Bordeaux and Burgundy to understand what fine wine tasted like. Then they challenged each other when tasting their own wines together in the cellar. Warren Winiarksi, Mike Grgich, Tim Mondavi, his brother Michael Mondavi, Tim's daughter Carissa, Paul Hobbes, and Zelma Long all got their start at Robert Mondavi. Winiarski went on to found Stag's Leap Wine Cellars and made the award-winning red in the 1976 Paris Tasting (see box, p. 164). Mike Grgich became winemaker for Chateau Montelena, which was bought and rebooted by Jim Barrett in 1972. Montelena's wine topped the whites in the 1976 Paris Tasting. Michael Mondavi founded his family winery. Tim Mondavi partnered with Robert, sister Marcia, and his children to launch Continuum. Carissa now helps lead it. Paul Hobbes founded wine projects worldwide. Zelma Long, regarded as one of the North Coast's first women winemakers, went on to expand SIMI and become a mentor for many young winemakers. She helped lead Chandon Estates, then launched her own Long Vineyards, before helping to build Vilafonte, some of the finest red wine in South Africa. Together, they became part of the new rise of Napa Valley, the basis of one of the world's leading wine regions.

The 1976 Paris Tasting

In the mid-1970s, friends Patricia Gallagher, from the United States, and Steven Spurrier of Britain (but living in France), joined forces with a specific plan. Spurrier believed wines from the US west coast were worth enjoying, but the legacy of France and its history of success made others skeptical. So Spurrier traveled to California to taste and select wines worth sharing.

Spurrier and Gallagher brought the wines back to Paris and in May of that year staged a tasting event, pouring the California wines alongside the best of France. Its purpose was to celebrate the American Bicentennial. France, after all, was the United States' first European ally.

Nine well-acknowledged French wine experts tasted French and California examples side-by-side with the specific wines hidden from the experts until all were tasted. Each expert scored the wines as they went. Spurrier tabulated them. And in the end it was revealed that California had earned top scores for both reds and whites. The results were shocking. Skeptical French connoisseurs hadn't expected California to be at all comparable to France, let alone besting French wines.

When it came to revealing the results, Warren Winiarski's 1973 S.L.V. Estate Cabernet Sauvignon from his Stag's Leap Wine Cellars took the lead in reds. For whites, it was Chateau Montelena 1973 Chardonnay, made from a blend of Napa Valley and Sonoma County fruit.

For Winiarksi, the importance of the event was less in the specific wines that won, but in demonstrating that California had found its way. The state's best stood up to the classics with the three tenets Winiarski saw as the foundation of the world's best wines: richness, ripeness, and restraint. Many wines have two of these, but according to Winiarksi the best have all three.

The 1976 Paris Tasting delivered unexpected results, but not until an article on the event was published did it have impact. George Taber was the only journalist to attend the Paris Tasting. Others turned the occasion down thinking everyone knew French wine would win, so as the only writer there, Taber got the scoop. Taber's write-up, a mere four paragraphs on the side of a page, appeared in *Time* magazine in early June. The editor appended the title "Judgment of Paris" as a cheeky allusion to the Ancient Greek myth. It was the only announcement of the tasting at the time. Yet, the modest coverage caused a sensation.

California wines had bested the French. The results since are considered controversial, indicating more that California wines held up alongside French wines rather than *won* a contest. Regardless, the news assuaged American insecurity and exposed the myth of French superiority. They also helped accelerate the rise of Napa Valley wine. The growth, well established during the 1970s, was suddenly urgent. As the end of the decade approached, the need for winemakers stretched the region's resources. Young people with almost no experience were given head winemaking jobs, and the region started the steady rise to becoming one of the best-known in the world.

Winiarski kept on with his efforts at Stag's Leap Wine Cellars. Winemaker Miljenko Mike Grgich, who made the Chateau Montelena wine, immediately partnered with Austin Hills, of the Hills Bros. coffee family, and launched Grgich Hills winery. It continues today. A bottle of each of the winning wines, from Stag's Leap Wine Cellars and Chateau Montelena, are now part of the Smithsonian exhibit, "101 objects that made America."

Beyond simply wine, Winiarksi helped create the first agricultural preserve in the country in Napa Valley. After selling Stag's Leap Wine Cellars he used the money to do extensive philanthropic work. During his tenure at Stag's Leap Wine Cellars, Winiarski was considered one of the great mentors of the region, helping to build the experience of a host of the next generation's pivotal leaders in the region. They include Françoise Peschon, who helped launch Araujo, Michael Silacci of Opus One, John Williams of Frog's Leap, Bob Sessions, known for his work at Hanzell, Dick Ward of Saintsbury, Paul Hobbes, Steve Matthiasson, and others.

Phylloxera reborn

Cabernet's status as Napa Valley's king grape didn't take hold until the late 1990s. Until then, Petite Sirah and Zinfandel each occupied a larger planted area. But the choice to plant *vinifera* on what turned out to be an unreliable rootstock, AXR1, throughout much of the 1970s and into the 1980s, created a crisis of reinvention for the region.

AXR1 is vulnerable to phylloxera, counter to expectations. Between 50 and 70 percent of Napa Valley relied on the rootstock, and in the early 1980s, phylloxera was rediscovered in the region. Initially its presence was either denied or ignored – surely the louse could not be decimating vineyards again. It was. The 1990s became a period of massive replanting and by 1997, Cabernet was Napa Valley's new dominant grape.

The one advantage with phylloxera this time, was that people knew how to resolve it. AXR1 had failed but they could turn to another rootstock proven to work. Vineyards established on the rootstock St. George, for example, were untroubled by phylloxera. The solution, new rootstock and replanting, would just cost money, and plenty of time. The AXR1 crisis was not confined to vineyards. Businesses closed, unable to afford the cost of replant and some were sold, mostly at reduced value. Others made a different decision.

Robert Mondavi led a host of wine investments at the time. The eponymous winery founded in 1966 was making 700,000 cases a year by 1992. In 1979, Mondavi opened his Lodi venture, Woodbridge. It made 3 million cases in 1992. Mondavi launched a collection of wines made from vineyards in the Central Coast called Coast Selections. Coast Selections made 60,000 in 1992. In 1985, Mondavi purchased Vichon, which made 40,000 in 1992. Mondavi also co-owned Opus One in partnership with Baron Philippe de Rothschild of Chateau Mouton Rothschild. In 1990 they'd purchased Byron in Santa Barbara County to accompany the Tepusquet Vineyard they already owned nearby. He also had 50:50 partnerships for wineries in Europe, Argentina, and Australia. By the time phylloxera reappeared, Mondavi investments stretched worldwide and the length of California. Being forced to replant at such scale meant changing the business plan. In 1993, Mondavi went public. The initial public offering of shares on Wall Street brought the business an estimated $50–60 million, the same estimated amount that Mondavi spent on eliminating phylloxera. Phylloxera led to an economic shift in Napa that laid the groundwork for the region to go corporate. After Mondavi went public it pushed others to rethink investments as well.

The French Paradox described in the 1991 CBS *60 Minutes* segment with Morley Safer helped assuage the problem. California wine had been losing sales throughout the 1980s. Safer unexpectedly turned that around. As sales increased across the country, it became worth re-invigorating Napa wine. Andy Beckstoffer bought his portion of To Kalon vineyards, one of Napa's most coveted sites, in 1993. He'd just bought Georges III in 1988, after phylloxera was recognized in the region. Both vineyards were prime sources for what was once the best Cabernet in Napa Valley, the Beaulieu Vineyards Georges de la Tour Private Reserve. In 1994, Beckstoffer co-founded the Rutherford Dust Society, a collaborative vintners' group essentially marketing the quality of red wines from the Rutherford AVA, another investment in the future of Napa Valley.

The French Paradox

In the 1990s, French research transformed American wine culture. In 1991, Canadian–American broadcaster Morley Safer offered a news segment on the CBS station television show *60 Minutes*. In it, he looked at the studies done by Serge Renaud of the University of Bordeaux. The studies showed that the French have lower instances of cardiovascular disease, and lower mortality related to heart conditions than Americans. The results were attributed to their Mediterranean diet. It's a diet that includes less meat and more vegetables than the typical American diet, hearty doses of olive oil, and moderate consumption of red wine.

Safer described how the resveratrol uniquely in red wine seemed to reduce cholesterol and so also heart disease. American wine drinking boomed. After a decade of declining sales in the 1980s, the health consciousness of that decade erupted into a wine-positive buying spree. The change in demand arguably launched the California wine industry. It built its foundations and grew in the 1970s, but in the 1990s it skyrocketed.

Since Safer's program was broadcast, the French Paradox has proved controversial. The data still shows lower incidents of heart disease for those eating a Mediterranean diet, but the reason seems unclear. Not everyone believes red wine is the solution.

The horror of phylloxera reborn led to a reworking of the region with already hard-won experience. Diseased vineyards were pulled out and burned. Riesling, Chenin Blanc, Petite Sirah, Pinot Noir, Chardonnay, Charbono, and Zinfandel, each planted like a patchwork quilt of row directions, vine spacings, and training methods, became the new vineyards of Cabernet Sauvignon planned with the newest viticultural knowledge.

Outside investors began arriving. Phylloxera losses brought the changing of the guard with new owners entering the region. At the same time, Robert Parker's attention was turning from Bordeaux to California. Wealthy wine lovers responded. Another reason to invest in Napa Valley. Amid the 1990s phylloxera crisis cult wineries were born: Bryant Family, Dalla Valle, Screaming Eagle, Abreu, Araujo, Harlan all emerged. The growth of this new kind of winery – niche, luxury, almost mysterious – brought new intrigue to the region as well. Outside money helped fund local infrastructure. Visitor tastings began to evolve. The country aesthetic of the bootstrapping family winery, where wine was sampled on an upturned barrel with the winemaker in muddy boots was disappearing. The 1990s launched the rise of the elegant destination tasting room. Napa Valley wineries drew a new kind of tourism, with people traveling to drink wine. It provided another means for the region to thrive.

Growing conditions

The Napa Valley appellation was the first recognized American Viticultural Area (AVA) in California, certified in 1981. The AVA encompasses the Mayacamas mountains up the county line (with Sonoma along its ridgeline), all the way down its slopes, through the valley floor and up again to the slopes of the Vaca Range on the eastern side. It includes, in other words, the geographical place called Napa Valley. But, importantly, the Napa Valley AVA also includes areas east of the Vaca Range that historically contributed to the reputation of Napa Valley wine. Growers towards Lake Berryessa, in the Chiles Valley, and its surroundings are included in the appellation.

Today, the region hosts 17 nested AVAs. Appellations spread across the valley floor from the Mayacamas to the Vaca, denoting growing areas surrounding villages the length of Napa Valley. On the valley floor, starting in the north, are Calistoga, St. Helena, Rutherford, Oakville, Yountville, and Oak Knoll District. And in the south just above San Pablo Bay is Los Carneros, which is shared with Sonoma County. Through the mountain ridges there are others. Along the top of the Mayacamas, north to south, these are Diamond Mountain District, Spring Mountain District, and Mount Veeder. This western side of the valley reaches peak elevations of 2,600 feet (792 meters). On the eastern side, peaks reach similar elevations. Through the tops of the Vaca Range are Howell Mountain, Atlas Peak, and, in the very south, Wild Horse Valley. Between Atlas Peak and Howell Mountain there also stands Pritchard Hill, certainly its own unique growing area, featuring famed wineries such as OVID, Continuum, Bryant, Colgin, Brand, and first of them all Chappellet – but all without a Pritchard Hill AVA. Along the slopes of the Vaca are Stags Leap District, Coombsville, and newest of them all, Crystal Springs.

The soils of Napa are profoundly varied. The Vaca and Mayacamas have formed through the forced folding of the landscape by the intersection of the Pacific and North American plates. Their impact has also caused volcanic activity, evidence of which can be found in the parent materials (read: bedrock) of the entire region. Volcanic stones have been pulled out of the mountains by flooding, creeks, rivers, and landslides then circulated in alluvial fans. Close to the river, pebbles mix with sands and clay and soils are deeper. In the mountains, shallow soils perch vines on bedrock with little water and lower vigor. To the east, andesite and volcanic tuff appear. In the west are more volcanic soils but also bands of sedimentary uplift. And in one unique territory of Napa Valley, the second generation of the Harlan venture, son Will Harlan and winemaker Cory Empting, grow vines on metamorphic stone for their winery Promontory.

Napa Valley shares the state's Mediterranean climate. It is coolest in the south near San Pablo Bay. Disease pressure from fog is also higher in this area. Through Carneros, temperatures are cool enough for Pinot Noir and Chardonnay, which are also grown in pockets of Coombsville. Going north, temperatures steadily increase. Through much of the Vaca Range afternoon breezes bring relief from the heat. In the west, vines are shaded during the afternoon. Flavors are darker and generally riper in the east due to the afternoon sun. The highest daytime temperatures appear in Calistoga, followed quickly by a sharp temperature drop, with cool nights. The highest overall temperature accumulation occurs in St. Helena, a steady temperature arc over the course of the day.

Regenerative farming plays an important role in some of the top wines of Napa Valley. Grgich Hills and Spottswoode lead efforts and Joseph Phelps has begun adopting these methods. Biodynamic farming has a long history in the region: Quintessa, Raymond, Robert Sinskey, and Neal all farm biodynamically. Frog's Leap, Turley, Tres Sabores, Corison, and Long Meadow Ranch work organically.

Napa County AVAs

Carneros (1983): straddles Sonoma and Napa Counties
Main varieties: Chardonnay, Pinot Noir, Merlot, Syrah

Napa Valley (1981): the first AVA in California, and its most famous
Main varieties: Cabernet Sauvignon, Chardonnay, Zinfandel

Howell Mountain (1983): the place where the first Cabernet was planted in Napa Valley
Main varieties: Cabernet Sauvignon, Merlot, Zinfandel, Chardonnay, Viognier

Stags Leap District (1989): the origin of the award-winning red from the 1976 Paris Tasting
Main varieties: Cabernet Sauvignon, Merlot, Sauvignon Blanc

Mount Veeder (1990): the foothills of the Mayacamas
Main varieties: Cabernet Sauvignon, Chardonnay, Merlot, Zinfandel

Atlas Peak (1992): rugged slopes on the western side of Napa Valley
Main varieties: Cabernet Sauvignon

Oakville (1993): home to some of the most renowned Cabernet in Napa Valley
Main varieties: Cabernet Sauvignon, Merlot, Sauvignon Blanc, Cabernet Franc

Rutherford (1993): made famous by André Tchelistcheff's love for its soils
Main varieties: Cabernet Sauvignon, Sauvignon Blanc, Merlot, Cabernet Franc

Spring Mountain District (1993): the location of the first Napa Valley winery devoted to white wine
Main varieties: Cabernet Sauvignon, Merlot, Cabernet Franc, Chardonnay, Zinfandel

St. Helena (1995): the northern stretch of the famed Mayacamas bench
Main varieties: Cabernet Sauvignon, Cabernet Franc, Merlot, Syrah, Zinfandel, Sauvignon Blanc

Chiles Valley (1999): tucked into hills east of the Vaca Mountains, near Lake Berryessa
Main varieties: Cabernet Sauvignon, Merlot, Cabernet Franc, Zinfandel

Yountville (1999): the area where the first *vinifera* was planted in Napa Valley
Main varieties: Cabernet Sauvignon, Merlot

Diamond Mountain District (2001): a rugged mountain district above Calistoga
Main varieties: Cabernet Sauvignon, Cabernet Franc

Oak Knoll District (2004): in a cooler section in the south surrounding the town of Napa
Main varieties: Cabernet Sauvignon, Chardonnay, Merlot, Sauvignon Blanc, Riesling

Calistoga (2010): the highest peak daytime temperatures in Napa Valley, with cold nights
Main varieties: Cabernet Sauvignon, Zinfandel, Syrah, Petite Sirah

Coombsville (2011): volcanic soils inundated with cooling winds off the San Pablo Bay
Main varieties: Cabernet Sauvignon, Chardonnay, Merlot, Syrah, Pinot Noir

Crystal Springs (2024): on the foothills of Howell Mountain
Main varieties: Cabernet Sauvignon

Wild Horse Valley (1988): continues into Solano County
Main varieties: Pinot Noir, Chardonnay

Wineries

Brown Estate

Napa

www.brownestate.com

Founded in 1996 with a first vintage of Zinfandel, Brown Estate has been a celebrated member of the biggest Zinfandel association in the world, Zinfandel Advocates & Producers (ZAP). They released the first vintages of their wine at a ZAP event and have since accumulated almost 30 years of winemaking experience. The estate stands in the eastern hills of Napa Valley, but to make the wines more readily available, they now have a Napa-based tasting room as well as a sister brand, House of Brown. The Brown family, and their winery, are recognized for incredible hospitality and a long-term following. Brown also has a commitment to fostering greater diversity in the wine industry and is one of the first Black-owned wineries in the state. Besides Zinfandel, they also make Cabernet Sauvignon, Merlot, and Petite Sirah. The wines are generous, plush, and filled with mouth-watering flavor. Part of the success of Brown Estate is their talent for making visitors feel welcome, whether already familiar with wine or just starting to explore.

Cathiard

St. Helena

www.cathiardvineyard.com

Bring Smith Haut-Lafite to Napa Valley and the wines are guaranteed to be of high quality and exciting. Since purchasing the property in 2020 (a difficult year), Florence and Daniel Cathiard along with head winemaker Justine Labbé have moved to regenerative farming, restored the wetlands, moved to year-round cover crops, upgraded the winery building, restored the original home of Louis M. Martini that remains on site, and delivered a quality of wine that recaptures the reasons to grow red wine in Napa Valley. Even this early in the project, the wines of Cathiard are already among the best in the region.

Chappellet

St. Helena

chappellet.com

In 1967, Don Chappellet climbed his way up the craggy slopes of Pritchard Hill and established Chappellet vineyards. Chappellet was among the first to plant vines there, and their success drew producers who now make some of the most coveted wines of the region. The red wines are known for their concentration, with a resinous quality; they express the brooding nature of Pritchard Hill and have the structure for aging. The Chenin Blanc remains an international favorite, a hat-tip to the history of white wines in Napa Valley.

Continuum

St. Helena

www.continuumestate.com

The Mondavi family has now enjoyed more than 100 years in California wine. Cesare and Rosa migrated from Italy and started their family in Minnesota. As the grape business began to gain currency during Prohibition, the Mondavis moved to Lodi to become brokers to home winemakers on the eastern seaboard. At the end of Prohibition, the family invested in a winemaking facility, Sunny St. Helena, making affordable, bulk wines for the newly emerging wine market. A few years later, they updated to fine wine, purchasing Charles Krug winery, which was led initially by brothers Robert and Peter. The foundations of the Mondavi family have led to generations of winemakers. Peter continued to lead Charles Krug, and now his children and grandchildren do. Robert's son, Michael, founded his own eponymous winery. After leaving Robert Mondavi winery in the early 2000s, Robert, son Tim and daughter Marcia built Continuum. It reflects a further iteration, a recognition based in the family's long tenure in California wine while retaining focus. Today, Tim's children are also owners and part of the project. Continuum makes two incredible wines. The flagship cuvée, bearing the name Continuum, offers a well-honed expression of the craggy, exposed, chapparal-covered Pritchard Hill area. Tim's lifelong experience in wine is brought to bear capturing finesse, transparency, and focus in a distinctly mountainous site. The result is an impressive and impressively ageworthy bottle based on Cabernet Sauvignon. Through the winery, visitors can also enjoy their second label wine, a more affordable, well-crafted Cabernet blend called Novicium.

Corison

St. Helena

www.corison.com

Launched in the late 1980s to make only 100 percent Cabernet Sauvignon, Corison remains one of the most elegant, savory, and beautiful examples of the variety in the region. Cathy Corison arrived in Napa Valley in the 1970s, becoming one of its first women winemakers and helping to launch the booming need for winemakers in the region following the impact of the 1976 Paris Tasting. During the Parker era of full-throttle wines, Corison stayed her course: she has never had Cabernet over 14 percent alcohol. Her Napa Valley cuvée relies on vineyards within 1.5 miles (2.4 kilometers) of each other and within that distance from the winery and her home as well. She also makes two single-vineyard wines, the historic Kronos, an incredible wine from older vines, and the more recent Sunbasket. In addition to Corison, there is also the Helios Cabernet Franc, Corazon Cabernet Rosé, and Riesling, all worth trying.

DiCostanzo

Oakville

mdcwines.com

A partnership between husband and wife team Massimo and Erin, DiCostanzo wines delivers single vineyard Cabernets with great ageworthiness from sites in both Napa Valley and Sonoma County. Their wine from Coombsville is one of the most detailed, fresh, and refined Cabernets at its price point in the region. They also make a blend of Cabernet under the more affordable bottling, DiCo. Massimo DiCostanzo gained winemaking experience apprenticing with Adi Badenhorst in South Africa, and Tiganello in Italy, among others. He then returned to Napa to help make wine at OVID, Screaming Eagle, and alongside famed winemaker Philippe Melka. This combination of experiences gave DiCostanzo a commitment to lighter touch winemaking, natural acidity, and savory poise.

Domaine Carneros

Napa

www.domainecarneros.com

An enormous affirmation from Champagne, Domaine Carneros is the investment of Taittinger in the Carneros section of Napa Valley. Founded in 1987, Domaine Carneros celebrated the longest tenured sparkling winemaker in the United States through founding maven, Eileen Crane and her 42 years. The estate sparkling wine offers the flavorful charm of the San Pablo part of Napa Valley. Their top cuvée, Le Rêve, their late disgorged, and their single vineyard wines are among the best in the region.

Dominus

Yountville

www.dominusestate.com

Christian Moueix had to sneak away from Bordeaux to establish his beloved Dominus. He managed to plant his vineyard on the site of the very first vines planted by the region's first settler outsider, George Yount, before California was even a state. With winemaker Tod Mostero in the cellar, Dominus has delivered a vision that's remained constant since its founding: to capture the intrinsic character of where the wines are grown. They are thought-provoking, decisive, and satisfying. The vines are dry farmed organically with techniques adapted to the unique nature of the site. Dominus also undertakes habitat and creek restoration, and encourages biodiversity throughout the soils, vineyards, and surrounding property.

Eisele

Calistoga

eiselevineyard.com

The historic Eisele vineyard remains one of the most coveted sources of historic wine in Napa Valley. In the 1970s, Ridge, Conn Creek, and Joseph Phelps made vintages from it that sell for thousands on the resale market. Their bottles are dwindling. Soon after, Bart and Daphne Araujo purchased the site, renamed it Araujo and farmed it biodynamically. Winemaker Françoise Peschon made clear the quality possible from the rare alluvial fan on the east side of the valley. The wines reveal the rocky, savory character of the vineyard while capturing the finesse and acidity possible but less often seen from Calistoga. The name, Eisele, has since been restored, and the vineyard is owned by Artemis Domaines and François Pinault. The new team has brought a honed refinement, transparency, and detail to the wines that is thrilling.

Etxea

Napa

etxeawines.com

In 2014, Luisa Bonachea and Ryan Pass decided to start a winery partnership. Ryan already had a decade of experience as a winemaker and viticulturist. Luisa was well established in trademark law, working with brands in the industry. Together, they devoted their project to Albariño from the schist- and limestone-ridden slopes of Rorick Heritage Vineyard in Calaveras County, and Cabernet Franc from vineyards throughout northern California. The result is mouth-watering, refreshing, flavorful wines that are highly drinkable and delicious.

Favia

Napa

www.faviawines.com

The extensive viticultural experience of Annie Favia, and the winemaking of Andy Erickson culminated in the creation of Favia. This husband and wife partnership delivers single-site Cabernet Sauvignon from vineyards throughout Napa Valley. Their own Coombsville estate is one of the most exciting examples. The natural acidity of the area combines with rocky, mineral length for a wine that's finessed, mouthwatering and savory. It's 100 percent Cabernet and a celebration of the structure and character that made it one of the favorites of wine lovers worldwide.

Frog's Leap

Rutherford

www.frogsleap.com

Organic vintner Frog's Leap was founded on the friendship of John Williams and Larry Turley (who went on to found his eponymous winery also farming organically). The Frog's Leap ethos has been staunchly based in organic and dry farming since its beginnings, and Williams remains a proponent of both. The original Frog's Leap estate vineyard, started with Williams' former wife, Julie Johnson, was the first to be organically certified in Napa Valley. Today, that vineyard continues to be farmed organically by Johnson for her Tres Sabores wines. Their son Rory Williams now works with both wineries, helping to farm Tres Sabores, while also helping to lead the vineyards and winemaking at Frog's Leap. The red wines remain textural, satisfying, and ageworthy. Each of the Zinfandel, Merlot, and Cabernet is a standout, in a classic, verging on rustic, but still finessed style.

Grgich Hills

Rutherford

www.grgich.com

The biggest biodynamic vineyard farmer in Napa Valley for a time, Grgich Hills now farms vineyards under the Regenerative Organic Certification. The farming and winemaking are led by Ivo Jimenez who reports that the health of the vines, and their resilience against both mildew and wildfire smoke, have increased since farming regeneratively. As the soil health has increased, so has the density of the cuticles on the vines and berries. Mike Grgich co-founded Grgich Hills with Austin Hills soon after the success of his 1973 Chateau Montelena Chardonnay in the 1976 Paris Tasting. Since 2006, the winery has relied entirely on their estate vineyards. The same year they also converted the winery to solar power.

Harlan

Oakville

www.harlanestate.com

Few people have had as much impact on the development of Napa Valley as Bill Harlan. Influenced by the commitment to community and collaboration demonstrated by Robert Mondavi, Harlan continues to share his Meadowood Resort property for some of the region's premium events. In 1984, he purchased a 240-acre (97-hectare) property on the steep slopes in the Mayacamas west of Oakville. Forty acres (16 hectares) of it are planted to a mix of Bordeaux varieties, primarily Cabernet Sauvignon. Bill Levy serves as founding winemaker and has mentored Cory Empting as winemaker for decades. The multi-generational perspective of Harlan includes not only the Harlan family but also the mentorship established through each essential position in the business from vineyards to winemaking, business, and sales. Harlan remains one of the founding cult wines of Napa Valley and is highly ageworthy. In 2020, the Harlan team successfully transitioned most of the vineyard to dry farming. The result has been earlier harvest, which has helped them avoid the impacts of smoke exposure. It also led to them evolving the wine's style to a fresher, redder fruited Cabernet blend with mouth-watering acidity and good ageworthiness.

Heitz

St. Helena

heitzcellar.com

Joe Heitz made a name for his winery when he partnered with Tom and Martha May to make a single vineyard designate wine from their Martha's Vineyard. Older vintages of Heitz Martha's Vineyard remain among the best finds of historic wines from the region. It ages for decades, with tannins the texture of cashmere. The aromatics are distinctive but it's the texture I can never get enough of. Today, Heitz relies primarily on Cabernet grown from its historic Trailside vineyard. The team has worked to intimately study the site's geology and adjust farming to match its specific needs. Geologist Brenna Quigley has helped identify the soil variation of the vineyard, which now guides the vinification choices in the winery. The wines continue to be made in the 1898-era stone building Heitz purchased to start his winery in the 1960s.

Inglenook

Rutherford

inglenook.com

Launched in 1879 by a Finnish sea captain, Gustave Niebaum, Inglenook winery was one of the founding wineries of Napa Valley, and then one of the new pioneers after the end of Prohibition. Niebaum sailed the world, including to my home, Alaska, before settling near Rutherford seeking to make wine to rival the best of Europe. The founder's great-nephew, John Daniel, Jr., restarted the winery following Prohibition and made what many regard as among the first great wines of Napa Valley. In the mid-1970s, Francis Ford Coppola purchased the property, first restoring the winery, then also purchasing its adjacent vineyard, thereby reuniting the historic estate. In 2011, he also repurchased the name Inglenook which is unified once again with the property.

Joseph Phelps

St. Helena

www.josephphelps.com

Joseph Phelps made a name for himself creating wines from some of the finest sites in Napa Valley. During the tenure of winemaker Walter Schug, Phelps bottled the first single-variety Syrah in California as well as its first proprietary Bordeaux-style red blend. He also made the first house wines, two Zinfandels, for Chez Panisse in Berkeley. Phelps was also one of the first to venture into the far western side of Sonoma County near the shores of the Pacific in the Freestone area of the West Sonoma Coast. There he began growing Pinot Noir. Today, Joseph Phelps practices regenerative farming techniques and is a leader in Fish Friendly Farming.

Keplinger

Napa

keplingerwines.com

Helen Keplinger has worked with some of the most famed wineries of Napa Valley: Bryant Family, Kenzo, Grace Family Vineyards. In 2006 and 2008, Keplinger and her husband, D. J. Warner also started Keplinger wines and then Vermillion, respectively. Vermillion offers Rhône wines sourced from the North Coast and Sierra Foothills. Keplinger serves as Helen's higher end devotion to wine. These are single-vineyard expressions of Rhône varieites from high elevation sites in the Sierra Foothills and Sonoma County. The wines are savory, juicy, thoughtful, finessed, and fully flavored.

Louis M. Martini

St. Helena

www.louismartini.com

Immediately following the end of Prohibition, Louis M. Martini built his winery in St. Helena. It was an important commitment in the region during uncertain times. Martini's role as one of the new founders of the area after Prohibition extended not only into winemaking but also vineyards, establishing them in new regions, honing mountain viticulture, and making some of the first, most ageable wines in Napa's then-new era. Martini was one of the first to plant in the Carneros region. He also purchased an 1800s-era planting, the Monte Rosso vineyard at the top of Moon Mountain, and began making some of the early mountain wines of the North Coast. Today, Martini's original winery has been beautifully restored and offers a range of wines alongside an impressive food program.

MacDonald

Oakville

macdonaldvineyards.com

Brothers Alex and Graeme MacDonald are the third generation of the family farming one of the oldest remaining sections of To Kalon vineyard. Their grandfather was considered one of the best farmers in Napa Valley when Robert Mondavi started working with him to source fruit for Charles Krug. The site also introduced him to To Kalon, which Mondavi helped make famous. They make a single wine, a Cabernet Sauvignon from older vines in the rocky, well-draining bench of the Mayacamas mountains. The wine has modest production size and can be hard to purchase, but is textured, erudite, restrained while savory and well worth aging.

Matthiasson

Napa

www.matthiasson.com

Steve Matthiasson and Jill Klein moved to Napa Valley in the early 2000s and successfully launched a family farm. It's a rare accomplishment given the more recent cost of living in Napa Valley. Matthiasson got his start helping to create one of the first integrated pest management programs in the country, in Lodi. He then moved to Napa Valley to serve as a consulting viticulturist. He has worked with some of the most admired vineyards in the region, including Eisele in Calistoga, Stag's Leap Wine Cellars in the Stags Leap District, Spottswoode in St. Helena, Ashes & Diamonds in the town of Napa, and others. He also farms the vineyards from which he and Klein make wine. Matthiasson has become part of the siren call of New California as described by Jon Bonné. Their wines arrived on the lower side of alcohol, with resplendent acidity, and a decided foodiness to accompany a meal, especially the fresher side of California vegetables and cuisine. Their red and white blends are compelling. They've also invested in more unusual varieties, Schioppentino, Refosco, and Ribolla Gialla among them.

Mayacamas

Napa

www.mayacamas.com

One of Napa Valley's historic wineries, Mayacamas was built in the late 1800s in the elevated peaks of Mount Veeder. The site is surrounded by historic oaks and perched against a canyon looking towards San Pablo Bay. Originally growing Zinfandel, after Prohibition Mayacamas began to make classic expressions of mountain Cabernet, enticing Chardonnay, and Sauvignon Blanc. In 2013, the Schottenstein family purchased the estate and revitalized the vineyards and winery. They rely on organic farming, with some biodynamic practices incorporated. They continue to make their three main varieties, and now also offer them certified kosher.

Newfound

St. Helena

www.newfoundwines.com

Founded by husband and wife team Matt and Audra Naumann, Newfound combine estate farming in their Sierra Foothills vineyard with wines made from distinctive vineyards through the North Coast. With a primary focus on Rhône varieties and Carignan, they also make surprises from iconic sites such as Pinot Noir from Enz vineyard in Lime Kiln Valley, and Semillon from Yount Mill in Napa Valley. The wines are mouth-watering with mineral backbone and consistently reflect the character of the sites where they are grown.

Opus One

Oakville

www.opusonewinery.com

Founded as a partnership between Robert Mondavi and Baron Phillippe de Rothschild of Chateau Mouton Rothschild, on opening Opus One immediately became one of the destination wineries of Napa Valley. The investment from Rothschild was a testament both to the quality potential of Napa Valley, and the marketing acumen of Mondavi. His ability to showcase enthusiasm for his region's wines helped make it one of the best-known in the world. While 1979 was the first vintage made of Opus One, it was not released until 1984. The extended aging, close vine density, and combined experience of Bordeaux elite with the familiarity of Napa's best created one of the region's iconic red wines.

OVID

St. Helena

ovidnapavalley.com

Perched on the edge of Pritchard Hill, overlooking a canyon on the eastern side of Napa Valley, OVID remains one of the most dramatic and beautiful sites in the valley to visit. Organically farmed on shallow mountain soils at 1,349 feet (411 meters), OVID makes four distinct cuvées from the estate. Each wine carries the mark of its surrounding landscape: the resinous aromas of chaparral, the gunsmoke accent of iron-rich soils, and the resplendent, rugged depth of the sun-exposed Vaca mountains. The flagship wine, OVID, brings together Cabernet Sauvignon with Cabernet Franc, Merlot, and Petit Verdot. The blend changes by vintage to express the best of the year. Hexameter centers on Cabernet Franc, capturing the aromatics, darker fruit, and structure of the variety. They also make a single block expression of mountain Cabernet Sauvignon, and a mountain Syrah.

Promontory

Oakville

www.promontory.wine

Before purchasing the estate that became Harlan in the Mayacamas mountains of Oakville, Bill Harlan spotted the property he now owns, which he named The Territory. It's a 20-acre (8-hectare), rugged, largely hidden, mountain-edged bowl to the south of Harlan. The site is surrounded by forest and invisible from the busy parts of Napa Valley. The unique land formation provides some degree of shielding from the sun throughout the day. It's also the only section of Napa with vines growing entirely in metamorphic parent material. Son Will Harlan created and leads the Promontory project in partnership with long-time winemaker Cory Empting. The result is enticing, sophisticated, and impressive. Its first vintage was 2008, but the wine already appears beautifully ageworthy.

Quintessa

St. Helena

www.quintessa.com

Farmed biodynamically since its founding, Quintessa was established with a commitment to make a single wine that expresses the complex intricacy of the estate. The grove of historic native oaks was retained. Vines were planted around the already established trees, rather than clearing any for planting. Each section of the vineyard is farmed uniquely to its needs then vinified primarily in concrete, delivering a truer expression of the site it was grown in by reducing the impact of new oak. Consistently a Cabernet-based Bordeaux blend, Quintessa today is making the best wines in its history, and one of the top wines of the region.

Robert Biale

Napa
www.biale.com

The history of Napa Valley rests, like much of California, in head-trained, historic Zinfandel plantings. There are far fewer today since the rise of Cabernet Sauvignon. Those that remain survived because of their exceptional quality and thanks to a few passionate winemakers. Robert Biale is an important part of this effort. The winery makes wines from some of the most beloved, historic vineyards in Napa Valley, RW Moore from Coombsville and Dickerson in St. Helena, as two examples. Biale wines consistently show what made Zinfandel California's signature grape: ripeness, richness, and restraint. Zinfandel is more commonly remembered today by its stereotype: largesse, impact, and alcohol. Classic expressions of the variety deliver the profound influence of where it is grown, including its distinctive structure. Producers still committed to this cause reveal that in a state like California, Zinfandel at its best might be even more idiosyncratic and enticing than Pinot Noir. Robert Biale wines are a demonstration of this assertion.

Rombauer

St. Helena
www.rombauer.com

Rombauer helped bring consumer drive to white wine, specifically Chardonnay, for Napa Valley. Capturing the spirit of California, with its feeling of independence, sophistication, and pleasure, Rombauer became a defining wine in the US market. Koerner and Joan Rombauer started the winery in 1980, making wines from Napa Valley and Sonoma County. Though famous for Chardonnay, Rombauer also makes a collection of red wines, the most intriguing showcasing the mountain structure of the Sierra Foothills. In 2023, they also began making Pinot Noir from the Santa Lucia Highlands. Their first Chardonnays were made with fruit purchased from the Sangiocomo family of Sonoma; the same vineyards that helped launch the 1973 Chardonnay from Chateau Montelena that won top spot in the 1976 Paris Tasting. Building on the success of the Chardonnay, Rombauer began making Sauvignon Blanc in 2014. Rombauer Chardonnay continues to draw wine lovers from throughout the United States.

Schramsberg

Calistoga

www.schramsberg.com

Jacob Schram founded one of Napa Valley's first great wineries in 1862. A century later, the vineyards and winery still stood at their original home but needed significant restoration. Jack and Jamie Davies purchased the site and began restoring the property, while making some of California's first and best traditional method sparkling wines. They remain standout wines and represent the history of California integrated with modern skill. Among their most distinctive wines they make a late disgorged sparkling from Carneros fruit aged for 19 years on lees. The savory quality is profound. Schramsberg wines have enjoyed national recognition, poured by American presidents for the world's top dignitaries.

Screaming Eagle

Oakville

www.screamingeagle.com

Founded in 1986 by Napa local Jean Phillips, Screaming Eagle has become one of the most enigmatic wines in the region. The winery makes two cuvées, the Cabernet-centered Screaming Eagle, and the Merlot-based The Flight. The Merlot is distinctive, a standout made from vines planted in gravely soils on the lightly sloped plateau beside the winery. Heidi Peterson Barrett was Screaming Eagle's first winemaker, introduced to Phillips by her friends and neighbors at Dalla Valle. The first vintage earned a near-perfect score from Robert Parker, immediately making it one of the hardest to get wines in Napa Valley. It is considered one of the first cult wines of the region and is sold primarily through a highly allocated membership list.

Shafer

Napa
shafervineyards.com

In the 1970s, John Shafer purchased a steep-sloped site in the Stags Leap District on the eastern slopes of Napa Valley. The site had been farmed, making award-winning wine in the 1880s, until the advent of phylloxera in the region. John established Cabernet Sauvignon and the winemaking in the early years of Shafer estate. In 1983, John's son Doug Shafer entered the business, joined the following year by winemaker Elias Fernandez. Together, they created the winery's flagship, Hillside Select. Fernandez has been recognized by the White House as a leader in the nation's Latino community. His winemaking has earned him numerous accolades. Hillside Select demonstrates the sun exposure and rugged soils of the Stags Leap District with power held strong with poise. The wines are fully framed while also savory and full of mineral length.

Spottswoode

St. Helena
spottswoode.com

One of the unquestionable leaders in sustainability efforts in California, Spottswoode winery stands strongly behind the cause with a host of certifications to prove it, as well as thoroughgoing efforts to provide consumer education where appropriate. Founded in the early 1980s by Mary Novak, the winery now enjoys second-generation leadership from daughter Beth Novak Milliken. Spottswoode serves as a member of the International Wineries for Climate Action, working diligently to reduce its carbon footprint while helping others learn how to do the same. They earned the challenging B Corporation Certification that not only demonstrates commitment to environmental sustainability but also requires both business and employee commitments. They are certified organic, biodynamic, and regenerative organic, a member of the Napa Green sustainability program, and part of both 1% for the Planet and the Napa County Land Trust, preserving open spaces in the region. They are known for their ageworthy, plush while refined, structured Cabernet Sauvignon, and their Sauvignon Blancs are also standout wines. The harder to find Mary's Block remains one of the best in the region.

Petaluma Gap

As temperatures rise in the inland part of Sonoma and Marin Counties, cold air is pulled through a low point in the coastal mountains into the eastern reaches of the regions. The low point in the mountains is affectionately named the Petaluma Gap, thanks to its location directly west from the town of Petaluma.

The effect of the Petaluma Gap is an intensity of daily wind and fog exceeding that of surrounding areas. Long a winemaker's secret, vineyards within what is now the Petaluma Gap AVA were an important part of the structure and natural spice of wines from the rest of Sonoma. Few estate wineries exist in the area. Instead, most wines from the area are made by wineries outside the gap.

As local recognition for the area began to grow, local growers decided to study the region and identify its unique characteristics. The result is the Petaluma Gap AVA, the first in the country defined by its perimeter of wind. Through a battery of weather stations, the air flow of the region was studied and the areas consistently receiving winds of 8 miles per hour (13 kilometers per hour) or greater were demarcated as within the influence of the gap. Those became the outer limit of the AVA.

The presence of wind in the region is important. Studies have shown that winds of 8 miles per hour and above slow the respiration of vines, and thus help preserve their natural acidity. Ripening and flavor development continue while acidity remains. Wind exposure also tends to yield fewer and smaller berries, increasing the skin-to-juice ratio and thus also tannin presence in red wines. Wines from the Petaluma Gap carry a distinctive spice element that often leads people to believe that whole cluster fermentation has been used when wines have been fully destemmed. The Chardonnays offer a balance of rich texture and abundant acidity.

The Petaluma Gap primarily grows Pinot Noir and Chardonnay but the few Syrah vineyards from the area are thrilling. Keller Estate is one of the region's important estate wineries and has helped bring attention and understanding to the area. The Petaluma Gap extends into Sonoma County, where most of its wineries are found.

Stag's Leap Wine Cellars

Napa

www.stagsleapwinecellars.com

Founded at the start of the 1970s by Warren Winiarski, Stag's Leap Wine Cellars is famous for its top-spot finish in the 1976 Paris Tasting.

Winiarski's legacy goes much farther. He mentored many of the next generation of Napa's top winemakers, and helped establish the country's first agricultural preserve in the valley. After selling the winery, he used the money to establish extensive philanthropic projects to benefit winemaking students, the Smithsonian Museum, higher education, wine writers, and more. Today, the Antinori family of Italy owns the estate and continues to make the cuvées Winiarksi founded. They are in the process of converting and certifying the vineyards to regenerative organic farming, have integrated sheep into the vineyard, and are intentionally focused on fostering soil health.

The Vice

St. Helena

www.thevicewine.com

Born from the creativity of founders Malek and Torie Greenberg, The Vice represents a new generation of Napa wine. Combining skilled winemaking with clever labels and marketing, the wines' eye-catching presentation attracts a broader, playful audience while delivering the depth and flavor Napa is known for. The Vice makes a series of small-batch wines released as their own idiosyncratic bottlings, while also anchoring the portfolio in recognizable cuvées available in most vintages. Malek originates from Morocco and served as a Team USA athlete before moving to wine. Torie attended one of the top design schools in the country and brings her experience to the visionary labels. Together they are expanding the audience of luxury wine at a time when that's just the push Napa Valley needs.

MARIN AND SOLANO COUNTIES

Marin and Solano Counties are two of the lesser-known parts of the North Coast region. Still, they appreciate the ocean influence and volcanic mountain formations of the larger appellation.

Marin County remains underdeveloped with vines. Oceanside homes and large elevation climbs on road bike are more common. As a result, most wine geeks drive through the area to reach Sonoma or Mendocino to the north. But science fiction aficionados, rejoice! George Lucas of *Star Wars* fame has invested in Marin County and there grows wine to make his Skywalker wines. Marin includes one officially recognized AVA, the Petaluma Gap which also continues into Sonoma County.

Solano County connects to the higher elevation sections in the southeast of Napa County. Wild Horse Valley AVA continues into both counties. German immigrants settled the Solano area after the creation of the state. Before them, the Patwin, also known as the Suisin, resided through the region. While Solano is less known for vineyards than its neighboring counties, it provides a full gustatory experience for people visiting to taste wine. Vineyards are coupled with farms and olive groves allowing for farm stands, and tasting experiences coupled with locally grown and made foods.

Marin County

Known for: Point Reyes, Muir Woods, Stinson Beach

Lesser-known strength: once the home of Julia Child, Janis Joplin, and Marilyn Monroe

Leader in: convincing sci-fi legend George Lucas to make wine

First peoples: Miwok, Pomo

First AVA: Petaluma Gap (also in Sonoma), 2017

Marin County AVA

Petaluma Gap (2017): defined by wind, overlaps Sonoma and Marin Counties

Main varieties: Pinot Noir, Chardonnay, Syrah

Solano County

First peoples: Patwin

First AVA: Suisun Valley, 1982, Solano County Green Valley, 1982

Number of AVAs (2024): 4 (2 overlapping other counties)

Solano County AVAs

Solano County Green Valley (1982): abuts the Wild Horse Valley AVA

Main varieties: Syrah, Pinot Noir, Chardonnay, Pinot Gris

Suisun Valley (1982): directly east of Solano County Green Valley AVA

Main varieties: Petite Sirah, a range of small varietal plantings

Wild Horse Valley (1988): continues into Napa County

Main varieties: Pinot Noir, Chardonnay

Winters Highlands (2023): in both Solano and Yolo Counties

Main varieties: Petite Sirah, Tempranillo, Zinfandel

8

THE SANTA CRUZ MOUNTAINS

California's first great sparkling wine came from the Santa Cruz Mountains, along with some its most influential winemakers, and possibly its most quixotic. It was the first AVA in the country to be defined by elevation *above* the fog-line. And it remains home to some of the most salient vineyards and winemakers today.

Look at any California map, and the Santa Cruz Mountains are right there, tucked along the Central Coast. But, when it comes to wine, they're entirely separate. When the Central Coast AVA was established in 1985, the Santa Cruz Mountains AVA had already enjoyed several years on its own. Considering the breadth, the reliance on the ocean and a series of transverse valleys, as well as a different range of soil types, the proposed Central Coast looked like a big contrast to the Santa Cruz Mountains. So wineries in the mountain AVA petitioned to be excluded. Today, the Santa Cruz Mountains continues to be its own geographical region, standing as its own legally recognized appellation, excluded from the much larger, surrounding Central Coast.

Founded in 1981, the Santa Cruz Mountains segregated themselves from the lower elevations wrapped around their base and created a region lifted above the valley floor. On the Pacific side, the Santa Cruz Mountains start at 400 feet (122 meters) in elevation. Along the eastern slopes, its lower limit stands at 800 feet (244 meters). Facing inland, the eastern side experiences a different relationship to the neighboring bodies of water, and thus the upper limit of the fog-line differs too.

The craggy peaks and folds of the Santa Cruz Mountains were lifted by the slippage of the North American and Pacific plates along the coast of California. The Farallon plate, now mostly subsumed under the continent, plays an important role here as well. Portions of it have been accreted along the coast, delivering uplifts of ancient seabed to the highest peaks of the mountains. Redwood forests cover the western sides of the mountains from the shore to the peaks. Most of the trees are considered second-growth, returned trees from after the clear cutting of the late 1800s. Even so, small sections of the forest include old-growth trees that are a few hundred years old. Redwoods create their own mesoclimates, surrounding themselves with humidity and an insulating effect that blocks cell phone service and other wireless technologies. The quiet from technology coupled with a feeling of resonance in the forest itself make the Santa Cruz Mountains one of the most refreshing and beautiful parts of California.

Most of the western side is a temperate rainforest that makes the redwoods possible. Mixed throughout are also some of California's other hallmark trees: live (evergreen) oak, madrone, bay laurel, and black oak. Sections of Douglas fir swirl throughout. These cooler, western slopes welcome a host of deer, the odd black bear, and occasionally cougars or coyote. The western side also includes one of the state's most unusual environments, the sandhills habitat. Here, forests emerge from sands uplifted from ancient seafloor. One of the state's most widely spread trees, the Ponderosa pine, grows near one of its rarest, the Santa Cruz cypress, which is found only in these sands.

Santa Cruz Mountains AVA (1981)

Known for: Ridge Monte Bello, pre-industrial winemaking

Lesser-known strength: some of the most winding winery roads in California

Leader in: heritage selections for Chardonnay and Pinot Noir, influential winemakers

First peoples: Awaswas

First vines: 1870

First AVA: Santa Cruz Mountains, 1981

Number of AVAs: 2

Planted vineyard area (2024): 1,600 acres (647 hectares)

The eastern slopes of the mountains are drier. Here, the highest peaks of the mountains are found, covered in a mix of chaparral, scrub oak, and manzanita. The soils are more exposed, the slopes and peaks more influenced by overhead sun. Rattlesnakes living in the area mean it's best to wear close-toed shoes. A rain shadow shields the eastern side of the mountains from precipitation. So while much of the west of the AVA lifts through temperate rainforest receiving around 50 inches (127 centimeters) a year, in the east rainfall hits around half that at 25 inches (64 centimeters) annually.

The elevation of the Santa Cruz Mountains creates significant temperature differences throughout the area. As air descends from the mountain heights, down their slopes, and into the valley floor, vineyards are cooled significantly, causing an important difference in temperature between the peak of daytime and the coolest night-time readings. Cold air settling into the valleys at night brings a risk of frost early in the growing season. Areas above the inversion layer experience generally more even temperatures between day and night, and snow accumulation rather than freeze over winter.

The highest peak in the Santa Cruz Mountains, Loma Prieta, reaches 3,786 feet (1,154 meters). Vineyards begin at the lowest elevations of the AVA and reach up to 2,700 feet (823 meters). The region's most famous site, Monte Bello, sits between 1,300 and 2,700 feet (396–823 meters). Though the AVA covers around 480,000 acres (194,249 hectares), only 1,600 acres (647 hectares) are planted to vines. The rugged terrain limits further planting. The sheer size of the region compared to its planted area offers an independence and privacy wineries of the area regularly mention. If you can't see a neighbor's vineyard, their farming and harvest decisions are less likely to impact yours.

Prudy Foxx, Santa Cruz Mountains viticulture

Ocean fog, ocean wind, the rainforest. Bay fog, limestone soils, steep slope vineyards. The challenges of the Santa Cruz Mountains mean successful wineries depend on viticulturists that understand how to find balance between the tension points of disease pressure, sun exposure, shallow soils, smaller canopies to the vines, smaller berries, vine stress, and varied access to water.

Viticulturist Prudy Foxx began her vineyard career alongside Randall Grahm at his Bonny Doon, where she worked for almost 10 years, starting in the winery's founding year. Later, she founded Foxx Viticulture with her geologist husband, Mark.

Known for its shallow topsoil and challenging soil types, many derived from limestone or serpentine, in the Santa Cruz Mountains Foxx believes a focus on microbial life of the soils is essential. Foxx works to build vineyard health through plant resilience, while respecting the need for occasional water supplementation (even if not regular irrigation) during droughts or in marginal climates. Though mountains are known to be among the most challenging environments to practice organic or biodynamic viticulture, Foxx relies on both for the sake of not only improving soil health but also expanding one's view of vine health beyond the plant itself to its broader environment.

In describing the needs of vineyards in the Santa Cruz Mountains, Foxx emphasizes the region's high variability and ecological diversity. As she describes, near the top of the mountains, there is more rock. Down the slopes eroded colluvial soils eventually mix with alluvial soils carried from the creeks and rivers. In the sandhill areas, vineyards are planted in exactly that, sand.

Proximity to the ocean increases maritime influence, but elevation also decreases it. The higher elevation areas gain more radiant heat, while the lower elevations are chilled by fog and ocean winds. The western ridges are generally uplifted seabed pulled upwards on the Pacific plate. Here, rocks and soils derived from mudstone, siltstone, and sandstone, sometimes with fossilized bands, proliferate.

In contrast, the eastern side, known as the Saratoga Hills, is warmer, with lower disease pressure, more radiant heat, and often enough warmth to ripen Cabernet Sauvignon. But throughout the mountains, the aspect of the slope (or the direction it is facing) determines its ripening potential most of all. A slope facing away from the sun, even on the warmer eastern side of the mountains, might struggle to ripen Pinot Noir.

Foxx has also helped bring influential winemakers from outside the Santa Cruz Mountains into the region, thus widening the influence and acclaim of the region. Arnot-Robert and Ceritas of Sonoma County have both worked with Trout Gulch Vineyards. Pax Mahle sourced fruit from Woodruff in Coralitos. Matthiasson wines of Napa Valley made Syrah from Zayante, a vineyard deeply cooled by Pacific influence. Lioco makes wine from Saveria. All are vineyards farmed organically by Foxx. She has also worked with Santa Cruz Mountains based wineries Big Basin, La Rochelle, Thomas Fogarty, Beauregard, Loma Prieta, and Soquel, among others.

HISTORY

Spanish settlers pushed their way into the cool, coastal redwoods along the northern shores of Monterey Bay and built in what is now the city of Santa Cruz. Around 1791, Franciscan monks established the *Exaltacion de la Santa Cruz*, the region's mission. Before the end of the century, they had also planted *vinifera*. The cooler temperature of the region made ripening a struggle. Some believe that, even so, the neophytes of the mission were able to make small quantities of Angelica but other evidence suggests the vines never adequately ripened. After missions were secularized in the 1830s, the site at Santa Cruz was likely abandoned. The reduced yields and underripe fruit seem to have made farming there difficult. Even so, new settlers to the area in the 1860s spotted the abandoned, feral growing *vinifera* and were inspired to plant vineyards of their own.

The area of the Santa Cruz Mission sits along the river in what today is known as Harvey West Park. Because the area falls below the 400 feet (122 meters) restriction of the Santa Cruz Mountains AVA, the vines of the mission that helped establish the town of Santa Cruz are outside the boundaries of the appellation. The first vines within the now-recognized growing region were planted in the 1860s by the entrepreneurial and forward-minded brothers Jarvis. Together, John and George purchased their Rancho San Andreas in the hills over what is today the town of Scotts Valley. There, they began the first plantings of vines in the area to make significant impact. Their 300 acres (121 hectares) of grapevines demonstrated that up the slopes temperatures were warm enough to grow wine. Within a decade, commercial winegrowing was established in the region. Immigrants from throughout Europe established more parcels alongside fruit trees, and a new local industry began.

By the 1880s, grapes proliferated into the eastern slopes of the mountains as well. Overlooking the Santa Clara Valley, the Santa Cruz Mountains offered proximity to town life on more affordable mountain land. Attorney Emmet Rixford established La Questa winery in the area in 1884, above what today is the town of Woodside. Vineyards dotted the forest around La Questa, but Rixford was the only one to make wine at volume. La Questa gained favor from consumers and won important awards of the time. By the time of the 1915 world's fair in San Francisco, La Questa had become one of California's celebrated producers. Five years later, its wine production was closed by Prohibition.

In the late 1800s, Paul Masson began his career in California wine, first working as apprentice to Charles Le Franc who founded Almaden in the Santa Clara Valley. In the late 1890s, Masson brought new vine cuttings of Pinot Noir and Chardonnay from France. They would become the foundation of his own Paul Masson Mountain Winery in the hills overlooking Saratoga. Focused on sparkling wine, Masson became known as the 'Champagne King of California' (the style was still named after its French counterpart at the time), and his wines became a coveted staple for Hollywood celebrities, internationally renowned singers, musicians, artists, and socialites. His regular parties drew glitterati from everywhere within travel distance. One of them led to the famed story of actress Anna Held "taking a bath in Champagne" and gave him and his winery a notorious reputation.

Masson's vineyard in the Santa Cruz Mountains founded some of the state's first important heritage selections of Chardonnay and Pinot Noir. He went on to become the mentor and friend of Martin Ray, the temperamental, isolationist proponent first of Paul Masson Champagne Co., then his own Martin Ray vineyard and winery, today known as Mount Eden. Ray was one of the first boutique winemakers of California post-Prohibition to gain a sort of icon status with young, aspiring winemakers across the United States. The best of his wines was considered among the best wines ever made. The rest of his wines weren't. The inconsistency in quality seemed to come from Ray's own pre-industrial winemaking taking an extreme approach – not only minimal sulfur, but probably also minimal hygiene. Even so, he drew a fan base still talked about in American wine today. Among his admirers was Warren Winiarski, who had his first vineyard job in California under the guidance of Ray. After a week, Ray sent Winiarski back home to Chicago saying he was too independent to work for Ray. Every story describes Ray as even more independent than Winiarski.

Regardless, the legacy of Martin Ray provides a link between the historical legacy of Masson's first vines, and the classic wines of California today. The Mount Eden selections of Chardonnay especially, as well as Pinot Noir, are among the most admired of California's heritage selections. While the experience Winiarski had with Ray might not have been ideal, it nevertheless taught Winiarksi his winegrowing dream was possible. Winiarski went on to become founding winemaker for Robert Mondavi of Napa Valley in 1966, and then to found Stag's Leap Wine Cellars, going on to make the wine that won the Paris Tasting of 1976 (see box, p. 164) and to mentor many of Napa Valley's most celebrated winemakers (see p. 165).

SINCE THE 1960S

In 1961, winemaker David Bruce moved to the Santa Cruz Mountains, planted vineyards, and established a winery under his own name. Pinot Noir was rare in the state at the time. Although Masson had established it, and Ray had gone on to expand it, few others had invested in giving it attention. Bruce entered his wine career set on Pinot Noir, planting it close to 2,100 feet (640 meters) above the town of Los Gatos, though admittedly he first planted Cabernet Sauvignon. As he settled into the world of Pinot Noir, Bruce became among the first in the state to use whole-berry fermentation (also known as carbonic maceration) for red wines, and to soften his whites by intentionally using secondary, malolactic fermentation (MLF). Many whites at the time relied on such high sulfur levels, they didn't go through MLF. While Bruce was not the only winemaker to rely on small-barrel fermentation for his wines, he was among the very first to use spinning, rotary tanks, and to also fruit tread for extended skin contact before fermentation. Bruce is regarded as a Californian pioneer of modern winemaking, especially in Burgundian varieties.

While Prohibition closed La Questa, Rixford's influence goes beyond the winery. In the 1880s, he wrote a manual for the cellar side of fine wine. His writing in *The Wine Press and the Cellar* became the guidebook Paul Draper used to learn winemaking, first while running a winery in Chile, and then after his return to California and move to the Santa Cruz Mountains to become head winemaker of Ridge. Rixford's book became the basis for the 'pre-industrial' winemaking Ridge continues to define itself by today.

Draper, in turn, became a legend of California wine, working with one of its historic vineyards of the Santa Cruz Mountains, and some of its highest elevation. The Monte Bello site of Ridge reaches up to 2,700 feet (823 meters). In the 1960s, the property was purchased by a group of friends from the Santa Clara Valley. While enjoying the site for weekends, they discovered the remnants of a vineyard established by Osea Perrone in 1885. There, Perrone had also built his Monte Bello winery; both the vines and the cellar were dug into the area's limestone. The wine of Ridge Monte Bello today is considered one of the top Cabernet-based blends in the world.

At the Ridge Monte Bello winery he helped make famous, Paul Draper opens another historically important California wine, the Hanzell 1965 Pinot Noir. Hanzell founding winemaker Brad Webb played an important role in advancing winemaking in general as well as specifically for Pinot Noir and Chardonnay in the state. Drinking Webb's 1965 Pinot with Paul Draper was one of my top memorable California wine experiences

Down the road and a hilltop over from Ridge Monte Bello, the historic Mount Eden, formerly known as Martin Ray winery, stands perched at 2,000 feet (610 meters) overlooking what today is Silicon Valley. Mount Eden was home to California's first professional woman winemaker, Merry Edwards, who began at the site in 1974. Gaining a reputation for her skill at Mount Eden, Edwards went on to lead one of California's most historically admired wineries, Matanzas Creek, where she helped grow the reputation of wine from the Russian River Valley, before launching her own eponymous brand in the 1990s.

The leadership and influence of Santa Cruz winemakers continues. In 1982, Randall Grahm began his winemaking venture, Bonny Doon,

Randall Grahm founded his Bonny Doon winery in 1982, bringing attention to first the wines of Santa Cruz County, and then California more broadly. Today, he farms his experimental vineyard, Popelouchum (seen here), in San Benito County, and also makes wines for The Language of Yes in partnership with Gallo

and bonded the winery in 1983. Though he famously began his effort with the aim of making one of the world's great Pinot Noirs, Grahm became the country's first "Rhone Ranger." Devoted to Syrah, then a blend of varieties from the Southern Rhône, and today also Grenache Noir, Blanc, and Gris, Grahm's career was guided by the grapes of his French counterpart region, the Rhône. His work proliferated not only a love for Rhône wines, but the recognition California could grow them well.

However, his influence on California and even the world of wine goes far beyond varietal types. He was among the first to turn to screw cap closures for premium wines, in revulsion at the unreliability of cork, as well as one of the first in the country to include ingredient listing on his labels. At the height of his fame, Grahm was one of two people in American wine known beyond the confines of the wine industry, recognized by everyday consumers. The other was Robert Parker. Grahm brought popularity and affordability to California wine through his bigger production labels, Big House and Pacific Rim. He's also been one of the state's great innovators, creating an endless list of experiments leading to his discovery of numerous techniques that have deepened the quality of his wines.

Among these, his use of lees aging might be paramount. Though winemakers the world over rely on aging their wine on the sediment settled in the wine barrel after fermentation, Grahm did so before many others in California, and went on to develop new ways to generate even greater lees exposure. One example is a wooden aging vessel he came to call his lees hotel. Inside, he'd built a stack of alternating shelves that allowed lees to settle at every level of the wine rather than only the bottom of the tank as is most common.

Like Ridge Monte Bello, one of the greatest attributes during Grahm's tenure with Bonny Doon was the profound ageability of his wines. Several years ago, I enjoyed the first wine Grahm ever made, a 1982 claret from Mendocino fruit. The wine, then almost forty years old, was still vibrant. Grahm has since sold Bonny Doon and is now focused on his Popelouchum project in San Juan Bautista. At Popelouchum, Grahm's experimental efforts continue. There he has created a seedling propagation project, growing vines from seeds of first Serine (a relation to Syrah from the northern Rhône), and next Tibouren (an obscure variety from where France meets Italy). When he started the project numerous wine experts told him it would never work. Today he has acres of seedling-bred vines bearing fruit.

Up the length of the Santa Cruz Mountains, the San Andreas denotes the boundary of the North American plate, hosting Ridge and Mount Eden, and the Pacific plate where Rhys was established in 2004, and down the road from it, Thomas Fogarty in 1978. Here, vineyards stand at around 1,500 feet (457 meters), planted in a mix of shale, clay, alluvial sands, and occasionally limestone.

When Kevin Harvey chose to invest in developing several vineyards on the western side of the Santa Cruz Mountains for his Rhys Vineyards it brought international attention to the region. Considered one of the most admired Pinot Noir projects in North America, Rhys has helped demonstrate even further the complexity and nuance possible from the region through organic and biodynamic farming. Rhys was founded in collaboration with winemaker and viticulturist Jason Jardine. Though Jardine had already established himself as a respected manager in top Willamette Valley sites, and before that in Lodi, his experience at Rhys propelled him and the winery to a new level of expertise. Jardine went on to bring his experience in mountain vineyards and holistic farming to Flowers in the Sonoma Coastal Mountains. Today, he leads Hanzell in the Moon Mountain district of Sonoma.

Santa Cruz Mountains AVAs

Santa Cruz Mountains (1981): first AVA in the country to be defined by elevation
Main varieties: Cabernet Sauvignon, Pinot Noir, Chardonnay

Ben Lomond (1988): nested within the Santa Cruz Mountains on its western side
Main varieties: Pinot Noir, Zinfandel, Cabernet Sauvignon, Chardonnay

Wineries

Birichino

Santa Cruz
www.birichino.com

Since 2008, Alex Krause and John Locke have combined forces while sourcing fruit from throughout the Santa Cruz Mountains as well as other parts of the Central Coast to make structurally driven, delicious, sometimes almost quirky wines. Their range of varietal wines lavishes diversity and choice upon the consumer. Pinot Noir, Riesling, and Chardonnay – oh, and Cinsault, Chenin Blanc, Vermentino, Mourvèdre, Carignan, Cabernet Sauvignon, and others – give away the enthusiasm Locke and Krause have for making wine. They are exploratory, adaptable, playful, and committed.

Bonny Doon

Santa Margarita
www.bonnydoonvineyard.com

A leader in the Rhône wine movement, Randall Grahm founded Bonny Doon in 1982 and bonded the winery the following year. He started it to make Pinot Noir, but ended up making racy claret-style wines first, turning later to Syrah and a blend akin to older styles of Châteauneuf-du-Pape. Grahm's marketing acumen sometimes distracts from his talented winemaking. He often anticipated the future of California wine before anyone believed him – like being a mythological Cassandra for the hippies and winemakers of California. Over time, Bonny Doon experimented widely: pure Grenache, Picpoul, sparkling Furmint, apple cider. Today, War Room Cellars owns Bonny Doon and continues to rely on its history of creative labels and boundary-pushing innovation to make wine. Grahm on the other hand makes small quantities of wine from his vineyard, Popelouchum, in San Benito County, as well as Tibouren rosé, a Grenache and Syrah, through the brand The Language of Yes, in collaboration with Gallo.

Mount Eden

Saratoga

www.mounteden.com

One of California's historically important vineyards, Mount Eden has vines that connect back to the establishment of fine wine in California. Founder Martin Ray established Pinot Noir and Chardonnay first, then slightly later added Cabernet Sauvignon. Standing at around 2,000 feet (610 meters) at the end of a thoroughly rugged mountain road, the vineyard remains not as eccentric as its founder, but unique. Winemaker Jeffrey Paterson has guided the wines since the late 1970s, bringing Mount Eden into the modern age while preserving its classic approach. Recently the wine team expanded its production into a more affordable sister winery with the same varieties, Domaine Eden.

Rhys

Los Gatos

rhysvineyards.com

Committed to cooler climates, mountain vineyards, and regenerative farming, Rhys has invested in plantings primarily in the Santa Cruz Mountains, more recently in Anderson Valley, and previously in Sonoma County. The wines are edgy and textural, an exploration of what it means to make distinctive wines from sites on the threshold of ripening. Rhys has commanded a devoted following. The wines are erudite, thought-provoking, and savory.

Ridge

Cupertino

www.ridgewine.com

Paul Draper has mentored and inspired generations of winemakers since his start at Ridge in 1969, both within California and far beyond. His wines and his influence are mentioned worldwide. The winery was founded in 1959 by three Stanford scientists, then named Ridge in 1962. Monte Bello is considered one of the best Bordeaux-blend wines

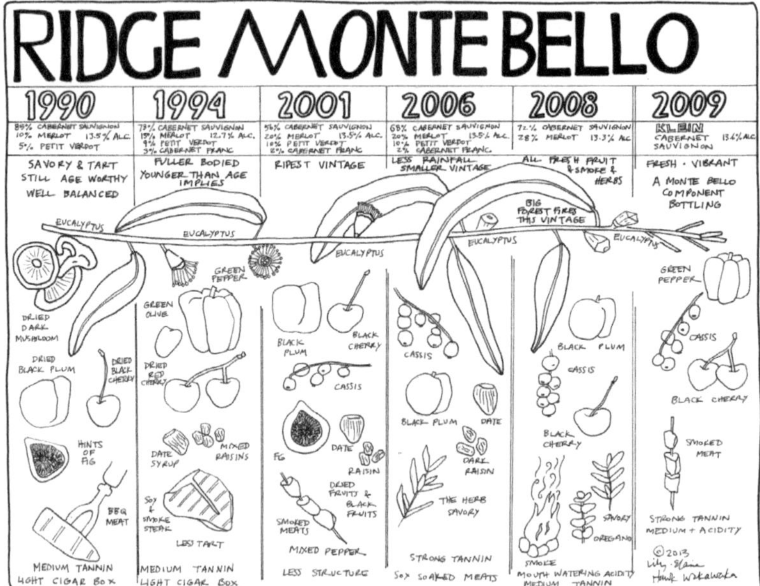

Ridge's flagship red blend, Monte Bello, is considered one of the best red wines in the world. Grown up the steep slopes of the Santa Cruz Mountains, it has become one of the iconic wines of California, made from a Cabernet Sauvignon-based blend along with varying amounts of Merlot, Cabernet Franc, and Petit Verdot

in the world. The winery has brought great success to Zinfandel from historic vineyards as well, including some down the slopes of Monte Bello. In the mid-1990s, they built a second winery at Lytton Springs in Dry Creek Valley. Having already worked with the Lytton Springs site in Sonoma for decades, the business invested and has made another destination for itself in the North Coast. Around the corner on the western side of Alexander Valley, Ridge farms one of California's most important historic vineyards, Geyserville. It makes one of the best aged wines in California.

Map 4: The Santa Cruz Mountains and the Central Coast

9

THE CENTRAL COAST

The northern stretch of California's Central Coast AVA wraps around the San Francisco Bay including both San Francisco and Oakland. It then continues south to Santa Barbara, covering 250 miles (402 kilometers) of coastline. The region includes 10 counties of California and almost 1.5 million total acres (607,000 hectares). It also includes incredible cultural richness and variety, spawning a range of pivotal innovations far beyond wine.

The Martini and Pisco Punch cocktails were both invented in San Francisco in the latter part of the nineteenth century. Less than a hundred years later, across the bay in Oakland, a bartender made the first Mai Tai. The Buena Vista Café, located near the last cable car stop for Fisherman's Wharf, didn't invent the Irish Coffee, but it did serve the first in the United States, a decade after the first Mai Tai. Foods created in the Central Coast region include sourdough bread, the fish stew cioppino, frozen popsicles, Jack cheese, taquitos, and ranch dressing. Santa Barbara County has its own style of barbecue, which relies on dry-rubbed tri-tip grilled over pieces of local wood, the red oak, served alongside *pinquito* beans. They're pink and native to the area. Below the southern stretch of the San Francisco Bay, the first Apple computer, MacBook, and iPhone were all created. Slot machines started not far away. An Italian immigrant invented the squeegee, a German settler the blue jean.

When it comes to wine, the Central Coast made contributions just as significant. Santa Clara Valley welcomed the state's first Cabernet Sauvignon in the 1850s. Around the bend, Livermore established the state's first Chardonnay. The varieties went on to become the two most planted wine grapes in California. La Cresta Blanca in Livermore won the first significant international award for wine of California, the Grand Prize at the 1889 Paris Exposition. After Prohibition, the Wente family's vineyards, not far from La Cresta Blanca, provided cuttings that became the source for some of the world's great Chardonnay, including around 80 percent of the plantings in California. They were also one of the first in the world to varietally label Chardonnay.

Today, the Central Coast includes the Rhône-style wines that took over Paso Robles, the old-vine field blends Contra Costa maintains, the first winery to earn the Regenerative Organic Certification, and Monarch tractor, the world's first fully electric, automatic, smart tractor. It's also home to around 100,000 planted acres (40,469 hectares), more than half of which are planted to Chardonnay. Within the Central Coast, there are around 40 smaller AVAs.

While not the largest AVA in the state (that honor belongs to the San Joaquin Valley), the Central Coast might be California's most varied. Growing conditions pivot and change thanks to the rugged folding mountain ridges, the transverse valleys, and the cold Pacific Ocean. In fact, the common thread in the Central Coast is as simple as coastline. Its entire length is bordered by the cold California Current, part of the larger North Pacific Gyre that churns cold water from the north along the west coast of the US, pushing it towards the equator. With it move weather changes, blistering fog, and cooling winds. Only one county in the Central Coast doesn't touch the coastline. San Benito was originally part of Monterey to its west, which is set alongside the Monterey Bay, but the residents of the inland mountains demanded their independence and won their own county. Visiting and seeing the isolation of its inland mountains, the unconventional feeling of the area makes sense.

The ocean has other influences. The folding coastal mountains create a series of valleys that open to the Pacific and move inland. Salinas Valley, Edna Valley, Santa Maria Valley, and Santa Ynez Valley are four of them. What they share is powerful afternoon winds. As inland temperatures rise, pulling the air upwards, air from the coast is pulled in to fill the vacuum created on the valley floor. That coastal air is chilling and powerful, lengthening the growing seasons in the valleys, lowering yields, and elevating the acidity (and sometimes also tannin). The Edna and Santa Maria valleys share the longest growing season in North America: budbreak can happen in February while harvest might wait until October or November.

Along the Central Coast's southernmost stretch, in the county of Santa Barbara, the coastal mountains change directions. Instead of bordering the coast in a north–south orientation, they turn east–west, fully exposing the county to the ocean. The series of parallel east–west ridges throughout the area are known as the Transverse Ranges due to the way they cross the area, rather than border it. They're among the few of this orientation the entire length of North and South America. The Santa Lucia Range, a stretch of coastal mountains from Monterey through to San Luis Obispo, includes the AVA's highest peaks, the tallest reaching to 5,856 feet (1,785 meters).

The San Andreas fault carves the length of the Central Coast. San Luis Obispo County sits entirely on the Pacific plate. Other growing regions of California along the coastal shores straddle multiple tectonic plates. Within San Luis Obispo, Paso Robles has become a destination for wine in the last few decades. It used to be a stop to get gas halfway between Los Angeles and San Francisco. But the success of its wineries changed that, bringing top restaurants and travel lodging needed for wine tourism.

From the area, Tablas Creek imported each of the approved varieties of Châteauneuf-du-Pape, including the really obscure ones. When they established 0.5 acres (0.2 hectares) of Picardan they increased the world's plantings by 50 percent. Bonny's Doon's Randall Grahm started the Rhône craze alongside Bob Lindquist and Steve Edmonds in the early 1980s. Grahm took Syrah from Paso Robles and planted it in a cooler stretch along Monterey Bay. Since then, Saxum, Ledge, Villa Creek, MAHI, Denner, Epoch, and many others in Paso Robles have followed suit.

The winds of the Salinas Valley elevate the natural acidity of wines grown in the area. They've helped create one of the best sparkling wines in the state. The Caracciolo family started the venture with Michel Salgues, the founding winemaker for Roederer's estate in Mendocino. Caraccioli sparkling wine became his retirement project.

Central Coast AVA (1985)

Includes: at least 40 nested AVAs

Does *not* include: Santa Cruz Mountains AVA

Includes at least part of 10 counties: Alameda, Contra Costa, Monterey, San Benito, San Francisco, San Luis Obispo, San Mateo, Santa Barbara, Santa Clara, Santa Cruz

Main varieties: varied

Wineries along the Central Coast are leading the charge in alternative packaging. Tablas Creek sells premium wine in the 3-liter bag-in-box format. Bonny Doon (whose headquarters moved to Santa Margarita after Grahm sold to the company WarRoom) now "bottles" rosé in paper bottles (they actually work). They essentially work like bag-in-box but are bag-in-bottle with a screw cap instead of a tap.

The sizeable O'Neill vintners, located in the Central Coast, has created and demonstrated new solutions for broader sustainability efforts. Certified as a True Zero winery, O'Neill diverts more than 99 percent of its production waste from landfills through reduction, reuse, and recycle efforts. O'Neill also hosts the largest vermicomposter, or worm farm, in the state. Earthworms help convert organic waste into usable compost.

Mixed amid the grapevines, the Central Coast also celebrates newer initiatives to expand the audience for wine. The state of California has recognized Tara Gomez, of Camins2Dreams in Santa Barbara County, as the first Native American winemaker. In San Luis Obispo, Dream

Tara Gomez harvesting in the Santa Ynez Valley for winery Camins2Dreams

Big Darling provides funding and mentorship for women newer to the wine industry. Talley Vineyards of Arroyo Grande makes a special PRIDE cuvée. It serves as a fundraiser for The Gala Pride & Diversity Center, which provides support and services for LGBTQIA communities throughout the Central Coast.

Given its sheer expanse and complexity, I have grouped the Central Coast into four areas (though these are not official subdivisions) to make it more understandable. From north to south these are: around San Francisco Bay, near Monterey Bay, the San Luis Obispo region, and the county of Santa Barbara.

AROUND SAN FRANCISCO BAY

San Francisco Bay area

Known for: planting the first Cabernet Sauvignon and Chardonnay in the state

Lesser-known strength: the Bay influence, moderating temperatures, increase in fog and cold winds

Leader in: experimental urban vineyards and wineries

First peoples: Ohlone

There wasn't a city of San Francisco until after the gold rush. Los Angeles was California's population and financial center until the wealth of the state came from the Sierra Foothills and the communities built to serve mining. Gold created a population boom far beyond the miners themselves. Adventure seekers hoping for riches moved into the northern part of California, first moving to the rugged slopes of the Sierra Nevada Range. Soon, other entrepreneurs followed. The services gold seekers needed built the foundations of the California economy. As miners struck it rich or failed, they moved into other counties. Lake County, Mendocino, the Central Valley, and areas around San Francisco Bay all became communities spurred by gold.

The timing was providential for wine. Sonoma became one of the first fine wine centers of California. Its fruit was shipped to San Francisco then sent by railway almost anywhere else in the state or country. South of the bay, the Santa Clara Valley became the other fine wine center of California. There, the first white wines emerged, as well as a new level of quality in sparkling wine, and the first coveted Cabernet Sauvignon in the state.

The earthquake and subsequent fire of 1906 dampened production. The eastern side of the bay became one of the biggest wine production areas after wineries in San Francisco were destroyed. The change was the first vestiges of an urban wine center that continues in Berkeley and Oakland today. Broc, Dashe, and Donkey & Goat, as a few examples, have all made homes tucked amidst city dwellings.

Historic vineyards still celebrated today also emerged. The oldest vineyards in the state, in Amador County, were planted during the gold rush and still provide fruit. The highest concentration of old-vine, head-trained, and own-rooted plantings in the state stands in Lodi. The area grew during the gold rush, then succeeded with the railroad. It also became one of the country's most important grape providers for home winemakers during Prohibition. Both the Sierra Foothills and Lodi sit outside the Central Coast but their success spurred that of areas further west. Across the California Delta, Contra Costa became an equally important old-vine center for California. Like Lodi, the sandy soils meant vines weren't impacted by phylloxera. Some of those historic vineyards still exist in Contra Costa, and it was recently recognized as an official

The sandy soils of the Contra Costa area near San Francisco Bay as well as the Mokelumne River area in Lodi prevent phylloxera from damaging the regions' vines. The two areas have a high concentration of older vineyards planted on their own roots as a result

AVA as a result. But vineyards are also regularly being pulled to make room for houses.

The Livermore Valley helped establish the state's first truly great white wines. Among those historic plantings, the now fifth- and sixth-generation Wente family became one of the best-known sources of Chardonnay in the world. They generally provided vine material to leading wineries, Stony Hill and Martini included, throughout the state and then to UC Davis. Virus-treated selections of Wente Chardonnay have made their way to Oregon and Washington, as well as down to New Zealand, Argentina, and Australia, amongst others. Today, like the Santa Clara Valley, Livermore looks more like an urban than wine center but some of its most important vineyards remain.

There's even wine in San Francisco itself. In 2012, the *San Francisco Chronicle* revealed it had planted a vineyard on its rooftop. Annie Favia, one of the state's great viticulturists, based in Napa Valley, helped train the vines and consulted on farming – the method of establishing a healthy rooting system on a rooftop hasn't been written about in viticultural textbooks. In 2012, the vines provided two clusters of Friulano. I must admit I haven't heard more since.

On the southern side of San Francisco, an urban vineyard climbs a hillside overlooking Interstate 280. The vines grow alongside a garden, both on public land. Christopher Renfro got permission from the city to maintain both and has used them to connect to the surrounding community, offering produce from the garden and weekly meals cooked on site. The goal is to return farming to the people, to help urban dwellers have an agrarian experience in the city. But Renfro used the vineyard to go even further. He reached out to North Coast viticulturist Steve Matthiasson, and UC Davis researcher Beth Forrestel to create the Two Eighty Project. The program provides viticultural apprenticeships for Black, Indigenous, and people of color newer to wine learning from some of the most respected producers in Napa Valley before they then go on to make wine on their own. The project has gained international funding to offer paid apprenticeships, and spawned a series of smaller production wineries launched by program graduates.

Renfro also rediscovered one of the state's more unusual sites, a gentleman's vineyard at Filoli Farm to the south of San Francisco. Filoli was planted as a sort of vine archive, or library, proliferating more than 100 different cultivars both *vinifera* and otherwise. Some have still not been

The 2024 apprentices of the Two Eighty Project. Christopher Renfro sits front left holding flowers

identified, their origins lost to history. Renfro maintains the garden and makes an intriguing wine from the collection annually.

The entire San Francisco Bay area fills daily with the cooling influence of the Pacific Ocean. Summers in the region are colder than many winters elsewhere in the state. But the city's great secret is its beauty and warmth from late September through October, after schools around the world have opened again and most tourists have gone home. The fog in the area is so prevalent that residents have named it. Karl the Fog rolls daily through the region between April and October, most eager to fill the area between June and August. In the past, Karl the Fog had his own Twitter (before it was X) and Instagram accounts. Time-lapse videos of Karl pouring over the coastal mountains into the bay really are worth watching.

The San Francisco Bay and Pacific Ocean provide two significant maritime influences for the extended region, keeping temperatures relatively even while also cooler than more inland areas. The eastern side of the bay is bordered by a modest mountain range that provides a shield from maritime influence, allowing daytime temperatures to increase in Livermore and Contra Costa. The southern portions of the bay through Santa Clara are also more removed from direct ocean influence and so enjoy warmer daytime temperatures.

Contra Costa gains significant warmth and direct sun exposure during the day. Its flatter topography and sandy soils allow greater warming influence on the vines as well. Vineyards closer to the bay like in the El Cerrito area are cooled by the Delta breezes and tend to grow wines that are brighter.

The topography of the Santa Clara Valley varies. Vineyards are no longer found along the valley floor where Silicon Valley grows instead. However, the surrounding hillsides do include some steep vineyards, albeit at much smaller concentration than historically.

San Francisco Bay AVA (1999): overlaps at least part of eight counties
Main varieties: varied

Contra Costa

Contra Costa AVAs

Contra Costa (2024): high quality old-vine field blends threatened by urban expansion
Main varieties: Zinfandel, Carignan, Mourvèdre

Lamorinda (2016): full of various planting conditions and grape types
Main varieties: varied

Contra Costa was considered a winemakers' secret for decades. The wealth of old-vine, head-trained, and own-rooted vineyards offers a flavor intensity and textural structure distinctive from other parts of California. The sandy soils mean phylloxera didn't affect the area. Some of the vines still grow from a pre-phylloxera era.

It sits along the cooling influence and water source of the California Delta, further increasing the distinctiveness of the area. Three Wine Company from Matt Cline, and Cline Family Vineyards from Fred Cline were among the first contemporary wineries to truly root their wines in Contra Costa. Ridge Vineyards made wines from the area decades ago and recently returned to sourcing fruit from the area. Other members of the Historic Vineyard Society have also brought significant attention to the region. Neyers, Turley, Bedrock, Enfield, and Extradimensional Wine Co. Yeah! have all turned heads with wines from the area. Turley's

Tegan Passalacqua now makes a Contra Costa Red Wine for his brand Sandlands from the area's Mataro and Carignan, the two varieties he feels most define the region's history.

In 2024, the acclaim these producers have brought to Contra Costa's old vines finally paid off with recognition from the Alcohol and Tobacco Trade and Tax Bureau (TTB) with an official AVA. Celebrated vineyards in the area hover around the Antioch, Oakley, and Brentwood areas. Evangelho is its best known.

Santa Clara Valley

Santa Clara had a vision for white wine. Paul Masson brought some of the most coveted cuttings of Chardonnay to the region. His work then inspired others in the region, some of whom moved up the slopes into the Santa Cruz Mountains.

The western flank of Santa Clara County rises into the Santa Cruz Mountains AVA, which is not part of the Central Coast AVA (see p. 188). Historically vines were spread across the valley into the eastern (less steep) slopes as well. Today, vineyards are harder to find. Santa Clara Valley is home today to one of the major technology centers of the world. Silicon Valley has created more computer designs and innovations than it has wine. But its success has also changed California wine. Vineyards and wineries throughout the state are now owned by those who made money first in Silicon Valley. Its technology has spurred closer examination and better technological understanding of the needs of vineyards as well as winemaking.

Estates that do still grow wine in the region hover around the towns of Gilroy and Morgan Hill. Gilroy is otherwise known for its wealth of garlic plantings. Morgan Hill, known for its horse shows, is also considered a mushroom capital: the region is the number one producer of fresh mushrooms on the continent.

When it comes to wine, there are wineries through the area worth visiting. Most sell wines direct through their tasting room. The region sits at the intersection of modest ocean influence with inland temperatures and so grows a range of Mediterranean varieties.

Santa Clara County AVAs

Santa Clara Valley (1989): tucked between Santa Cruz Mountains in the west and the Diablo range to the east

Main varieties: Cabernet Sauvignon, Chardonnay, Zinfandel

San Ysidro District (1990): set along the Pajaro River, separates Santa Cruz and Monterey Counties
Main varieties: Chardonnay, Pinot Noir, Merlot

Pacheco Pass (1984): overlapping both Santa Clara and San Benito counties
Main varieties: Chardonnay, Merlot, Zinfandel

Alameda County, Livermore

The sweeping freeways of Livermore sit in easier view than its vineyards. But grabbing an exit from the freeway to turn up the Livermore Valley means visiting one of the state's historically most important regions. White wine was launched in Livermore. Fine wine made advancements here too. Two of California's most valuable varieties, Chardonnay and Cabernet Sauvignon, found the early stages of acclaim here.

In the 1880s, the story goes, James Concannon invested in property within the Livermore Valley and began importing Cabernet Sauvignon cuttings from Chateau Margaux. His vineyard became the first important source block for the variety in the state. The second generation of Concannons managed to keep making smaller quantities of wine throughout Prohibition, an effort so rare that bottles of Concannon wine from the period are now on display in the Smithsonian. In the 1960s, the third generation joined forces with UC Davis to catalog the Concannon clone of Cabernet as part of California's heritage, as well as to create several virus-treated selections of the Concannon clone. Today, the results of the UC Davis collaboration mean around three-quarters of California's planted Cabernet reaches back to the original vines from Concannon.

Almost simultaneously with Concannon, Carl Wente planted vines nearby. His son, Ernest Wente, planted 4 acres (1.62 hectares) of Chardonnay from clonal material first established by Charles Wetmore at his nearby Cresta Blanca winery. According to Wetmore, his vines originate in Meursault. A devotee of the variety, Ernest traveled to a then-famed nursery in Montpellier, in southern France, and brought back more Chardonnay cuttings to plant in 1912. Wetmore's selection from Meursault blended in the field with Ernest's from Montpellier.

During Prohibition, almost all white wine vineyards all but disappeared from California. In Livermore, Ernest was licensed to make both sacramental and medicinal wine throughout the 1920s, saving the family winery through Prohibition. Thanks to his love of Chardonnay, he also chose to maintain his original vineyards. By 1933, Wente was one of only two sites growing the cultivar in the state.

As the California wine industry rebuilt itself, vintners slowly regained interest in white wine grapes. In 1936, Ernest bottled the first varietally labeled Chardonnay in the United States. At the end of the 1940s, Stony Hill in St. Helena was gifted cuttings of Chardonnay from Wente. A few years later, Stony Hill gave cuttings from their Chardonnay to Hanzell in Sonoma. Louis Martini also obtained cuttings from Stony Hill. He propagated his Chardonnay first in Carneros, and then Haynes Vineyard nearby in Coombsville, both chilly areas in the southern Napa Valley. In the 1960s, Ernest's son Karl established some of the first vineyards south in Arroyo Seco of Monterey County with cuttings from Ernest's beloved vines.

UC Davis also took cuttings of Wente Chardonnay. The university took Ernest's vines through multiple virus treatments, creating several clonal selections derived from Wente vines. These moved into Oregon as selection 108, into Argentina where it became the Mendoza clone, and into Australia where it is celebrated as Gingin. Gingin, Mendoza, selection 108, and even the original Wente have diverged from each other over time and are today regarded as distinct clones of Chardonnay (selection 108 is only known as that in Oregon and is actually a combination of clone 4 and clone 5, heat treated from Wente). Even so, the three expressions each originate from the work at UC Davis,[1] which depends on Ernest's vines at Wente. Today, Chardonnay stands as California's most planted wine grape. It competes for the top with Cabernet Sauvignon, the two occasionally trading between first and second in their standing. Around three-quarters of the Chardonnay grown today in California reaches back to the first plantings at Wente.

Livermore was also the first region to varietally label Petite Sirah. Prior to Concannon's varietal bottling in 1964, the variety had only been used for blending, primarily in Zinfandel field blends. Today, Cabernet Sauvignon is the most planted variety in Livermore, followed by Chardonnay, then Petite Sirah. The area hosts primarily gravels and sandy loam soils with around 5,000 acres (2,023 hectares), and 50 wineries.

Alameda County AVA

Livermore Valley (1982): one of the founding regions of fine white wine in the state

Main varieties: Cabernet Sauvignon, Petite Sirah, Grenache

The Monarch tractor is the first commercially released electric, driver-optional, smart tractor in the world. It was developed and tested in Livermore in California's Central Coast. Here, driving one of the first commercially released tractors at the Monarch production warehouse

Monarch tractor

Since the introduction in 1973 of Roundup, made from the herbicide glyphosate, the population of Monarch butterflies has been reduced by 90 percent. They are today considered an endangered species. But Monarchs are also an indicator species. The size of their populations indicates changes impacting the health of the environment. As Monarch butterflies visibly decline, other pollinators like bees, wasps, moths, and other butterflies also decline. Yet, 35 percent of the world food supply depends on pollinator species.

In 2016, Carlo Mondavi of RAEN winery on the West Sonoma Coast and Continuum on Pritchard Hill of Napa Valley launched the Monarch Challenge.

They asked wineries in Napa and Sonoma to shift from the use of glyphosate to other under-vine weed-management approaches to reduce the use of the herbicide, while also improving the relationship with the soil, the basis of organic farming. Launching the Monarch Challenge led to Mondavi's realization that the main barrier to farmers switching to organic farming rests in the cost increase. The use of herbicides is often lower, and their use is less time-consuming.

In 2018, Mondavi met with three technology professionals. Praveen Penmetsa built a career at the intersection of robotics, energy, mobility, and engineering then realized, through visits to some of the world's remote farming regions, that farmers needed assistance from more useful and affordable technology. Mark Schwager built the Tesla-Gigafactory from concept to fruition through his expertise in manufacturing, logistics, and operations planning. Zachary Omohundro applied his knowledge of integrated, intelligent electro-mechanical systems to robotics, electric vehicles, energy storage, and autonomous cars. Mondavi provided his experience in farming with a commitment to organics and familiarity with wine producers around the world.

Together, the four founded the Monarch tractor initiative, named for Mondavi's original inspiration, the Monarch butterfly. In 2022, the MK-V, the first production units of Monarch were released. The MK-V is the world's first fully electric, driver-optional, smart tractor. It uses GPS-guidance systems to operate the tractor autonomously. Remote monitoring assesses field operations and captures and analyzes data during tractor operation. When its electricity requirement is fulfilled by renewable energy sources, the tractor eliminates that portion of a vineyard's carbon footprint.

The Monarch also functions as a generator battery for electrical needs in the field, or during power outages. Monarch tractor was designed and built in Livermore. Developmental models were utilized for farming at Wente winery. The first MK-V units were purchased by Constellation and used at To Kalon vineyard in Oakville.

NEAR MONTEREY BAY

Monterey

Writer John Steinbeck made Monterey famous. The Salinas Valley, just east of Monterey Bay, remains one of California's most vibrant agricultural centers. Rows of tomatoes, strawberries, lettuce and cabbage, broccoli, cauliflower, celery, and artichokes cover the valley floor. The history of farm work in Salinas Valley and fisheries in Monterey Bay fed several of Steinbeck's novels. *Of Mice and Men* examines the hardships of farming the Salinas Valley during the Great Depression. *East of Eden* follows several generations to witness how the region evolved over time. One of his first novels, *To a God Unknown* also takes place in Salinas Valley in the end of the 1800s. Later in his writing, Steinbeck moves closer to the coast. *Tortilla Flat* steps into a then-impoverished neighborhood near Monterey Bay. *Cannery Row* followed by *Sweet Thursday* move to the edge of the bay to follow the decline and evolution of the farming industry there. Steinbeck's non-fiction book *Their Blood is Strong* examines the struggle of migrant farmworkers especially in the Salinas Valley and is considered a precursor to *The Grapes of Wrath*. The closest Steinbeck got to writing about wine was in the drinking habits of the characters in *Tortilla Flat*. His lack of interest makes sense. The first commercial vineyards didn't arrive in the area until the 1960s. It was a long wait from the mission vines planted in the region at Mission Nuestra Señora de la Soledad in the late 1700s.

Known for: sea life and bird watching, Monterey Bay Aquarium, John Steinbeck, Chardonnay
Lesser-known strengths: prior home of Salvador Dali, Jimi Hendrix, Joan Baez, and Doris Day
Leader in: Chardonnay, Pinot Noir, San Bernabe – the biggest vineyard in North America
First peoples: Ohlone

It was A. J. Winkler of UC Davis who spurred vineyard plantings in Monterey County. In 1960, he recommended several cooler climate areas he thought would be successful for wine. One was the elevated benchlands and slopes on the western side of the Salinas River. Within a few years, new vineyards were being established. By 1970, 2,500 acres (1,012 hectares) were already established. One of the first was Wente, alongside Mirassou. In the 1970s, plantings expanded throughout Monterey County. J. Lohr helped establish Arroyo Seco. Scheid put roots near Greenfield in the Salinas Valley. Hahn moved into Santa Lucia Highlands. In 1982, the county was awarded its first AVAs, Carmel Valley and Chalone. Today, Monterey County includes nine significant AVAs and 45,915 acres (18,581 hectares) of vineyards.

The northwest portion of the county borders Monterey Bay. Just offshore, the Monterey Canyon turns out to be one of the deepest underwater canyons in the western Pacific. Its depth compares to that of the Grand Canyon in the southwestern United States. The Monterey Canyon plays an important role in the region's winegrowing. Its depths fill with cold ocean water carried by the California Current from the north, making this part of Monterey even colder than other sections of the state's coastline. Every afternoon it triggers powerful winds, which are pulled into Monterey Bay, cooling off growing conditions through the region, extended the growing season, and slowing down ripening.

The AVAs closest to Monterey Bay are its coolest. Santa Lucia Highlands receives some of the most direct impact from the wind. The winds off the bay cool the entire length of the Monterey AVA through the Salinas Valley, bringing winds over Arroyo Seco. Further south in the valley, San Lucas, Hames Valley, and San Bernabe feel the winds to a much lesser degree and as a result are warmer. Santa Lucia Highlands sits on an alluvial fan with small gravels. The gravels increase in size into Arroyo Seco. Further south, San Lucas and Hames carry significant portions of shaley loam, while San Bernabe enjoys mostly sand.

Outside of Monterey Valley, Carmel Valley, San Antonio Valley, and Chalone are warmer. San Antonio nests into a basin at high elevation within the Santa Lucia mountains. Its highest point reaches 2,350 feet (716 meters). Carmel Valley goes even higher, to 2,762 feet (842 meters). Chalone stands at around 1,800 feet (549 meters) and features limestone and decomposed granite.

Chardonnay covers more than half the plantings in Monterey County, distantly followed by Pinot Noir, then Cabernet Sauvignon.

Monterey County AVAs

Monterey (1984): runs along the Salinas River, includes several nested AVAs
Main varieties: Chardonnay, Pinot Noir

Carmel Valley (1982): sheltered from the Pacific Ocean, tucked within the coastal mountains
Main varieties: Cabernet Sauvignon, Merlot

Santa Lucia Highlands (1992): impressive winds and relative aridity produce Pinot with power
Main varieties: Pinot Noir, Chardonnay, Syrah

Arroyo Seco (1983): "dry creek," named due to the rain shadow produced by the mountains
Main varieties: Chardonnay, Riesling, Zinfandel, Merlot

San Antonio Valley (2007): impressively complex soils and warm summer temperatures
Main varieties: hearty reds

San Bernabe (2004): one of the biggest vineyards in the world, a monopole AVA
Main varieties: varied

San Lucas (1987): borders San Bernabe and crosses the Salinas River
Main varieties: varied

Hames Valley (1994): nested within the southern stretch of the Monterey AVA
Main varieties: hearty reds

Gabilan Mountains (2022): overlaps San Benito and Monterey Counties
Main varieties: Pinot Noir, Chardonnay, Syrah, Grenache

Wineries

Caraccioli

Carmel-by-the-Sea

www.caracciolicellars.com

Using all estate fruit since 2015, Caraccioli makes some of the most refined and balanced sparkling wine in California. Their Escolle Vineyard sits in the northern foothills of the Santa Lucia Highlands, exposed to both morning fog and afternoon winds. The combination keeps natural acidity elevated, making it an ideal foundation for sparkling wine. These wines are made with Chardonnay and Pinot Noir, while the Caraccioli family also grows Gamay and Syrah. They make still wines from their vineyard and sell fruit to other producers in California. The sparkling wines are all made in the traditional method and deliver deliciousness with ageworthiness that is impressive.

Pisoni

Gonzales

www.pisonivineyards.com

Gary Pisoni planted his first vineyards in the Santa Lucia Highlands in

1982. The adjacent valley is more known for row crops, which served as the economic base of the region for decades. But Pisoni decided to climb the benchlands and slopes above the Salinas River to plant *vinifera* instead. To increase interest in his grapes in what was then still a relatively unknown region, Pisoni made wine from his vineyard and sent bottles of it to winemakers throughout the state. The effort worked. Pisoni became a source for top Pinot Noir, Chardonnay, and Syrah of the Santa Lucia Highlands, and his planted area kept expanding. In 1998, Pisoni partnered with his sons Mark and Jeff to make Pisoni Estate wines. Pisoni has also sold fruit to celebrated Pinot Noir wineries Patz & Hall, Peter Michael, Testarossa, Siduri, ROAR, and Paul Lato. The Pisoni family also makes wine for their sister labels, Lucia and Lucy.

ROAR

Castroville

www.roarwines.com

Rosella and Gary Franscioni began planting their estate vineyards in the mid-1990s in partnership with the Pisoni family. The two families co-own the Soberanes and Garys' vineyards, and the Franscionis own their own Rosella's and Sierra Mar sites. Sons Adam and Nick lead ROAR with father Gary to grow and make Pinot Noir, Chardonnay, Syrah, and Viognier from their Santa Lucia Highlands vineyards. While the Santa Lucia Highlands are best known for Pinot Noir, ROAR and their vineyards are a demonstration of how enticing Syrah from the region can be.

Scheid

Soledad

www.scheidfamilywines.com

Two generations lead Scheid family wines, located in Monterey County. They offer a simple mission for the winery, "bringing people together to enjoy and savor life." Their portfolio offers affordable wines with quality that exceeds their price point while farming with the goal to continually hone and improve their approach. The winery relies on a wind turbine that generates all of the power needed for winery operations as well as enough energy fed into the local electrical grid to power more than 100 homes. They were the first vineyard in the country certified with GLOBALG.A.P. standards to ensure health, safety, and welfare for their agricultural workers. And they also provide scholarship money to students in the area for furthering their education. At Scheid they compost all of their grape pomace and organic materials produced from harvest and fermentation, and use a water recycling system that captures water used in the winery to be reused in vineyards when irrigation is needed.

Talbott

Carmel-by-the-Sea
www.talbottvineyards.com

The Talbott family planted one of the first vineyards in the now-famed Santa Lucia Highland in 1972. A decade later they followed with vines in Carmel Valley. In 1994, they added a third vineyard, making it possible to make wines entirely from estate vineyards. Known particularly for its Chardonnay, Talbott also makes Pinot Noir, and both have won significant national and international awards. The wines express the persistent natural acidity of the region with a concentration of flavor also expressive of the region.

San Benito County

San Benito County feels remote. Its mountains swirl with the golden grasses that make the Central Coast of California seem both so barren and so beautiful. The population of the county is relatively small, 68,175 as of 2023. The area's claims to fame are several successful professional athletes. Dave Tipton played professional football for the Seattle Seahawks, New York Giants, and San Diego Chargers during the 1970s. For two decades, in the 1920s and '30s, Charlie Root played professional baseball in the American Major League and still holds the club record for career wins. In the same period, Babe Hollingbery became one of the most successful football coaches in the history of what is now Washington State University.

Beside a few prominent sports stars, San Benito County remains a little obscure. In vines, it grows 2,464 acres (997 hectares). That planted area includes a little less than 30 percent each of Pinot Noir and Chardonnay, followed third by Cabernet Sauvignon, and small percentages of other varieties. In the north, Pacheco Pass AVA crosses into San Benito County but sits primarily in its northern neighbor, Santa Clara. It falls outside the Gabilan Mountains that carry the other AVAs of the county, hosts mainly alluvial soils, and is swept regularly by cooling breezes from the ocean. Inside the Gabilan Mountains, the warmth of the AVAs depends on proximity to the ocean. Mount Harlan and Cienega Valley are the coolest. Chalone (which extends into Monterey

County) can be, but temperatures vary so that in some vintages it is much warmer. San Benito varies too. Furthest east, Paicines is warmest, though still considered moderate.

History

Though San Benito County is relatively unknown in wine, it has fostered some of the state's most idiosyncratic vintners, and even more obscure varieties, that have played a pivotal role in the history of California wine.

In the 1950s, Almaden moved south from its original home in Santa Clara Valley to the Gabilan Mountains of San Benito. There they planted 2,000 acres (809 hectares), in its time one of the largest vineyards known, hidden in a high-elevation valley. The site was devoted to Chardonnay and Cabernet Sauvignon in equal measure, with the addition of a host of oddball varieties meant to give complexity to the winery's high-volume jug wines. In that location, a hundred years before, the first vineyards had been established on the property by Theophile Vache. In 2016, it became the site of Eden Rift, a Pinot Noir and Chardonnay valley in the foothills of the Gabilan Mountains. As in much of the area, vines grow in a mix of limestone soils and granitic sand.

Ron Siletto helped keep San Benito wine going after Almaden sold in the 1980s. He'd served as president of the company, so as its vineyards in the region declined, Siletto started buying land of his own. As a grower, Siletto provided unknown varieties such as Cabernet Pfeffer (not related to Cabernet Sauvignon), Pinot St. George, Frappato, and Corvino to winemakers in love with idiosyncratic varieties. These included Kenneth Volk, who played a significant role in the growing acclaim of Central Coast wines, and Bryan Harrington, whose small production winery carried a modest while staunch fan base throughout its tenure.

Nearby, Wirz vineyard is now farmed by its fourth generation. The site was planted in the early 1900s to a mix of Mourtaou, Palomino, Mission, Carignan, Riesling, Mourvèdre, and Zinfandel, some of which remain from the original planting. Randall Grahm was drawn to the area first by Wirz Riesling, which he used to make his brand Pacific Rim, well known in the 1990s. In 2011, Grahm bought his own property nearby and started his passion project, Popelouchum. Today, Cathy Corison makes a compelling Riesling for her Corazón project.

Not far from Wirz, Bob Enz started farming the Lime Kiln area in the 1960s. His site has become a coveted secret for winemakers seeking standout fruit from distant locations. Ian Brand, Benevolent Neglect, Birichino, Newfound, and Terah all make wine from Enz fruit. Other distinctive vineyards can also be found. Ken Gimelli grows the largest-known planting of Cabernet Pfeffer in the world, in a site planted in 1908, measuring 11 acres (4 hectares).

When it comes to pushing the possibilities of holistic farming, Paicines might be going the farthest in the state. There, Kelly Mulville has elevated the canopy and fruit zone of the vineyard to make year-round sheep possible. The site remains untilled. The sheep turn out to be enriching the soils and returning native ground cover even more than expected. No chemicals are used. The native insects and animals have returned in unexpected numbers. Over sixty species of birds can be found as well, including the endangered tricolor blackbird, which had been a rare sighting in California before the last few years at Paicines. Since 2017, the ranch has grown Assyrtiko, Grenache, Picpoul, and Syrah. In 2020, it expanded into Cinsault and Carignan. A host of winemakers eager for regeneratively farmed fruit now make wine from the Paicines. They include A Tribute to Grace, Camins2Dreams, Margins, and Terah.

It was also in San Benito County that two of California's most historically import Pinot producers got their start. Richard Graf of Chalone and Josh Jensen of Calera both planted in the limestone soils of the area's mountains. Their devotion to crafting the best of Pinot Noir changed what the industry thought possible for the variety, and drew attention to the potential of limestone in the area.

San Benito County AVAs

Pacheco Pass (1984): overlapping both Santa Clara and San Benito Counties
Main varieties: Chardonnay, Merlot, Zinfandel

San Benito (1987): includes several other nested AVAs
Main varieties: Chardonnay, Cabernet Sauvignon

Chalone (1982): in both San Benito and Monterey Counties
Main varieties: Pinot Noir, Chardonnay

Cienega Valley (1982): on the western side of the Gabilan Range
Main varieties: Riesling, Pinot Noir, Chardonnay

Lime Kiln Valley (1982): within the southern part of the Cienega Valley AVA
Main varieties: Mourvèdre, Zinfandel

Mount Harlan (1990): Pinot Noir grown in limestone at high elevation
Main varieties: Pinot Noir, Chardonnay, Aligoté

Pacines (1982): previously for bulk wine, today home to regenerative farming
Main varieties: Chardonnay, Cabernet Sauvignon

Gabilan Mountains (2022): overlaps San Benito and Monterey Counties
Main varieties: Pinot Noir, Chardonnay, Syrah, Grenache

Wineries

Calera

Hollister

www.calerawine.com

In the late 1960s and early 1970s, Josh Jensen worked with mentors in Burgundy to learn winemaking and vineyards. As the story goes, his mentors emphasized that the best Pinot Noir must be grown in limestone soils. Jensen returned to California determined to make the highest quality Pinot possible in the state and thus spent two years searching for his ideal site to grow wine, especially one resplendent with limestone. At 2,200 feet (671 meters), Jensen established Calera near Mount Harlan in San Benito County. The mountain-top vineyard was rooted into uplifted seabed, limestone in one of California's highest elevation and windblown vineyards. While growing his Calera vineyards, Jensen sourced fruit from other sites to make wine. His first vintage was 1,000 cases of Zinfandel in 1975. In 1982, he expanded his plantings to Viognier before many growers had the variety in the state. And in 1984 he added Chardonnay. In 1998, Jensen became one of the first vintners in California to also grow Aligoté. Calera wines became an international benchmark for California Pinot Noir enjoyed at tastings alongside other Pinot Noirs considered the most distinctive in the world. Today, Jensen, who passed away in 2022, is regarded as one of the pioneers of each of these varieties in California. Calera continues to make Chardonnay, Pinot Noir, Viognier, and small quantities of Aligoté worth seeking out.

Eden Rift

Hollister

edenrift.com

Sitting in the Cienega Valley AVA, Eden Rift was first planted in 1849 and has been continuously producing wine ever since. Vineyards grow alongside the San Andreas fault and a short distance from Mount Harlan, where Calera planted vines. According to the winery, some of the first Pinot Noir in California was planted at the site in the 1860s, though it was lost to Prohibition as hearty varieties took its place. The site is primarily known for Pinot Noir and Chardonnay but the varied terrain allows it to grow Pinot Gris and to foster 1906-planted, own-rooted and head-trained Zinfandel as well.

Popelouchum

San Juan Bautista

www.popelouchum.com

Still a mostly experimental vineyard growing small quantities of its established varieties, Popelouchum has captured attention beyond its volume. Part of its acclaim originates from the man who founded it, Randall Grahm, one of the state's first Rhone Rangers. But it is also dependent on the risks and trials Grahm is willing to bring to the site. Grahm was one of the first in California to establish Tibouren. He's a regular proponent of Cinsault. He makes racy, cool-climate Grenache, sometimes a sparkling Furmint, and one of the best Grenache Blanc–Grenache Gris blends I have tried anywhere. Each of these grow on site but he also has a small planting of his beloved Pinot Noir planted at close spacing on a steep slope tucked out of site in the hills of the property. Most intriguingly, Grahm has also established a seedlings project with shoots grown from the seeds of Serine, as well as those grown from Tibouren. The vines have begun delivering fruit so we will soon hear more about the resulting wine.

Terah

Morgan Hill

www.terahwineco.com

With training in culinary arts, hospitality, viticulture, and winemaking, Terah Bajjalieh brings formidable experience to her Terah Wine Co. and consulting work. Terah works exclusively with organic and biodynamic vineyards from sites she admires that deliver structured, vibrant wines. The wines tend to capture a fresh complex of aromatics and flavors with earthiness and mineral lines expressive of where they are grown. Bajjalieh's take on iconic California vineyards brings fresh insight to the sites as her more minimal approach in the cellar reveals lip-smacking flavors and stimulating verve while still expressing a signature of their place recognizable from a history of wines from the sites. Terah works with a range of varieties including Verdejo and Aligoté (two of my favorites), old-vine Grenache, and Barbera. When she finds a vineyard she wants to work with she will sometimes add other varieties as well.

THE SAN LUIS OBISPO REGION

Known for: Paso Robles, Chardonnay, Morro Bay

Lesser-known strengths: the Madonna Inn, Bubblegum Alley, Hearst San Simeon state park

Leader in: commitment to reaching carbon neutrality; has numerous e-charging car stations

First peoples: Te'po'ta'ahl, Michumash

San Luis Obispo

San Luis Obispo County grows with a split personality. Along its coastal areas the climate is cold, verging on too foggy to ripen fruit. The wines are lean, almost ethereal, and mouth-watering Pinot Noir, Chardonnay, Vermentino, and Gruner Veltliner. In a few pockets there is Syrah. In 2022, it was recognized with its own San Luis Obispo Coast (SLO Coast) AVA.

At its north end, Arroyo Grande enjoys some of the wind rushing down Salinas Valley. Here, Chardonnay and Pinot Noir do well with a little less structural tenacity and still plenty of flavors. Talley made a name for itself investing in this part of the region. Long-time Talley winemaker, Eric Johnson, has spent his career crafting mouth-watering and elegant Pinot Noir and Chardonnay from the estate's vineyards. He and his wife also own their own brand, Ann Albert. Made in Arroyo Grande, it primarily sources fruit from Santa Barbara County to the south. Arroyo Grande also sits within the SLO Coast AVA.

To the southwest, Edna Valley was founded on Chardonnay in the 1970s. Chamisal and Paragon were among the first to put vines there. Its influence from the Pacific Ocean makes it one of the coolest, longest growing seasons in California. More recently, Raj Parr has established himself at Phelan Farms near Cambria, which is known more as an oceanside town. Parr's plantings sit at the edge of ripening, full of juiciness and length with a lighter touch. He now relies on regenerative farming techniques for the vineyard.

Other projects in the SLO Coast tend to be a bit further south than the isolated, cold-weather area of Cambria. Pinot Noir and Chardonnay as well as other less common cool-climate varieties, and even Gewurztraminer and Syrah, do well here.

The SLO Coast area includes around 3,800 acres (1,538 hectares) and stretches from San Simeon at its north to Santa Maria at its south. San Simeon also includes the eccentric home of the eccentric publisher William Randolph Hearst, complete with now-wild zebra, lavish pools, extensive grounds, and a difficult road to reach it. Santa Maria overlaps San Luis Obispo and the northern part of Santa Barbara County.

Uniquely, San Luis Obispo also includes the warmer conditions of inland areas tucked in among the coastal mountains and further east through the plains. York Mountains welcomes direct sun exposure in close proximity to the Pacific and pushes against the western side of the larger Paso Robles.

San Luis Obispo County AVAs (not including Paso Robles)

San Luis Obispo Coast, or SLO Coast (2022): includes the coolest, coastal parts of San Luis Obispo
Main varieties: Pinot Noir, Chardonnay, Syrah

Arroyo Grande (1990): a cool section of the San Luis Obispo close to the Pacific
Main varieties: Chardonnay, Pinot Noir, Syrah, Grenache

Edna Valley (1982): first vines planted were Chardonnay, close to the Pacific
Main varieties: Chardonnay, Pinot Noir, Syrah

Wineries

Phelan Farms

Cambria

phelanfarm.com

The vineyards at Phelan Farms were planted by Greg Phelan, beginning in 2007, but the Phelan family had been on the site since the 1850s. Tucked along the curving valley of Steiner Creek, the vines are swept through the day by rotating shadows cast by the surrounding hillsides. Close to the Pacific near Cambria on the coastal side of San Luis Obispo, the growing conditions are cool enough to leave most varieties unripe. Phelan established Pinot Noir and Chardonnay in the vineyard, which eventually drew Raj Parr to the site as a fruit source for the edgier side of his Sandhi wine. In 2019, Parr began working more directly with the site, beginning to graft over some of its established vines to rarer varieties, especially from the French Alps and Jura. Today, Parr leads the farming and winemaking through his winery Phelan Farms. Parr explores regenerative techniques, including integrated sheep, composts, and non-chemical vineyard treatments, and a light-touch, no sulfur approach to the winemaking. The best of the wines are compelling, transparent, and mouth-wateringly enjoyable. My first tasting of the wines from barrel and tank before they were ever bottled was an exciting glimpse at their potential. Parr's shift from more classicist wines from Chardonnay, Syrah, and Pinot Noir to more distinctive expressions of Mencia, Trousseau, Mondeuse, Gamay, Poulsard, and Savagnin while still working with Chardonnay, small portions of Pinot Noir, and varying blends of any of these has ignited enthusiasm for them in California and for California wine.

Scar of the Sea and Lady of the Sunshine

San Luis Obispo
www.scaroftheseawines.com
www.ladyofthesunshinewines.com

In 2010, Mike Giugni founded Scar of the Sea, making Pinot Noir, Chardonnay, and cider from San Luis Obispo and the coastal side of Santa Maria Valley. In 2017, Gina Hildebrand founded Lady of the Sunshine to deliver fresh wines with a focus on transparency to place. Mark and Gina biodynamically farm their vineyard together in San Luis Obispo. They also run their two wineries side-by-side. Scar of the Sea makes an impressive lineup of Chardonnays while also sourcing fruit from other vineyards for mouth-watering Syrah, old-vine Zinfandel, and a surprising Palomino made with genuinely old vines from California. Lady of the Sunshine wines are decidedly aromatic with multiple examples of Sauvignon Blanc, a Gewurztraminer, Sauvignon Blanc–Chardonnay blend, Pinot Noir, Gamay, and even a non-vintage solera of white wine steeped with foraged botanicals from the surrounding area.

Verdad and Lindquist Family

Arroyo Grande
www.verdadandlindquistfamilywines.com

Biodynamically and organically farmed, Verdad and Lindquist have been pioneers in their farming methods as much as the varieties they choose to farm. Bob Lindquist was one of the original Rhone Rangers, making his first Qupé Syrah in 1982. It quickly became a staple of the Chez Panisse wine list in Berkeley. Lindquist has made impressive wines from other Rhône varieties as well, first through Qupé and now through his Lindquist Family wines. Louisa Sawyer Lindquist founded Verdad in 2000 with a focus on Spanish varieties, most especially Albariño, Graciano, Tempranillo, and Garnacha, both red and rosé. The couple have also made small lots of Pinot Noir and Chardonnay for their collaborative brand, Sawyer Lindquist. The first planting of Albariño in California was established by Brian Babcock on the far western side of the Santa Rita Hills in 1992. The site was too cold and so Louisa and Bob took cuttings from Babcock to plant their own in 1996. In 2000, Verdad made the first varietal Albariño in California.

York Mountain

York Mountain sits 8 miles (13 kilometers) from the ocean, at between 1,500 and 1,700 feet (457–518 meters). The first winery in the area was established in the 1880s, on the site that today hosts Epoch. The proximity to the ocean, and exposure to maritime winds, make York Mountain cooler than its neighboring Paso Robles. It also receives more rain. The shallow topsoils and uncovered bedrock make it hard to access groundwater, so residents rely on rain as a water source. Sandstone with pockets of limestone and veins of sandy clay loam covers the area. The area grows primarily Cabernet Sauvignon and Grenache, but do not sleep on any Clairette Blanche you find from the area: it's fantastic.

York Mountain AVAs

York Mountain (1983): adjacent to, but outside of, the Paso Robles AVA
Main varieties: Grenache, Syrah, Cabernet Sauvignon

Epoch Estate

Templeton
epochwines.com

The York Mountain home of Epoch Estate rises from the site of the region's first winery, York Mountain, built in 1882. Epoch was founded

The women of York Mountain. From left: Vailia From of Desperada, Jordan Fiorentini of Epoch, Hilary Yount of The Royal Nonesuch

in 2004 with the purchase of the historic Paderewski Vineyard in the Willow Creek District of nearby Paso Robles. In 2010, they began revitalizing the old York Mountain property, and moved their winery onto the site in 2016. In the Templeton Gap AVA, they also own Catapult Vineyard. Together, these three sites are the basis of Epoch wines. They grow a mix of Bordeaux and Rhône varieties, vinifying them both as blends and as varietal wines, as well as Tempranillo and Zinfandel. Winemaker Jordan Fiorentini has easily become one of the Central Coast's most respected winemakers, bringing a classicism and finesse to the intensity of the wines. She relies especially on concrete for fermentation and aging. The vineyards are farmed biodynamically. Fiorentini also creates what she calls Vinpressions®, visual tasting impressions drawn to convey the feeling and texture of the wines.

The Royal Nonesuch Farm

Templeton

www.royalnonesuchfarm.com

At 1,800 feet (549 meters), overlooking the Pacific Ocean, Hillary and Anthony Yount grow their Royal Nonesuch Farm in the York Mountain District of San Luis Obispo County. The site is rocky, with a mix of limestone and sand and traces of clay. Topsoil is shallow, vineyards steep, but rainfall is substantial enough that they dry farm. They grow Grenache, Graciano, Syrah, and Clairette Blanche with exciting texture and energy washing through deliciousness. They bring together whole bunch fermentation with spontaneous fermentation mostly in concrete as well as older, large-format oak barrels. The production size remains small, but the distinctiveness and life of the wines make them worth seeking out.

Graciano masquerades as Mourvèdre

In 2002, a new selection of Mourvèdre, labeled by one of its Spanish names, Monastrell, became available from a nursery in the Central Coast of California. Producers noticed it ripened earlier than other selections of Mourvèdre, and it offered both more acidity and a more interesting range of flavors. Due to the success of Rhône wines at the time, the Monastrell was planted throughout York Mountain, Paso Robles, the Sierra Foothills, and into Santa Barbara County. Ninety percent of the Mourvèdre sold from the nursery was the new selection of Monastrell.

In 2018, the nursery sent a letter to its Monastrell-buying clients all the way back to 2002. The Monastrell had been mislabeled. They might have arrived instead labeled Morastell. Or, the vines could have arrived with an obscure Spanish name, Monastrell Menudo or Monastrell Verdadero. Any of the three could easily be misread as Monastrell, which is the same as Mourvèdre. But each of the three instead refer to Graciano, a variety not found in France's Rhône Valley at all.

By 2018, Graciano had become a favorite element in many of the state's Rhône blends, most especially in Paso Robles. Most producers growing the mixed-up grape decided to rely on the freedom of California wine to retain it as part of their flagship blends. Since 2018, California has enjoyed the almost-new variety. Numerous producers, including Saxum, a Tribute to Grace, Camins2Dreams, Ferdinand, Epoch, Villa Creek, Field Recordings, Paix Sur Terre, and Hope Family all celebrate the benefits of Graciano without giving up their Monastrell.

Paso Robles

Once barely a stop between Los Angeles and San Francisco, Paso Robles has grown into one of California's wine destinations. Vines arrived in the late 1700s via the mission trail stop at Santa Margarita in the region's south. At the end of the eighteenth century, a second mission emerged in the region, a short distance north of what is today Paso Robles. By the 1820s, the area was known as Paso Robles.

Head-trained Zinfandel plantings arrived in the 1880s. One of the first was in what today is known as Ueberroth Vineyard, farmed organically and made by Turley. In today's Willow Creek District, Frank Pesenti planted Zinfandel in the 1920s. Vines that remain today are also farmed by Turley.

In 1921, Ignacy Paderewski planted Zinfandel, Petite Sirah, and Mission on a significant ranch within today's Willow Creek District. Just two years before, he had helped Poland gain independence, serving as its first prime minister and signing the Treaty of Versailles, the document that ended World War I. He'd first found Paso Robles just before the start of the war and returned to enjoy the region for his health while resuming his career as a composer and concert pianist. Shortly afterwards Benito Dusi established his Paso Robles Zinfandel, a site made famous by its long-term relationship with Ridge Vineyards, the only wine they make from Paso Robles. Plantings grew slowly in the region. Dante Dusi planted in the Templeton Gap area in 1945. Dr. Stanley Hoffman established a high elevation estate in the western hills of the Adelaida District. There, in the 1960s, he planted what today is the oldest Pinot Noir and Chardonnay site of the Central Coast, known as the HMR Vineyard.

In 1973, Gary Eberle planted his Estrella Vineyard and became known as the pioneer of Paso. The Syrah he brought from France and established there became one of California's most important selections of the variety, today accounting for two-thirds of the state's plantings. His fame, and that of the region, expanded when Ronald Reagan selected a case of the 1980 Eberle Cabernet Sauvignon to take on a diplomatic mission to China. Bottled in 1983, it was also the first wine ever to showcase the Paso Robles AVA and was selected by UC Davis as one of the top Cabernets in California history at the university's seventy-fifth anniversary celebration.

Incredibly, Gary and his wife Marcy Eberle opened a bottle of the 1980 Eberle Cabernet to drink with me in 2016. He explained it was one of the last 12 he had. We sat together for the afternoon slowly sipping the wine, which proved to be impressively aged, with a refreshing lightening from time and still fresh fruits spun with rose cream and lengthy acidity. It remains for me an example of how beautiful Paso red wine can be, and how ageworthy.

Eberle's acclaim brought increasing attention to Paso Robles. Kenneth Volk launched his eponymous winery in the early 1980s. J. Lohr committed to a Paso location in 1988. In the same year, the first high-volume winery, Meridian, moved in. And by 1990, Rhône varieties were pushing into the region. Tablas Creek was founded in 1989, and the same year John Alban established his estate. The Smith family established Rhône varieties in their James Berry Vineyard in 1990. It is the site that later birthed Saxum. The Rhône movement was developing a second wave.

The Rhone Rangers

In 1973, winemaker Walter Schug made the first fully varietal Syrah in California for the winery Joseph Phelps. The fruit was sourced from a vineyard farmed by the Christian Brothers originally planted in Napa Valley in the 1950s. The Brothers used their Syrah in blends and never made Syrah on its own. Gary Eberle established Syrah in his Estrella Vineyard of Paso Robles in 1975. His cuttings originated from France and became one of California's most important selections, the Estrella clone.

Randall Grahm received cuttings of the Estrella clone and planted them in his original Bonny Doon vineyard, tucked into the cool side of the Santa Cruz Mountains in 1980. Within a few years, the site was lost to Pierce's disease and Bonny Doon took a turn towards purchasing fruit for a Rhône-style blend of Grenache, Syrah, and Mourvèdre that became the winery's flagship, Le Cigare Volant.

Devoted to Syrah, Bob Lindquist launched Qupé in 1982; Steve Edmonds of Edmonds St. John made his first Syrah in 1986. Lindquist, Edmonds, and Grahm met regularly to share insights gained into making Rhône wines. Few others joined them. It led Grahm to joke he was something like the Lone Ranger of Rhône wine. In 1989, Grahm was featured on the cover of *Wine Spectator* wearing a Lone Ranger costume (complete with mask) with the words "The Rhône Ranger" beside him.

Soon, a second wave of winemakers entered California. In 1986, Sean Thackery began making his top wine, Orion, with what he thought was a field blend of primarily Syrah at Rossi Ranch of St. Helena. In 2000 UC Davis researchers, including Professor Carole Meredith, toured the site and realized it wasn't Syrah but a field blend of multiple grape types. Thackery turned to calling it a proprietary red blend, without specifying it any more than that. He began his wine, Taurus, in 1988, a Mourvèdre from Oakville, and fell in love with Rhône varieties. He also eschewed organizations and never joined the Rhone Rangers. Even so, his influence expanded interest in Rhône-style wines.

In 1989, John Alban launched his estate for only Rhône-style wines. The same year, the Haas and Perrin families partnered to bring the experience of Chateau Beaucastel to an American location. Lindquist helped them scout sites until they settled in the coastal mountains of Paso Robles to launch Tablas Creek. Together they imported the varieties from Châteauneuf-du-Pape, in the southern Rhône, providing new vine material for the country.

The same year, the Rhone Rangers formed a non-profit, bringing together producers from all over the country. For a wine to be named a Rhone Ranger bottle it had to include at least 75 percent Rhône varieties. 1994 marked a turning point. Robert Parker recognized Sine Qua Non (SQN) 1994 Queen of Spades Syrah with 95 points, the first American Rhône-style wine to receive such acclaim.

Plantings steadily, persistently increased, until in 2002 Syrah became one of California's top seven grapes. Parker wrote an article about the quality of the Rhone Rangers. Planted area for Rhône grapes expanded. Local chapters of the Rhone Rangers formed.

Then, in 2004, the movie *Sideways* came out. It tanked interest in Merlot as it expanded Pinot Noir to California's third most planted variety. Syrah descended to 3 percent of grape plantings in California. Pinot Noir still sits at 18 percent. Today, the Rhone Rangers includes around 100 winery members from across the United States.

Growing conditions

The western hills of Paso Robles are cooler than common expectation. The heat of the eastern side of the region, and the bolder wines that grow there have established a reputation for the area of heat and largesse. And in its easternmost reaches, such characteristics are true of Paso Robles. But the coastal range of Paso Robles fills with morning fog and ocean breezes, creating numerous microclimates and micro growing areas. Rainfall is also higher as you go further west through the mountains. In the westernmost side of the Adelaida District, vineyards are dry farmed.

The Willow Creek District has become an epicenter for the Rhône movement. James Berry, one of its most admired vineyards for Rhône winemakers, helped spread interest far into the North Coast as winemakers from the region traveled south to buy fruit. The Templeton Gap, also in the western mountains, is a low spot in the coastal range that cold wind pours through. Vineyards in the route of the gap enjoy lower temperatures and plenty of air flow.

The highest peaks in these coastal mountains reach 2,070 feet (631 meters). They include banks of calcareous soils – uplifted limestone – adjacent to stretches of silicious soils. These sands tend to be warmer than the calcium-rich bands, and often host heartier reds like Cabernet. The opportunity to taste wines from each side-by-side is rare, but an interesting example is a bottle of Ledge Vineyards, from silicious soils, alongside a bottle of Saxum James Berry, grown in calcareous earth. Mark Adams of Ledge made wine alongside Justin Smith of Ledge for decades and uses the same winemaking approach for Ledge. The two sites grow at the same elevation about a mile (1.6 kilometers) apart yet show incredibly different character.

Paso Robles warms increasingly heading east. The Paso Robles Highlands in the southeastern corner of the region is its warmest. If you're ever near Paso Robles in spring, the highlands area fills with a shockingly dominant and beautiful spread of multicolored wildflowers. Seeing them in the area remains one of my most vibrant memories. French Camp Vineyards, the largest single vineyard in Paso, is here. It consists of 2,000 acres (809 hectares) with more than 20 varieties. Due to its size, wineries from throughout the Central Coast make wines from the site. Shandon Hills and Shell Creek are two of the other significant plantings in the area. Neither has its own brand and instead provide fruit to winemakers.

The Geneseo District, San Miguel District, and San Juan Creek are also in warmer areas. San Juan features an alluvial bench of well-draining soils and enough warmth to grow Cabernet, Zinfandel, and Petite Sirah. San Miguel features the significant diurnal swing (35–40°F/1.7–4.4°C) much of Paso Robles enjoys. Through a series of terraces, the area climbs up to 1,600 feet (488 meters) and grows mainly Zinfandel. The Geneseo District is known primarily for Cabernet Sauvignon.

In the southernmost stretch of the region is Santa Margarita. The ranch was originally planted by Robert Mondavi in his quest to build an empire for affordable wines further south. The property is varied enough to grow a wide span of varieties. In the coolest spots there are also fossils and calcareous stones. Here, they grow Chardonnay. In the warmer areas there appears more sand, and Cabernet as well as other hearty reds does well. The Santa Margarita area sits on the crest of what locals affectionately call the grade, a steep slope of highway over La Cuesta Pass from the town of San Luis Obispo up to Paso Robles. The grade pulls cool air off the ocean below and blows it through Santa Margarita.

This diversity of growing conditions earned Paso Robles 11 nested AVAs in 2014.

Geography

Unlike other parts of coastal California, Paso Robles falls entirely on the Pacific plate, displaying soils of its uplifted seabed. The San Andreas is in the region's east, making it one of the few growing regions along the coast not bifurcated by the fault. The town of Paso Robles sits at 740 feet (226 meters), tucked against the coastal mountains to its west. The area rolls over grassy plains heading east until reaching the La Panza hills and climbing to 1,200 feet (366 meters) elevation.

The Salinas River begins here in Paso Robles. The Los Machos Hills in the Los Padres National Forest are its starting point. From there it flows north, through the Salinas Valley and out into Monterey Bay. Its movement from south to north has earned it the moniker the upside-down river. Its river valley has created the largest basin in the Central Coast, home to one of California's most important produce-growing regions. In Paso Robles, the Salinas River is the principal water source for area farms. It also provides an important wildlife corridor and habitat for a range of fish, reptiles, songbirds, raptors, and other creatures.

Paso Robles AVAs

Paso Robles (1983): incredibly varied range of microclimates
Main varieties: Cabernet Sauvignon, Syrah, Merlot, Zinfandel

Adelaida District (2014): Paso Robles' westernmost AVA
Main varieties: Bordeaux and Rhône reds

Creston District (2014): east of the Templeton District
Main varieties: Bordeaux reds

El Pomar District (2014): a comparatively cool area for Bordeaux reds
Main varieties: Bordeaux reds

Paso Robles Estrella District (2014): the location of some of the first, important Syrah vineyards in the state
Main varieties: Cabernet Sauvignon, Petit Verdot, Merlot, Syrah

Paso Robles Geneseo District (2014): established by German immigrants in the 1880s
Main varieties: Cabernet Sauvignon

Paso Robles Highlands (2014): a higher elevation, and warmer portion, in the eastern side of Paso
Main varieties: varied

Paso Robles Willow Creek District (2014): home to some of the most coveted Rhône wines in California
Main varieties: Rhône reds, Zinfandel

San Juan Creek (2014): a series of old terraces full of alluvial soils
Main varieties: Cabernet Sauvignon, Zinfandel, Petite Sirah

San Miguel District (2014): named for a Spanish mission in the area
Main varieties: Zinfandel, Sangiovese, Aglianico

Santa Margarita District (2014): the southernmost part of Paso, originally planted by Mondavi
Main varieties: Cabernet Sauvignon, Sauvignon Blanc, Chardonnay

Templeton District (2014): influenced by the winds pushing through the Templeton Gap
Main varieties: Zinfandel, Rhône reds, Mataro

Wineries

Adelaida

Paso Robles
www.adelaida.com

Tucked into the western district of Adelaida in Paso Robles, Adelaida vineyards and winery farms at a range of elevations with a commitment to organic viticulture. They have also converted significant portions of the site to dry farming. Across the portfolio, Adelaida wines bring restraint to the unctuous flavors of Paso Robles. The Adelaida property includes the historic HMR vineyard, planted by Dr. Stanley Hoffman with Pinot Noir and Chardonnay in the mid-1960s. As surprising as it might seem to grow these varieties in a region better known for heat, the location of HMR reveals the complexity of Paso Robles. It sits on an arching slope facing west towards the ocean that creates something like a wind funnel and cool pocket that is ideal for cooler climate varieties. HMR grows the oldest plantings of the two varieties in the Central Coast. Adelaida also makes a range of other wines: Cabernet Sauvignon, Zinfandel, Syrah, and a host of Rhône varieties. The complex topography makes them possible.

Ancient Peaks

Santa Margarita
ancientpeaks.com

Besides creating wines impressively priced to overdeliver quality in their category, Ancient Peaks also has a zipline near its winery that is a lot more fun than I had expected. There are opportunities for some of the more straightforward (read: no bull riding) ranching activities, picnics, and a café offering food worth seeking for a casual meal. Ancient Peaks wines are grown in a historic vineyard established by Robert Mondavi and today led by three local families also connected to ranching. The site is impressively varied. In cooler pockets the soils are dominated by fossilized shells and chalky soils. Slightly east, more gravelly loam appears. The variation and size of the property mean it can reasonably grow Cabernet Sauvignon in warmer pockets and Riesling or Chardonnay in cooler ones. The wines are crowd-pleasing with finesse, texture, and backbone, an excellent option when you want your wine of poise with the flavor to please your nana.

J. Lohr

San Jose
www.jlohr.com

Founded by Jerry Lohr in the 1970s, J. Lohr is now led by the second generation of the family. Committed to sustainable farming in both Monterey and Paso Robles, J. Lohr produces reliable wines at multiple quality tiers, most more affordably priced than many other wineries in the state. Founder Jerry first established vineyards in the Arroyo Seco; a cool, windy, and rocky section of Salinas Valley. He has since also expanded into the Santa Lucia Highlands. There he grows Chardonnay and, more unusually, Valdiguié, which is consistently charming and delicious. In the 1980s, he established vineyards in Paso Robles that are home to their more robust reds, especially Cabernet Sauvignon. Lohr's investment in both regions was ahead of the curve, and was one of the foundations for the success now seen in each.

L'Aventure

Paso Robles

www.aventurewine.com

Beginning his winery Domaine de Courteillac in Bordeaux in the early 1980s, Stephen Asseo became known as an almost rebellious winemaker with impeccable skills. His interests over time reached beyond the traditions circumscribed by French appellation laws. In the 1990s, he set out on a quest to make wine outside the restrictions, looking to regions around the world as possibilities. After traveling Napa Valley, Argentina, Lebanon, and South Africa he found his way to Paso Robles. Today, his L'Aventure wines are made entirely from estate fruit on the west side of Paso Robles. Combining the craftsmanship still respected in France with the freedom allowed by California, Asseo blends Rhône varieties with those from Bordeaux. His Optimus cuvée brings Syrah, Cabernet Franc, and Petit Verdot into the blend offering both the mineral intensity and ripeness characteristic of his part of Paso Robles and the refinement that reveals his winemaking experiences.

MAHA

Paso Robles

www.villacreek.com

The production sizes are small, but the quality high with MAHA wines. Developed from the Villa Creek vineyard in the far western canyons of Paso Robles, MAHA is farmed biodynamically and crafted for energy and texture. As well as making the more familiar range of Rhône varieties, MAHA now also bottles a shockingly good Clairette Blanche, and makes sure wine lovers know they were the first in the region to do so. They are also experimenting with a sparkling version with good promise. Wines in the range are savory, energetic, and sophisticated with subtlety. Over time, founder–winemaker Cris Cherry has developed a more finessed and mouth-watering style that still delivers the incredible flavor of Paso Robles.

Saxum

Paso Robles

www.saxumvineyards.com

Saxum stands as one of California's few Rhône-based cult wines. The wines can be hard to get, and the winery hard to visit, but Justin Smith who owns and makes Saxum also advises some of the region's top vineyards. And from his own site he sells fruit to some of the other celebrated Rhône winemakers of California. The range of producers who have worked with Smith's vineyard demonstrates the quality of his farming and location. Saxum's own portfolio celebrates density, intensity, and power of presence. There is also soul, nuance, depth, and a lot of persistence. The combination is what established the cult following and helped bring acclaim to Paso Robles Rhône winemakers.

Tablas Creek

Paso Robles

tablascreek.com

Born in partnership between the Haas family of the United States and the Perrine family of Chateau du Beaucastel in the Rhône, Tablas Creek has become one of California's leaders of farming, sustainability, and the California Rhône movement. Founded in 1989, Tablas Creek successfully imported all the approved varieties of Châteauneuf-du-Pape,

Tablas Creek was the first in the world to earn the respected Regenerative Organic Certification for vineyard and winery. As part of their regenerative practices, they rely on sheep in the vineyard, here shown with their llama guard, Paco

including the really obscure ones. They are also among the few producers in California that grow Tannat and Vermentino. Tablas vineyards are planted in the cooler mountains of the Adelaida District, on the western side of Paso Robles, predominantly in calcareous soils that bring tension and texture. They've converted most of the winery to dry farming and helped create the Regenerative Organic Certification for vineyards and wineries. They were the first winery in the United States to gain the certification. Tablas Creek has also been a leader in other aspects of sustainability, including packaging. They now release multiple examples of premium wine in 3-liter bag-in-box containers. The approach reduces wine waste as well as the winery's carbon footprint.

SANTA BARBARA COUNTY

Known for: Pinot Noir, *Sideways*
Lesser-known strength: homes to Charlie Chaplin, Oprah Winfrey, and Ronald Reagan
Leader in: savory Chardonnay, Santa Maria BBQ, *pinquito* beans
First peoples: Michumash

Situated on the late-eighteenth-century mission trail, Santa Barbara of course includes a Spanish history. Misión de Santa Barbara was established in the 1780s and came to be known as Queen of the Missions for its beauty and that of its surroundings. It also provided the name for the surrounding city.

Prior to the Spanish period, Michumash (now known as the Chumash) lived throughout the region all the way south into Malibu and over the Santa Ynez Mountains into the cooler northern parts of what is now Santa Barbara County and the southern parts of San Luis Obispo. The rich resources of the extended area provided seafood, foraged plants, pine nuts, elderberries, chia, acorns, and wild meat as staples of their diet. Their skills at boat building, basket weaving, and stone cookware, along with the beautiful shells they gathered, wore, and traded were known throughout the extended region. Several bands of Chumash continue to inhabit the central and southern coasts of California. In 2024, they succeeded in lobbying the US government to create the third largest marine sanctuary in the country along 116 miles

(187 kilometers) of California coastline. It is known as the Chumash Heritage National Marine Sanctuary. Their history has also been recognized with the Chumash Painted Cave State Historic Park, an important part of Chumash cultural heritage that is still being preserved.

Though wine was established during the mission era, modern wine slowly began to emerge in the 1960s and 1970s. Successes began to appear in the 1980s, but the region didn't gain a sizeable reputation until the end of the twentieth century.

The cold, windy, and fog-ridden conditions of the region, as well as its low rainfall (9–17 inches/23–43 centimeters annually) meant agriculture never significantly featured in the region's economy. The slow growth of wine began when Uriel Nielsen planted on the benchlands of the Santa Maria Valley in 1964. Nielsen vineyard became the primary source for Santa Barbara Winery, situated near what is today the Funk Zone, along the water of the city. Santa Barbara Winery had been making wine from the Templeton area of Paso Robles until Nielsen offered a more local opportunity.

In 1971, Richard Sanford and Michael Benedict established the first winery in the Santa Rita Hills, in direct line with the winds from the ocean. Sanford had returned from Vietnam and sought a reclusive, agrarian life to center himself in the land and gain a new focus in farming. Some of Sanford's original vines remain as a coveted fruit source in the Sanford & Benedict vineyard, one of the extended region's most important heritage sites.

The success of Nielsen and Sanford & Benedict helped demonstrate that wine was possible in the cold of the region. Slowly, others followed. A year later, Tepusquet Vineyard was established by Louis Lucas and Dale Hampton, in the Santa Maria Valley. Soon after that, the Miller brothers, Stephen and Robert, established what would become their Bien Nacido Vineyard, one of the state's great quality fruit sources. It eventually became the home of Au Bon Climat, Qupé and eventually the Millers' own Bien Nacido winery. A few months after Napa Valley's recognition, the Santa Maria Valley was named an official AVA.

Zaca Mesa winery began in 1973, led by John Cushman. The experimental and collaborative approach Cushman fostered created the first generation of star winemakers in the region. Both Clendenen and Lindquist worked there before founding their own wineries. The winery is still affectionately called Zaca University. Ken Brown, Adam Tolmach, Jim Adelman, Scott Osborn, Lane Tanner, and others got their start at

Zaca Mesa before going on to become changemakers themselves in the region. The 1980s saw many of Zaca U's graduates founding their own wineries. And then another round of growth took hold in the 1990s, including the growth of the Santa Rita Hills. By 2001, it was its own recognized AVA (with the name shortened to Sta. Rita Hills to avoid conflicting with a winery of almost the same name); in 2004, it became famous because of the movie *Sideways*.

The Funk Zone

Tucked along Santa Barbara's waterfront, near the lower stretch of State Street, the Funk Zone is a once industrial area that has been converted to a bohemian collection of surf shops, cafés, and artist centers. A mere 13-block walking district, it has evolved to include destination restaurants and bars, breweries, and a host of tasting rooms.

The late Seth Kunin was the first to establish his Kunin tasting room in one of the surf shacks near the entrance of the area. He and his wife Magan eventually opened The Valley Project a little further into the Funk Zone with a focus on elevating understanding of the county's various AVAs. Their success helped build a winery draw for tasting rooms in the area.

Well-known wineries Melville, Margerum, J. Wilkes, and Blair Fox all pour wine in the area, alongside others. The first post-Prohibition distillery is also here, featuring local gin. Al and Terry Merrick's Channel Islands Surfboards, founded in 1969, is also found in the Funk Zone. Loquita offers Spanish tapas with fresh ingredients, or for street tacos worth seeking out, there is Mony's Mexican Food. With kids (or silly grown-ups), the Wolf Museum of Exploration & Innovation includes a dynamic mix of scientific learning and artistic play through interactive, experiential exhibits.

Growing conditions

Santa Barbara County faces the water on two sides. Its western and southern boundaries tuck against the Pacific Ocean. Its transverse mountains mark it as unique within California growing regions. They are the most significant transverse ranges in north or south America. While the rest of the state is bordered by north–south oriented mountains, those coastal mountains turn east at Santa Barbara County. The region sits fully exposed to the cold ocean influence.

The Sierra Madre Mountains rise from the county's east, the San Rafael to its north. Together, they create a rain shadow preventing northern storms from reaching the region and keeping annual precipitation at an average below 18.5 inches (47 centimeters). During the growing season, little rain falls, and only the largest rivers flow. The aridity of the region, combined with its daily winds, increases natural concentration in its wines and extends its growing season beyond those in northern California.

Ruben Solorzano, grape whisperer

Ruben Solorzano arrived in California in the late 1980s to join his older brothers as a seasonal worker farming vineyards in Santa Ynez Valley of Santa Barbara County. He was 19. Solorzano had been raised growing a mix of vegetables and other produce alongside his parents and grandparents in Jalisco, Mexico, before traveling north to California in time for pruning grapevines. He fell in love with the process almost immediately.

Solorzano's talent with grapevines has earned him legendary status in the wine community of Santa Barbara County. He's known as "the grape whisperer" by area growers. Solorzano brings such skill and work ethic to his efforts that soon after his arrival in California, winery owners sponsored him to remain in the region permanently and become a citizen.

Viticulturist Jeff Newton launched the region's top vineyard-management company, Coastal Vineyard Care, in 1983, relatively early in the growth of Santa Barbara County's vineyards. He is one of the highly regarded vineyard managers in the state, and his efforts have helped build Santa Barbara County wine. Recognizing Solorzano's talent, Newton made a bold move and offered Solorzano co-ownership in Coastal Vineyard Care in the 1990s, to ensure they could continue working together. The duo has also since partnered with Ben Metz and Mike Testa.

Solorzano has helped establish and convert vineyards throughout Santa Ynez Valley to organic viticulture. Working especially with Rhône varieties, Solorzano has expanded such plantings within the area as well. At Stolpman, he has built experimental plantings of Syrah that test viticultural training methods while also testing numerous clonal selections of the variety, with good success. These include sites that require pre-industrial methods – that is, no tractor and farming done entirely by hand. In 2008, Solorzano also began making his own wine, Hecho Por Ruben, which is well worth seeking out, joking that making his own wine is less expensive than buying it.

The coastal side of Santa Barbara County is profoundly cold. The vineyards closest to the ocean struggle to ripen and create lighter-bodied, ethereal wines when they do. The saying goes that temperatures climb by 1°F (0.6°C) per mile (1.6 kilometers) traveled east.

The western sides of the county feature Riesling, Chardonnay, and Pinot Noir. In its far east, Cabernet Sauvignon and Sauvignon Blanc are most common. Through the middle, it is Rhône varieties, especially Syrah, that are most celebrated.

Soils throughout the county also vary. Uplifted sand fills the western side of Santa Maria Valley, as well as the northwestern portion of Santa Rita Hills. Solomon Hills in the north and Melville in the south are planted entirely in sands. In the southwestern parts of Santa Rita Hills shale, chert, and diatomaceous earth (one of the least common parent materials for vines) can be found. Through Ballard Canyon calcareous soils appear, with sand in the west, mostly planted to Rhône varieties, especially Syrah. One AVA east, Los Olivos is slightly warmer and dominated by sandy loam. Syrah appears but so does Cabernet Sauvignon. All the way east, Happy Canyon features cobbles and clay loam growing Cabernet Sauvignon, Malbec, and Merlot.

Santa Barbara County AVAs

Santa Maria Valley (1981): in San Luis Obispo and Santa Barbara County, exposed to the Pacific
Main varieties: Chardonnay, Pinot Noir, Grenache, Syrah

Alisos Canyon (2020): cooled by the San Antonio Creek basin
Main varieties: Rhône reds, Tempranillo, Albariño

Santa Ynez Valley (1983): a windy valley on the southern end of Santa Barbara County
Main varieties: varied

Ballard Canyon (2013): limestone soils planted with Rhône varieties
Main varieties: Rhône varieties

Los Olivos District (2016): tucked between Ballard Canyon and Happy Canyon in the Santa Ynez Valley
Main varieties: Sauvignon Blanc, Cabernet Franc, Cabernet Sauvignon

Sta. Rita Hills (2001): one of the state's best-known Pinot Noir regions
Main varieties: Pinot Noir, Chardonnay, Syrah

Happy Canyon of Santa Barbara (2009): the warmer inland side of the county
Main varieties: Cabernet Sauvignon, Sauvignon Blanc

Wineries

A Tribute to Grace and Folded Hills

Los Alamos/Gaviota

gracewinecompany.com / foldedhills.com

Founded in 2007 by New Zealand-born winemaker Angela Osborne, A Tribute to Grace is a dedication to Osborne's love of Grenache. Each wine is made as a single vineyard, exploring the distinctive growing conditions of the state and their character expressed through one variety. The winery is named for Osborne's grandmother as well as one of her favorite attributes. Grace sources fruit from hallmark vineyards in Santa Barbara County and has also made Grenache from the Sierra Foothills.

Osborne served as the founding winemaker for Folded Hills from 2015 to 2020, a Rhône-based estate located in the middle of Santa Ynez Valley where it is just warm enough to ripen Grenache and Syrah but not as warm as some of the nested AVAs further east. Osborne returned to the winemaking helm of Folded Hills in 2024. The project is now led by an all-woman team with co-owner Kim Busche overseeing the business side. Folded Hills continues to center on Rhône varieties, growing Grenache, Syrah, Grenache Blanc, Marsanne, and Clairette Blanche. Folded Hills also grows row crops for local markets and raises heritage-breed animals.

Au Bon Climat

Santa Maria

aubonclimat.com

Founded in 1982 by Jim Clendenen and Adam Tolmach, Au Bon Climat quickly became a beacon for Pinot Noir and Chardonnay in the Central Coast, relying on barrel aging, malolactic fermentation and other methods now considered standard for the highest expression of the varieties. In 1990, Clendenen became sole proprietor. He also turned the winery space into a mentoring incubator for a host of younger winemakers who have since gone on to become leaders of the region. They include Bob Lindquist who founded Qupé and now leads Lindquist Family wine,

Jim Adelman, who is still the winemaker for Au Bon Climat, Gavin Chanin of Chanin wines, Paul Lato, Raj Parr, the Miller family of Bien Nacido, Joshua Klapper of Timbre, Marcel Giesen of New Zealand and innumerable others. He is widely regarded as having changed the quality and acclaim of Santa Barbara County and brought the region to international attention. His wines are beloved in France, the UK, Japan, Vietnam, and beyond. Au Bon Climat has continued to make supreme wines with Jim Adelman at the helm, and Clendenen's children Isabelle and Knox playing important roles in the winery. Although Au Bon Climat is best known for its Pinot and Chardonnay, it makes some of the best Nebbiolo in California. Hildegard, a blend of Aligoté, Pinot Blanc, and Pinot Gris is one of my favorite whites in California.

Bien Nacido

Santa Maria

biennacidoestate.com

Situated in the cool Santa Maria Valley, Bien Nacido is one of the Central Coast's most celebrated vineyards. The vineyard was established in 1970, and the original blocks of own-rooted vines from the time are still in production and highly desired by the region's winemakers. The Miller family has continued to expand the site, taking advantage of its highly varied terrain, elevations, and soil types. Today they grow not only Chardonnay and Pinot Noir, but also Syrah, head-trained Grenache, Viognier, Roussanne, Marsanne, Mourvèdre, Pinot Blanc, Petit Verdot, and Nebbiolo. Bien Nacido makes its wines from its estate vineyards under the winery names Bien Nacido and Solomon Hills. Solomon Hills makes Pinot Noir and Chardonnay, while Bien Nacido also bottles Grenache, Syrah, Viognier, and several blends.

Camins2Dreams and Kalawashaq' Cellars

Lompoc

camins2dreams.com

Co-founded by wives Tara Gomez and Mireia Taribó, Camins2Dreams brings together the distinctive winemaking perspectives of Taribó, whose experience comes from the mountains of her native Northern Spain, and Gomez who was born, raised, taught, and mentored in California's Central Coast. Gomez and Taribó worked together in Spain before returning to California in 2010 to launch the first tribally owned, grown, and made winery in the world, Kita. In its ten-year tenure, Kita became one of the most celebrated wines of California, gaining national attention and awards for both Gomez and the wines. In 2017, the duo started their own project, Camins2Dreams, focusing on mouth-watering, flavorful wines of the Santa Rita Hills. Initially focused on Syrah, their portfolio has expanded to include Albariño, Grüner Veltliner, Grenache, Mencia, Carignan, and occasional blends. Don't sleep on their next release of sparkling wine; it is consistently well crafted, fun, and delicious. Gomez has also relaunched her very first label, Kalawashaq' Cellars, which donates proceeds to non-profit organizations and is one of the first wines in North America to receive the Certified Authentic Made/Produced by American Indians seal. In its first return vintage, Kalawashaq' donated proceeds to The Cultural Conservancy, a Native-led non-profit in San Francisco dedicated to preserving and supporting Indigenous culture and traditions as well as ecosystems. Gomez is a member of the Santa Ynez Band of the Chumash and has been recognized by the California State Legislature as the first Native American winemaker.

Tara Gomez, Iris Rideau, and the Santa Ynez Valley

Tara Gomez, a member of the Santa Ynez Band of Chumash, was recognized by the California State Legislature as the first Native American winemaker. For her work in agriculture and winemaking, Gomez was also asked to be part of the International Indigenous Agribusiness Trade initiative to bring Indigenous people and their businesses together across international boundaries.

For ten years, she founded, built, and led the world's first and only fully tribally owned, grown, and made winery, Kita. In its decade-long lifespan, Gomez won numerous national awards, and her wines were featured in both television shows and national magazines. Her work brought attention to the Santa Ynez Band and heritage of the Chumash. Since 2017, Gomez and her wife Mireia Taribó have had their own winery, Camins2Dreams.

Also in the Santa Ynez Valley, Iris Rideau became the first Black woman to own and operate a commercial winery in the United States. She purchased her 23 acres (9 hectares) in 1995 and began planting Syrah, Mourvèdre, Marsanne, Roussanne, and Viognier imported from Southern France, an important part of California's Rhône movement. In 1997, she opened her winery and tasting room.

Rideau led her eponymous winery for twenty years. Throughout, she hired women winemakers, cellar crew, and tasting room staff, opening the door to more women and minorities gaining experience in the wine industry. She has also been a mentor to other Black women and women of color in wine since. The McBride Sisters, who make wine from both California and New Zealand, credit Rideau's legacy as part of the motivation they bring to their wine business.

Iris Rideau is a respected entrepreneur in Los Angeles for using her success to increase opportunity for other people of color in business. She went on to become the first Black woman in the United States to own a winery, located in Santa Ynez Valley under her name. Here seen with Dorothy Gaiter, multi-award winning journalist and news editor for the Wall Street Journal, *who with her husband John Brecher started the wine column "Tastings" for the journal. She is the first Black woman wine writer in the United States*

Jonata and The Hilt

Lompoc

www.jonata.com

Owned by the man behind Screaming Eagle, in Napa Valley, Jonata grows on the sandy side of Ballard Canyon and is reliant on a vineyard of varieties not always seen together. Jonata wines are primarily reds made from Cabernet Sauvignon, Cabernet Franc, Sangiovese, and Syrah, as well as some proprietary blends. Occasionally, Jonata also makes a white wine, though they make clear their primary focus is on reds. From the Santa Rita Hills, the same team makes Pinot Noir and

Chardonnay from their own steep-sloped vineyard under the winery name The Hilt. The wines are full of bright fruit as well as vibrant acidity and a sea-fresh, mineral tension that makes each sip satisfying.

Sandhi and Domaine de la Cote

Lompoc

domainedelacote.com

The influence of Raj Parr on the world of wine, most especially in California, is hard to overstate. He consistently credits Master Sommelier Larry Stone and Au Bon Climat lead, Jim Clendenen, as two of his most important mentors in wine. Parr made one of his first wines with Clendenen. In 2010, he founded Sandhi to focus on Chardonnay from Santa Barbara County's top heritage vineyards, and quickly added Pinot Noir. More recently Sandhi's reach has expanded to include vineyards further north along the coast of San Luis Obispo. Almost immediately, Sandhi drew national acclaim and brought a new generation of sommelier interest to Santa Barbara County wines. In 2013, Parr partnered with winemaker Sashi Moorman to start the estate-focused winery Domain de la Cote, using fruit from one of the more marginal sites in the Santa Rita Hills that is sectioned into five more or less adjacent vineyards. Portions grow in eroded diatomaceous earth: it's one of the few vineyards in the world planted in such soils. Many portions of Domaine de la Cote are planted at high density. Each vineyard has a different elevation, aspect, slope, and microclimate but all sites are only 7 miles (11 kilometers) from the Pacific.

Sine Qua Non

Oak View

www.sinequanon.com

Also known as SQN, Sine Qua Non was started by Manfred and Elaine Krankl as a small-production experiment intended to be poured in Los Angeles restaurant Campanile. Only a couple years after launch, an SQN 1994 Syrah, Queen of Spades, scored 95 points from Robert Parker, propelling the brand (and the Krankls) onto the world stage as one of California's niche cult wines. It was also the first time a Rhône-style wine from the United States had received such a score from Parker. Until 2021, SQN never made the same wine twice, crafting unique blends every year with equally unique labels featuring Manfred's art. The wines have become collector's items not only because of their distinctiveness year to year or the early praise from Robert Parker, but also because of a wine lover drive for SQN's gutsy, full force style. In 2021, SQN announced it would have to begin reusing cuvée names as it had become too difficult to register new names with the TTB. Though SQN primarily makes wines from Rhône varieties, especially Syrah and Grenache, in their estate vineyard they have also established Graciano, a variety they've worked with since 2017. Graciano spread across California at that time thanks to being mislabeled as Mourvèdre (see box, p. 230). By the time the mishap was realized its growers and winemakers realized they like the verve and brightening acidity of the variety and so many of California's Rhône winemakers now include Graciano.

Two Wolves

Santa Ynez

twowolveswine.com

Just west of the Happy Canyon of Santa Barbara AVA in the Santa Ynez Valley, Two Wolves makes smaller volume cuvées of Cabernet Franc, Cabernet Sauvignon, and Petit Verdot, along with Graciano, Syrah, Sauvignon Blanc, Semillon, a rosé, and occasional blends. Each is only a few hundred cases. The vineyard is organically farmed with a focus on water and energy conservation as well as promoting biodiversity in the vineyard and surrounding area. The project was founded by Alecia Moore (also known in some circles as P!nk) in 2013. Moore works with co-winemaker and director of operations Alison Thomson to craft Two Wolves wines. And unlike a lot of so-called celebrity projects, Moore trained in winemaking and most of all loves time in the vineyard. The wines of Two Wolves are structured, with invigorating acidity and profound flavor. They are distinctive within California. Thomson also makes her own brand, Lepiane, also worth seeking out, where she focuses on Italian varieties, primarily Barbera and Nebbiolo, as well as occasional blends.

10

THE CENTRAL VALLEY

California is carved by the vast Central Valley, which once formed an inland sea. The state's two largest rivers, the Sacramento from the north, the San Joaquin to the south, intersect in the middle, their union creating the California Delta. Each of the rivers forms its own valley, bearing the relative river's name; together they are known as the inland valleys or Central Valley. A continuous drive from the northern point of the Sacramento Valley at Redding, to the southern tip of the San Joaquin Valley at the Tehachapi Mountains takes almost ten hours. To cross the width takes four to six, depending on location. The sheer mass of this inland basin informs the weather for the entire state.

The Central Valley is surrounded by a ring of mountains. Along the east, the Sierra Nevada Range separates it from the stark continental climate of Nevada. To the north, the Cascade Range descends from British Columbia and the Pacific Northwest to end with Mt. Shasta and Lassen Peak. The southern rim is bordered by the Tehachapi, and to the west, along the Pacific, sits the Coastal Range.

As daytime temperatures rise over the course of the day, air along the floor of the inland valley lifts. The sheer mass of warm air rising creates something like a vacuum effect sucking new air in to fill in the valley floor. The vacuum pulls from the lone opening in the ring of mountains found at the California Delta. There, the low spot west of Lodi reaches through the Carquinez Strait into the San Francisco Bay and out through the gap to the ocean at the Golden Gate. The vacuum effect pulls air east off the cold California Current in the Pacific. There it rushes across the San Francisco and San Pablo Bays, through the Carquinez Strait, and into the Central Valley. Wind farms built through the area

Significant winds blowing into the Central Valleys through the Carquinez Strait and near the California Delta provide renewable energy via wind turbines for much of the Central Valley

provide zero-emissions power for parts of Central California. The entirety of the Central Valley enjoys the state's Mediterranean climate. Broadly speaking, cooler areas sit along the central area where the valley intersects the delta. Going south, the warmest areas tuck into the San Joaquin Valley.

The Sacramento and San Joaquin Valleys formed as the Farallon plate was subsumed (almost entirely) beneath the North American plate. The tectonic shift over time created a low spot that let in the ocean to form an inland sea. As evidence, California includes two deep seaports within the Central Valley, one next to the city of Sacramento, the other by Stockton. Remnants of the ancient sea are also found in the rocks and soils of the area. The slopes of the Diablo Range are filled with ancient marine sediments left as the inland sea receded. Much of the soil through other parts of the valley is also rich in salts and related minerals, creating challenges with excess irrigation.

The two large rivers descend from the Cascade Range near Mount Shasta into the Central Valley in its northern half, and out via the Sierra Nevada Range through the valley's southern section. The Sacramento Valley remains cooler than the San Joaquin, with more humidity and

Fog lifts from the oak groves of the Central Valley outside Lodi

a shorter growing season. As the river approaches the California Delta, grass and marshlands take hold. Here a history of rice and cotton farming took over the state. Peat soils can be found here. Outside the delta area, sediments along the rivers contain gravels and sand.

The heart of California pumps through the Sacramento–San Joaquin River Delta. It's a vast inland intersection of tributaries and estuaries connecting the inland watershed to the San Francisco Bay. Portions of the delta fall below sea level. The land has been reclaimed for farming as a series of islands and small peninsulas surrounded by levees and sloughs. The experience of traveling the delta by car or on foot includes looking up to see water channeled through troughs far above: traveling by boat one has to move slowly in case land reaches near the surface.

The portions of reclaimed land in the delta are encircled by levees that must be maintained and protected from erosion or leaking. Some of the delta islands have stopped being maintained and returned to the marsh. To keep the islands afloat, each has an attendant whose job is to walk the perimeter of each levee twice a day and perform repairs at the earliest sign of slippage. Even in some of California's greatest floods, these islands have continued to float below sea level. Here, vineyards are planted to grow vibrant white wines with unexpected freshness and length. Lighter reds can also be found. Elsewhere cattle and row crops are also growing.

The far western side of Lodi sits below sea level within the delta. Gravelly alluvial soils fill the area, and daily winds called the Delta breezes keep it cool. As the sun descends, the winds go down. Slightly north, the Clarksburg area sits along the Sacramento River within the delta as well. The region grows primarily white wines, especially Chenin Blanc or Pinot Gris, with smaller portions of Vermentino and Albariño. The majority of Lodi is to the east. While the Central Valleys are generally warmer growing areas for California, Lodi must be seen differently. Portions of it are warm, though nowhere hotter than northern Napa, but it sits within the Delta breezes. Properly speaking, Lodi stands at the intersection of the warm inland valley with the chilly delta. The combination gives nuance to its growing conditions.

Traveling south, the San Joaquin Valley is warmer as well as more arid. The Central Valley and State Water Projects transformed the growing potential of the area by channeling river water from the abundant north into reservoirs (as well as hydro-electric power stations) and irrigation centers in the warmer south. Even so, the Central Valley aquifers are among the most pumped underground water resources in the nation. They provide approximately 20 percent of the US groundwater demand.[1] Urban centers such as Fresno and Bakersfield play a significant role in water use, but farming the nation's food plays another. As water availability in the San Joaquin improved over the second half of the last century, agriculture grew. So did oil drilling, since the availability of water made cooling machinery for pumping possible. The combination of agricultural land and petroleum harvesting in the area raised the value of the San Joaquin. Over time, smaller family farms evaporated, to be replaced by larger growers and international oil companies.

In the second half of the 1800s, much of the Central Valley grew grain, both for eating and for pasture animals. In the 1900s, cotton came to dominate the region. In the mid-1900s cotton was such an important crop for California that the state held annual contests to crown a Cotton Queen who traveled the world hosting international dignitaries to sell California cotton.

With the arrival of Prohibition, the Central Valley became one of the state's most valuable grape growing areas. The planted area of the region doubled in order to supply home winemakers of the east, who needed fruit to make wine. During World War II, the region served as one of the country's most important raisin centers, providing shelf stable rations to last through the war. Today, the inland valley forms one of the nation's most important agricultural centers, providing a quarter of the nation's food supply while also exporting crops to other countries. Tree nuts, vegetables, berries, other row crops, cotton, rice, and grains all grow here. The Central Valley also grows at least 70 percent of California's wine.

SACRAMENTO VALLEY

Known for: the breadbasket of the nation, growing half of the state's grapes
Lesser-known strengths: the world's largest yoyo, walnuts, waterfowl wetlands
Leader in: rice, citrus, nuts, the California capital Sacramento
First peoples: Miwok, Maidu
First vines: 1850s
Planted vineyard area (2024): 23,834 acres (9,645 hectares)
Primary varieties: Chardonnay, Cabernet Sauvignon

Spanish explorers pushed into more accessible parts of the Sacramento Valley during the late eighteenth and early nineteenth centuries. But the area was distant from their other settlements. Indigenous peoples already settled throughout the area remained relatively undisturbed. As settlers occupied the western coast along Spain's mission trail, some Indigenous groups pushed east along the waterways, having to negotiate new territories or co-exist with tribes they had previously not been part of. California's Indigenous peoples had always formed evolving cultures and colonies, following trading practices throughout the extended region into numerous areas now considered parts of other states. Until the 1830s, after Mexico took Alta California from Spain, inland groups experienced change as a result of coastal peoples forcibly moving into their territory, but experienced less contact from European settlers.

Mexico's land management was different from Spain's. Seeking sustainable independence, Mexico needed to increase settlements and productivity of the land. To do so, they invited immigrants to become citizens, the most productive of whom were given land grants. Such parcels became means for extensive ranching, milling, and various forms of agriculture. A Swiss settlement established by John Sutter became the first outside settlement in the area, tucked into the hills of the Sierra Nevadas, overlooking the Sacramento Valley. Hoping to capitalize on the wealth of the forest, Sutter and a carpenter, James Marshall, built a sawmill powered by a tributary of the American River running through the foothills. During construction, the duo discovered gold. For almost a year, Sutter and Marshall enjoyed the solitude of a partnership gold mine. In 1849, their riches were discovered, and more than 80,000 fortune seekers rushed to the area.

The discovery of gold created the state of California and filled the Central Valley between the foothills and San Francisco with settlements providing hospitality and resources to outsiders hunting gold. The Maidu and Miwok were pushed from the area, and by the 1870s those who survived were moved to containment reservations. The influx of European and Chinese settlers moved the economic and agricultural centers of California to the north. The waterways of the Central Valley became the transportation system for moving gold, timber, food, and other resources out of the foothills, to the city. The railway served as the means to move it around the nation from there.

The same rivers also changed the landscape. Gold mining in the region was done by washing large amounts of water over sections of land to remove the lighter sediments and reveal comparatively heavier stones of gold. As sediments were washed from the foothills, they filled the riverbeds and estuaries in the Sacramento basin below. Granite sands became a formative part of the soils of the extended area. Salmon habitats were lost to the smoothing out of the riverbed. Once the largest spawning ground for wild salmon in the world, the San Francisco Bay area gained too much dirt to bring salmon home anymore. Food sources transitioned, with more of a reliance on farming.

Today, neither the Sacramento nor the larger Central Valley is considered an AVA. They instead act as geographical references within the state of California. Within the Sacramento Valley there are 17 AVAs, eight of which fall inside Lodi. Parts of Lodi stretch south into the San Joaquin Valley. The Clarksburg marshlands overlap the California Delta and Sacramento Valley, including two AVAs. Slightly north, Capay Valley and Dunnigan Hills sit entirely within the Sacramento Valley. Yolo and Sacramento Counties are entirely within the Sacramento Valley. Yuba and Solano Counties are only partially so. Solano (see chapter 7) lifts into the North Coast along Napa County on its western side. Yuba (see chapter 11) climbs part of the Sierra Foothills in its east. Capay Valley, Dunnigan Hills, and the Winters Highlands AVAs are each modestly planted and host few wineries. They each grow primarily hearty red varieties including Tempranillo, Cabernet Sauvignon, and Petite Sirah. Clarksburg and Merritt Island AVAs also fall at least partially in Yolo County but are mentioned below as part of the California Delta.

To put it simply, the areas further west and closer to the bay are cooler than those to the east. The western sections also bring more humidity, and the soils with more clay are also cool and slow ripening. Moving east, elevations increase towards the Sierra Foothills and soils become warmer and rockier.

Water throughout the region comes from winter snowmelt occurring in spring. Reservoirs have been built throughout the extended area to capture melting snow and prevent flooding. The snowmelt from the Sierra Foothills in the east and Cascade Range to the north both help fill aquifers in the area as well.

Yolo County AVAs

Capay Valley (2002): especially planted to Iberian varieties
Main varieties: Tempranillo, Cabernet Sauvignon, Petit Verdot

Dunnigan Hills (1983): less prone to frost than other parts of the Sacramento Valley
Main varieties: Cabernet Sauvignon, Merlot, Tempranillo

Winters Highlands (2023): in both Solano and Yolo Counties
Main varieties: Petite Sirah, Tempranillo, Zinfandel

CALIFORNIA DELTA

Known for: critical habitat for fish and other wildlife, bird watching, white wine
Lesser-known strengths: the Delta Queen, an 1890s paddle boat that's now a restaurant
Leader in: supplying fresh water to two-thirds of the state's population
First peoples: Miwok
First vines: 1860s

Sunset on the levees of the California Delta mean standing above vineyards in the area planted below sea level

The tributaries and estuaries of the California Delta provide some of the most fertile soils of California. The water confluence with fog and Delta breezes keeps the area cool.

Agriculture began in the area in the 1860s. Settlers removed rocks and drained water to access farmland. By the 1880s, it was one of the state's farming resources, providing alfalfa, corn, and cereal grains. In the 1960s, multi-generational families turned to tomatoes and wine grapes. Today, the fertile soils and water table still provide farmland for asparagus, corn, safflower (for oil), cereal grains, and alfalfa. On its eastern side, portions of the delta fall within the Lodi AVA and also grow wine.

In the 1960s, grapes moved into Clarksburg as well, which sits partially in the Sacramento Valley, verging into the delta. The afternoon breezes and wet soils keep Clarksburg cool, encouraging early-ripening white wine grapes in the area. Chenin Blanc has made a name for itself in Clarksburg, and the area is planted to Pinot Gris, Albariño, Chardonnay, and Vermentino as well. The Bogle family were the first to plant vines in Clarksburg. From the late 1800s, the Bogles farmed the area, initially growing row crops and corn. In 1968, the family began their move into wine, planting 20 acres (8 hectares) of Chardonnay and Petite Sirah. Within Clarksburg falls Merritt Island, properly speaking a peninsula surrounding by sloughs. Situated within the delta, Merritt Island is even cooler and grows almost entirely white wine varieties.

California Delta AVAs

Clarksburg (1984): continues into Sacramento, Solano, Yolo Counties
Planted area: 9,000 acres (3,642 hectares)
Main varieties: Albariño, Chenin Blanc, Chardonnay, Pinot Gris, Sauvignon Blanc, Verdelho

Merritt Island (1983): located in Yolo County
Planted area: 63 acres (25 hectares)
Main varieties: Chardonnay, Chenin Blanc

LODI

Known for: Zinfandel, one of the first sustainability programs in the world
Lesser-known strengths: Isenberg Crane Reserve with over 100 breeds of cranes
Leader in: the biggest diversity of grape varieties in California, the largest planted vineyard area in the country
First peoples: Miwok
First vines: 1850s

Vinifera entered the Lodi region in the 1850s, largely because of the expansion of the gold rush. The flatter lands and easily dug silt and sand along the turns of the Mokelumne River made it the heart of Lodi.

The highest concentration of family farms and head-trained vineyards rose along the Mokelumne. The granitic sands of the area prevented phylloxera taking hold. The result is some of the oldest, own rooted (i.e. ungrafted) vines in North America. It even includes what is likely the oldest surviving Cinsault vineyard in the world – ungrafted, dry-farmed, and head-trained – Bechthold Vineyard planted in 1886. Next door, the Royal Tee vineyard was planted in 1889 with a mix of Zinfandel, Petite Sirah, Mataro, Mission, and others.

Lodi enjoys the warmth and sun exposure of the Central Valley with temperatures mitigated by afternoon wind. Overall temperature accumulation for Lodi resembles that of the St. Helena area of Napa or the eastern side of Sonoma County. But Lodi's temperatures do not get as high during the day, or as cool at night, staying more even over the course of the day and night. The Delta breezes, as they are affectionately named, play a powerful role in the most exposed parts. Winds through the delta power much of the region through wind farms planted along the Montezuma Hills. Deep fog can fill the area. In winter it is sometimes so dense that cars slow on the corners of country roads, roll down the window, honk the horn and listen for other cars doing the same. In the growing season, fog stays throughout the morning, cooling soil temperatures before lifting with rising temperatures from the vines.

To Lodi's north is the city of Sacramento, the capital of California; in its south is Stockton. As an AVA, Lodi was recognized in 1986. By 2006, it was clear that the area held even more nuance, and seven nested AVAs were also recognized. The seven nested AVAs must be named on a wine label in conjunction with the larger regional name of Lodi. The Lodi AVA crosses both the Sacramento and San Joaquin Counties. The city of Lodi and, nearby, the region's historic center, the Mokelumne River, sit in San Joaquin County along the southern side of the appellation. Mokelumne River is the largest AVA, nested within the larger Lodi appellation.

Traveling northeast in the region through the other nested AVAs, elevation increases, soils become rockier, and temperatures generally increase. Clements Hills, Borden Ranch, and Sloughhouse each gain progressively in gravels. The elevation and rocky nature of the northeastern areas kept it as ranchland for a long time because it was difficult to grow crops or vineyards. Vineyards in the northeast tend to be both larger and younger than those in the older parts of the region, especially compared to Mokelumne River. The more western AVAs of Jahant, Cosumnes River, and Alta Mesa instead host more clay, clay loam, and in sections clay hardpan, making the soils cooler than those in the east as well.

In terms of planted area, the Lodi AVA is the largest, not only in California but in the entire country, with more than 100,000 acres (40,469 hectares), compared to the entire Central Coast AVA, which includes 90,300 total planted acres (36,543 hectares). The yields of Lodi exceed Washington and Oregon states combined, and Lodi provides around 20 percent of California's grape production per year.

The Mokelumne River remains Lodi's best-known AVA. The first outside settlers came from German and Italian immigrants who arrived in the second half of the 1800s. They grew the foods needed for a farming family, and as the distribution networks developed from the railroad, melons and Zinfandel became significant crops. Before the rise of seedless grapes (grown mainly in the south of the San Joaquin), Lodi also grew the most beloved table grapes in America, Flame Tokay.

Winegrowing expanded in Lodi during Prohibition, then became part of regional cooperative wineries, and later served as a crop for the Gallo family. When white Zinfandel erupted in success from the slopes of the Sierra Foothills Lodi became an important source for growing production. Arguably, the white Zin boom saved many of California's historic field blends. Almaden took over Lodi grapes, for a time, and then Robert Mondavi returned from Napa Valley to found Woodbridge, a more affordable brand grown in Lodi. The brand served two purposes: it helped elevate the level of farming and grape quality in Lodi, and it also provided financial support for the fine wine venture to the west.

Even with the wealth of vineyards, winemaking didn't properly begin in Lodi until the 1990s. Tim and Barbera Spencer of St. Amant played an important role in the region's transition from grape farming to wineries. Their vineyard further east died from phylloxera, so the Spencer family started buying fruit in Lodi. Spencer vinified single vineyard lots of wine and brought bottles of each site back to the relevant grower. It was the first time most had tasted wine from their vineyard. Over time, many of them started slowly building their own wineries.

St. Amant continues and is now run by the second generation, Stu Spencer. Other family wineries from local, multi-generational farming families today include Mettler, Harney Lane, Michael David, Klinker Brick, LangeTwins, Van Ruiten, Jessie's Grove, and Lucas. Lucas Winery played an important role in expanding the possibilities of the region. Lucas was one of the wineries that raised the case Granholm vs. Heald along with other modest, family-owned wineries seeking the ability to sell wine across state lines directly to consumers. It eventually reached the Supreme Court where, in 2005, the decision allowed selling wine outside the state directly to consumers. It was an important shift in the rigidity of the three-tier system. Prior to the ruling, wineries were required to enter the wholesale distribution system to deliver wine in another state. Some states continue to have their own restrictions on this question. As of 2024, Lodi has around 85 local wineries. Gallo continues to source fruit from the area, Bonterra buys from organically farmed sites in the region, and Trinchero family continues to make its Sutter Home wines with Lodi fruit. Sutter Home created the country's first commercial white Zinfandel, initially by accident while making wine with Darryl Corti of the Corti Brothers in Sacramento.

Brothers Louie, Jr. (left) and Joe Abba, third-generation farmers and grape growers in Lodi. They grew up farming and making wine with their father and grandparents. Louie made home wine from his family's old vine Zinfandel for over 70 years. Today, his son Phil Abba is a fourth-generation grape grower

Winemakers from other parts of California have helped elevate the region's reputation from a perception of bulk wine to demonstrating it also makes fine wine. Steve Matthiasson has made wine from Royal Tee. John Lockwood of Enfield buys Rhône varieties from the Abba family, a multi-generational Italian–German farming family in the region. Nathan Kandler's Precedent (he's also winemaker for Thomas Fogarty in the Santa Cruz Mountains) shone a light on the quality of Carignan in the area. Tegan Passalacqua started seeking out old vine vineyards to farm organically for Turley, then invested in the area himself, buying the old Kirschenmann vineyard and building his own Sandlands winery. Morgan Twain-Peterson soon followed, buying Katushas vineyard around the corner from Kirschenmann. Ridge vineyards has returned to making

Zinfandel from Lodi. Patrick Cappiello launched Monte Rio to make affordable wines from distinctive areas of California that over-deliver on quality. He significantly relies on Lodi fruit for the venture. His friend and mentor, Pax Mahle of Pax, sources Syrah from the area as well.

Numerous viticultural and mechanical innovations have occurred in Lodi as well. Its vineyards helped foster tractor innovations, including machine harvesters and leaf pullers. The region served as one of California's first integrated pest management viticultural areas, developing the means to fight vineyard pests using more holistic and non-chemical interventions such as hormonal confusion and birds of prey. More recently, Lodi vineyards have been developing the use of high-wire training methods that, with the right varieties, make the major steps of the growing season easier to manage and less expensive to farm, increase the quality possible with machine harvesting, and make the presence of farm animals within the site possible as the canopy is above the height of sheep or goats.

Wine writer Randy Caparoso has played an important role in the region's recent evolution. He formed the group Lodi Native, bringing together local winemakers to make Zinfandel from historic vineyards with no new oak, and no additions save sulfur. Fields Family, M2, McCay, St.

Vineyard workers harvesting wine grapes by hand from head-trained vineyards in Lodi

Amant, Maley Brothers, and Macchia all joined the effort. The project led to a glimpse of these old vine sites with minimal varnish, attracting a new generation of winemakers from outside the area to investigate Lodi fruit. Caparoso has also written on Lodi more than any other wine writer anywhere.

And the area is worth investigating. Lodi has the greatest varietal diversity of any region in California. Around 130 wine grape types grow in the region commercially. Some of the nation's largest collections of Austro-Hungarian (German and Austrian) varieties, as well as Iberian (Spanish and Portuguese) can be found in the region, alongside cultivars from the Rhône, Italy, and Greece, as well as the more expected Cabernet Sauvignon, Sauvignon Blanc, Chardonnay, and Pinot Gris.

Lodi AVAs

Lodi (1986): continues into Sacramento and San Joaquin Counties
Main varieties: Cabernet Sauvignon, Chardonnay, Zinfandel

Sloughhouse (2006): the highest elevation and most gravelly area of Lodi
Main varieties: varied

Borden Ranch (2006): volcanic hills and prairie mounds with some alluvial influence
Main varieties: varied

Alta Mesa (2006): a mesa of sandy and gravelly clay loam
Main varieties: varied

Cosumnes River (2006): wetlands and floodplains in Sacramento and San Joaquin Counties
Main varieties: Chardonnay, Sauvignon Blanc, Pinot Gris, Vermentino

Jahant (2006): rolling woodlands with sandy, alluvial soils in both Sacramento and San Joaquin Counties
Main varieties: varied

Clements Hills (2006): rolling alluvial hills with rocky clay loam
Main varieties: varied

Mokelumne River (2006): the heart of Lodi, its original population and vineyard center
Main varieties: Cabernet Sauvignon, Chardonnay, Zinfandel

Wineries

Bokisch Family

Lodi

bokischvineyards.com

Inspired by his childhood in Spain, Markus Bokisch, along with his wife Liz, established his first Lodi vineyards in 1998 with a host of Spanish cultivars. Bokisch was one of the first to import Albariño to California, and the first to plant it in the Lodi region. The cuttings Bokisch imported became the first of the variety in the state to be officially recognized and registered at UC Davis. Albariño was first made in Santa Barbara County by Verdad. The variety has had unexpected but impressive success in the region, where sun exposure delivers flavor development while the influence of the delta retains vibrant acidity. Prior to his Lodi venture, Bokisch served as a viticulturist for Joseph Phelps in Napa Valley. Upon moving east to Lodi, Bokisch began growing vineyards for other producers and provides some of the most distinctive quality fruit in the region. His winery is situated in the Clements Hills area, and he also farms other sites in the Mokelumne River and Borden Ranch areas of the region.

LangeTwins

Acampo

langetwins.com

The Lange family have farmed in Lodi since the 1870s. In the 1970s, they started growing fruit for Woodbridge, working directly with Robert Mondavi to develop quality fruit for the venture. The Lange brothers (twins as the name suggests) then founded their own viticultural business, operating vineyards throughout the extended area. The family farm stands in Jahant, where they have worked hard to restore natural habitats including a beautiful on-site lake that's become home to a wealth of birds, frogs, reptiles, and other wildlife that have all naturally found their way there. They also work with regional school programs to provide hands-on learning about water management, wildlife preservation, and science resources from the farm. They have helped preserve some of the old vine sites in the region by partnering with the grower families to make distinctive wine from their heritage plantings.

Kevin Phillips of Michael David Winery farms Bechthold vineyard organically, here seen with spring covercrop. Bechthold is likely the oldest Cinsault planting in the world, planted on its own roots in 1885. The sandy soils mean the area is not impacted by phylloxera

Michael David Winery

Lodi

michaeldavidwinery.com

Brothers Michael and David Phillips are the fifth generation of a farming family that made its way to Lodi in the 1860s. The brothers have developed one of the most successful independent and family-owned wineries in the region. They've created a destination winery on the west side of the region that brings together their wines, food, and activities to enjoy. The marketing brilliance of the brothers has attracted a sizeable following. Relying on clever names with the imagery to match, their cuvées like Freakshow or Earthquake have staunch supporters. Their brand 7 Deadly Zins relied entirely on Lodi-sourced fruit and provided a valuable market for growers in the area. It was so successful that the brand was sold in 2018. They also organically farm Bechthold vineyard, likely the oldest surviving Cinsault planting in the world, with vines dating from 1885.

Sandlands

Victor

sandlandsvineyards.com

I first met Tegan Passalacqua the day he'd been elevated to head winemaker and vineyard manager for Turley. He spent much of his time driving through lesser-known regions of northern California getting to know the older farming families of their areas. His talent for relationship building helped find some of Turley's best vineyards, which are each organically farmed. Passalacqua has since also started his own winery, Sandlands, making wines from historic regions as an expression of their heritage and what Passalacqua believes are the region-shaping varieties. Sandlands wines are small production with an international following. They are delicious, mouthwatering, ageworthy, and a reminder of how much California wine has accomplished.

SAN JOAQUIN VALLEY

Known for: providing a quarter of the nation's food
Lesser-known strengths: ghost lake, Tulare, which disappeared in 1890 and reappeared in 2023
Leader in: grows 70 percent of state wine, a major Amazon distribution center
First peoples: around 50 distinctive tribes known collectively as the Yokuts
First vines: 1850s
Planted vineyard area: 69,823 acres (28,256 hectares)
Main varieties: Chardonnay, Zinfandel, Cabernet Sauvignon, French Colombard, Rubired

In the 1960s, cotton was one of California's most important crops. Every year the industry held a contest for the annual Cotton Queen, who was given a collection of fine clothing, including ballroom gowns, and flown about the world meeting international dignitaries to encourage the purchase of California cotton. Cotton first arrived in the 1800s but became an important part of agriculture as American settlement took hold. It's still grown in the region, identified with sturdy fibers in the industry. But cotton is no longer California's most important crop.

Vinifera arrived in the San Joaquin with the gold rush; the first vines were planted in 1850. It marked the arrival of the first outside settlers too. Before the gold rush, Indigenous peoples of the San Joaquin Valley largely enjoyed their own regions. The Yokut spread through the valley, relying on various river systems, native flora, and wildlife for resources. The Yokut included around fifty individual groups, each with their own traditions and territory. As outside settlement expanded, Yokut dwindled. It wasn't until 1901 that newer residents of the San Joaquin realized they had a unique opportunity in wine. The venture grew during Prohibition, encouraged by the demand for grapes from home winemakers in the east. By 1927, 72 percent of California's wine grapes came from the San Joaquin Valley.

The region today is often dismissed as the home of bulk wine, but the value of wine in the area is more complex. Lodi at its northern side has long been home to some of the oldest ungrafted and head-trained vines in the state. And Madera, further south, has been developing its reputation for quality wine more recently. The San Joaquin doesn't even have the biggest vineyard in the state. That's found in Monterey. That said, the San Joaquin does provide a sizeable volume for wine. The fertile soils, warm temperatures, and importing of water support yields. The climate of most of the San Joaquin Valley is very warm to hot, as it is shielded from ocean influence by the coastal mountains. The further south, the warmer. The northern sections of the San Joaquin enjoy a hot Mediterranean climate but descending to the southern edge, the area is desert. Without the assistance of groundwater pumping and the State Water Project that diverts river water from the north, much of the region would be too dry to farm. South of Stockton, Bakersfield, and Fresno are the two most significant population centers in the San Joaquin. They are served by water from the San Joaquin, Kern, and Kings Rivers. Even so, the aquifers of the San Joaquin are one of the most pumped aquifer systems in the country.

Mixed within the vines of San Joaquin are some of the state's most unusual varieties. UC Davis researcher Harold Olmo bred a series of grape types during his career, meant to provide unique crops, and new viticultural advantages, for California and beyond. As part of his effort, he also worked on identifying and developing phylloxera-resistant rootstock. Many of Olmo's varieties were intended for table grapes, but he also bred grapes for juice, and for wine specifically. Among wine cultivars, Ruby Cabernet was his most successful. A cross between Carignan and Cabernet Sauvignon, the variety captures the aromatics of Cabernet with the vitality and bright acidity of Carignan. Like its Carignan parent, Ruby Cabernet holds its acidity, with moderate color concentration and impressive yields. The one criticism leveled at it is that it lacks tannin, but even that makes it ideal for approachable, crowd-pleasing wines. Ruby Cabernet mainly grows in the Central Valley of California but has also been established in other parts of the world, including South Africa, Argentina, and Australia.

Gallo

The largest winery in the world, Gallo has been led by four generations of family. Prior to Prohibition, brothers Ernest and Julio worked in their parents' grape-growing business in Modesto. During Prohibition, at age 17, Ernest traveled to Chicago to sell a railcar of grapes for winemakers in the east. Chicago was considered one of the toughest markets in the country at the time. His success demonstrated a sales acumen that guided the Gallo business for decades.

In 1933, Ernest and Julio launched their E&J Gallo business. As the story goes, they found an old winemaking pamphlet in the local library and taught themselves winemaking from it. Full of ambition and a strong work ethic, the brothers continued to teach themselves all aspects of the business. They realized the most secure way to succeed was to own almost every essential aspect of the business; they were the first in US wine to develop vertical integration. First they grew grapes and sold bulk wine to others. Then they made wine under their own brand. When that was stable, they started printing their own labels and established an in-house design firm. They created the first specialist wine salespeople in the United States, expanding into new regions only once they dominated those they had already entered. They started their own glass manufacturing, then bought the sand mines to make the glass. They founded their own transportation business. After success with jug wine they invested in fine wine vineyards.

Today, Gallo is one of the largest recyclers of glass in California. In 2024, they began a shift to partially electric glass furnaces to reduce emissions. Gallo has one of the most powerful marketing research groups, as well as viticultural and winemaking research programs, in the world. When Gallo starts a new marketing style or buys a wine brand it's a sign of trends to come.

Gallo is controversial for its sheer size. The wine industry favors the romance of small producers. Gallo has had moments of controversial history. The UFW boycotted Gallo in the 1970s, and again in 2005. Their glass bottles had production issues for several years. Ernest and Julio legally stopped their younger brother from using the Gallo name to sell his cheese, arguing it would create market confusion. However, Gallo has also consistently won awards for its treatment of employees, its sustainability initiatives, and its support of LGBTQIA communities. Anecdotally, employees say Gallo has offered significant support during family challenges and employee emergencies.

Today, Gallo remains family owned. Its portfolio includes more than 100 wine, beer, and spirits products. They export to 110 countries, and import from others, particularly Italy.

Even more successfully, Olmo bred a juice grape known as Rubired. The inter-specific hybrid was released in 1958, offering abundant yields even in hot temperatures, with pigmented juice delivering deep red color. Importantly, Rubired is a teinturier grape, which means its pulp has color rather than just its skins. Its grandparent, Alicante Bouschet, is teinturier as well. The difference with these sorts of grapes is the juice immediately shows red. In non-teinturier grapes used to make red wine, the color comes entirely from the skins. With those grapes, to make red wine, the skins must be left to soak in the juice during fermentation. As alcohol builds, it extracts color from the skins turning the wine red. With ancestry largely based in Portugal, Rubired was thought to be an option for sweet wines of California (the dominant part of the wine market when the cultivar was released). It was used for a time in some Port-style wines in Australia. A few producers, such as Monte Rio Cellars, make it as a varietal red wine.

Rubired also serves as a color addition for wines in vintages that produce less color. And there begins its most lucrative value. In the 1970s, a winery started experimenting with the color-enhancing qualities of Rubired and developed a juice concentrate called Mega Purple. It's an impressively dark, sweet liquid that became a game changer in the wine industry. As aesthetics matter when it comes to how US wine is sold, Mega Purple is used in wines meant for a consumer audience that wants darker color. It began a mega industry. The intensity of Mega Purple means very little is used for any wine. The process to make it is proprietary, developed by Constellation brands and now owned by Vie-Del. In 2023, Rubired represented more than 5 percent of all the wine grapes harvested in California, almost entirely planted in the San Joaquin Valley.

The San Joaquin Valley was also home to some of the biggest farm labor strikes in California history. The United Farm Workers, started by Larry Itliong, Cesar Chavez, and Dolores Huerte in the mid-1960s, successfully picketed and boycotted grape growers in the region to demand better working conditions. Prior to their efforts, farmworkers were the sole unprotected labor industry in the country. While the UFW did gain labor contracts with specific growers, their bigger success was changing state regulations to create labor protections, and to regulate deadly pesticides.

The San Joaquin's other lucrative industry is oil. Both the Sacramento and San Joaquin Valleys were once part of an inland sea. The marine sediments left when it receded deposited layers of shale, and underneath that, oil. The San Joaquin Valley is now considered the center of the

California oil industry, including more than 80 percent of the state's still-active oil wells.

Like its northern sibling, the Sacramento Valley, the San Joaquin Valley is not a recognized AVA. Instead, it includes several official appellations. The San Joaquin also crosses at least part of eight counties: San Joaquin, Stanislaus, Merced, Fresno, Madera, Kings, Tulare, and Kern.

San Joaquin Valley AVAs

(Not including Lodi, which is discussed earlier in this chapter)

San Joaquin County AVAs

River Junction (2001): a shallow bowl surrounded by rivers
Main varieties: Chardonnay, Cabernet Sauvignon

Tracy Hills (2006): drier and less foggy than surrounding areas
Main varieties: Cabernet Sauvignon, Chardonnay, Montepulciano

Madera County AVA

Madera (1984): overlaps both Madera and Fresno Counties
Main varieties: French Colombard, Chardonnay, Cabernet Sauvignon

Fresno County AVA

Squaw Valley-Miramonte (2015): in the Sierra Foothills but south of the Sierra Foothills AVA
Main varieties: varied

Stanislaus County AVAs

Salado Creek (2004): currently no wines are designated with this appellation
Main varieties: Cabernet Sauvignon, Syrah, Sauvignon Blanc

Paulsell Valley (2002): soils are mainly volcanic tuff, sometimes with cobbles
Main varieties: Cabernet Sauvignon, Petite Sirah, Petit Verdot

Diablo Grande (1998): originally established as a one-winery AVA, the winery no longer exists
Main varieties: Cabernet Sauvignon, Chardonnay

Kern County AVA

Tehachapi Mountains (2020): the very southern end of San Joaquin Valley
Main varieties: Zinfandel, Viognier, Tempranillo, Zinfandel

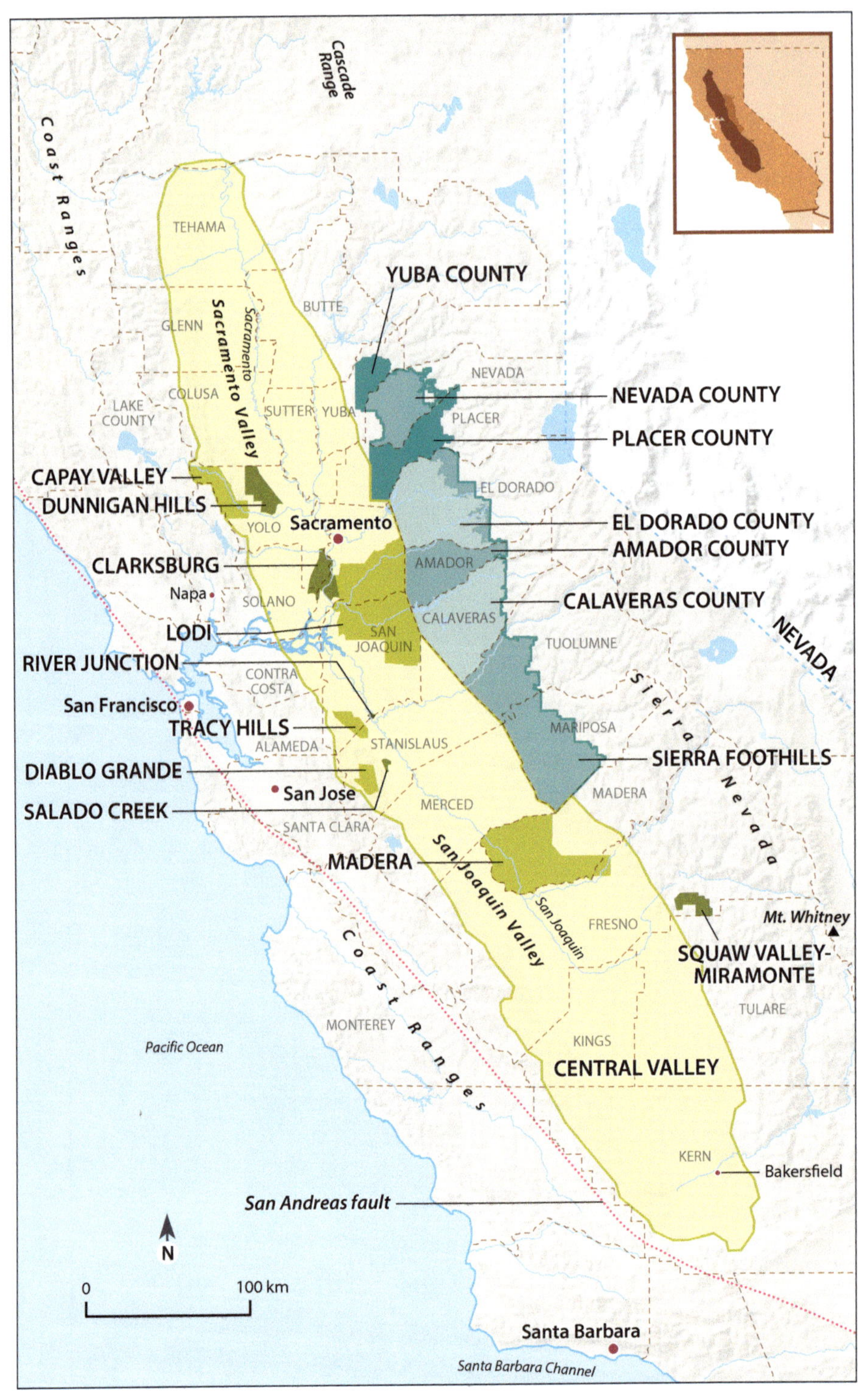

Map 5: The Sierra Foothills and the Central Valley

11

THE SIERRA FOOTHILLS

Like many wine regions, the Sierra Foothills experienced a series of rebirths. Settlers from outside only entered the region during the gold rush. The first, of course, was John Sutter, who built a sawmill with carpenter James Marshall and accidentally discovered gold. The duo got a year-long head start in mining before a rampant band of followers took over the region. Sutter's and Marshall's gold fortune changed California. Fortune seekers worldwide changed direction to hunt the Sierra Foothills, many of them staying in California whether they found gold or not.

The international influx also brought wine to the region. Into the early 1900s, vineyards grew in spots throughout the forest and steep slopes of the mountains. The rugged terrain limited development, but the granitic soils – full of iron in places, limestone in others – and the combination of warm temperatures with cold mountain air made wines to remember. It was the advent of wine in America. That meant experimenting with varieties and exploring ways of making wine. Advertisements of the time name cultivars rarely spoken of today. Native, the local name for the Mission grape, Green Hungarian, Muscatella, and even some hybrid varieties like Isabella and Catawba grew in the area.

Wine types were named by recognizable styles from Europe: Hock meaning in the style of a German white blend; Tokay, in the fashion of mouth-watering sweet wines; Sherry, Port, Burgundy, each a reference to a wine type California winemakers were trying to emulate. Varietal naming for wines didn't start until the 1930s, a California innovation, though the world's first iteration was in Alsace.

Known for: the gold rush, old vine Zinfandel
Lesser-known strength: oldest vineyard in North America
Leader in: boutique wineries, modestly sized vineyards, quality Zinfandel
First peoples: Yokut, Miwok, Nimi
Elevation: 500–6,000 feet (152–1,829 meters)
First vines: 1850
First AVA: Sierra Foothills, 1987
Primary varieties: Cabernet Sauvignon, Zinfandel, Syrah, Barbera

When Prohibition hit in the 1920s, vineyards in the foothills were largely decimated. Those in more settled areas on upland plateaus (rather than steep slopes) remained. But the low yields and difficult farming in most of the foothills meant selling grapes didn't fund the effort. As vines moved out, pears and other orchard fruits moved in.

The 1970s in California included a return to the hillsides and an expansion in mountain vineyards throughout the state. Cabernet Sauvignon climbed up the slopes of Napa Valley, then Sonoma County in the 1960s. Riesling had already gone up to elevation. But the examples of older vines in the Sierra Foothills brought winemakers back to that region as much as it drew others to older vineyards through the north and central coasts.

In the 1990s, like the rest of California, the Sierra Foothills stepped into brash, brawny, boozy Zinfandel. But with so much of the foothills known at the time for Zinfandel, the reputation stuck here and became a stereotype. Even so, around the same time other growers were establishing exciting new plantings of Rhône varieties. Terra Rouge and Easton wines became the new young guns of the region, showing off structural Syrah with resinous, wild aromas and flavors. Turley started making wine from Amador County in 1996, then bought land there in 2012. At its founding, Turley wines were extracted, generous, pumping with flavor, closer to the stereotype, but not today.

Since 2012 or so, the Foothills have been experiencing their most recent renovation. The idea of "no wimpy wines" is shifting into a recognition that size and ripeness can be delivered through poise rather than excess. Turley has refined its style and expanded its portfolio of unforgettable single vineyard Zinfandel. Interest in Rhône-style wine has expanded. Terra Rouge and Easton continue to offer their craggy, textured expression, also joined by the Clos Saron. And a new generation of winemakers has moved in with Rhône varieties. Matt and Audra Naumann at Newfound are reviving a previously neglected vineyard in El Dorado County. In Calaveras, Matthew Rorick of Forlorn Hope has taken on a similar, though larger, concern near the town of Murphy. Tegan Passalacqua of Sandlands (and Turley) has invested in Amador near the town of Volcano.

At the same time, others are rethinking not the varieties but the style of wines. Clos Saron in Yuba County has always preferred minimal winemaking, doing as little as necessary to showcase the site over the cellar. Soon after Clos Saron started, Hank Beckmeyer and Caroline Hoel settled in El Dorado County where they planted a few acres of vines: Tempranillo, Tannat, Syrah, Grenache, and Negro Amaro. But they also started sourcing fruit from mountain sites they admire to start their winery La Clarine Farm. The Petite Manseng from Fenaughty Vineyard in El Dorado is a wine I still think about years later. They too chose a minimal approach to winemaking, capturing a refreshing style that is gulpable and compelling. More recently, Frenchtown Farm in Yuba has taken a similar approach, making textural white blends, juicy Malbec and Cinsault, and an evolving range of red blends.

The combination, winemakers challenging the common approach to winemaking alongside others seeking known varieties but in an older, classic, mountain expression, has been transforming the perception of the Sierra Foothills.

GROWING CONDITIONS

The Sierra Foothills includes eight counties with Yuba at the northern border, and Mariposa the other end. Between sit the counties of Nevada, Placer, El Dorado, Amador, Calaveras, and Tuolumne. Amador is the region's best-known county. It grows the oldest *vinifera* in the country and some of the oldest Zinfandel in the state. El Dorado was plundered harder when it came to gold, but Amador retained its older vineyards. North Yuba developed later and gained one of the most unusual stories in California wine as a result: a religious cult and hidden commune growing some of the most attractive, ageworthy Cabernets in the history of the state.

Soils through the foothills are shallow and rocky. Granite runs the length, but schist and limestone also appear, as well as volcanic rocks, shale, and some unexpected finds along the way.

As the most easterly growing region in the state, and among its highest elevation as well, the Sierra Foothills fill with snow in winter. In summer, parts of them are profoundly dry. The winds and ocean influence reach all the way across the San Francisco Bay and California Delta into the slopes of El Dorado and Amador Counties. But to the south, Calaveras doesn't feel it, and by Mariposa, ocean influence isn't present at all. Even so, cold air settles, and the placement of the foothills along the flank of the Sierra Nevada Mountains brings downdrafts of cold air even in summer. The foothills remain part of the state's Mediterranean environment. Elevations in the Sierra Foothills appellation span between 500 and 3,500 feet (152–1,067 meters).

YUBA

Yuba County carries a modest wine history in terms of planted area and number of vineyards, but those that found their way to the area made significant impact. One of the most enigmatic and strange stories in California wine history comes from Renaissance Vineyards, which was established in the area in the 1970s. The site was established as one of the largest and longest lasting religious communes in California. Almost no members live on site today, yet it still exists, a lone member charged with its upkeep and selling the remaining wines. The height of quality for Renaissance wines was in the 1990s and early 2000s, when Gideon Beinstock made its wines. In 1999, Beinstock launched his own small-scale and impactful project, Clos Saron, downslope from Renaissance. More recently, Aaron and Cara Mockrish have taken over farming parts of the Renaissance vineyard as well as their own site down the road called Frenchtown Farms.

Yuba County AVA

North Yuba (1985): northernmost area in the Sierra Foothills

Main varieties: Cabernet Sauvignon, Syrah, Grenache, Roussanne

NEVADA AND PLACER

During the height of the gold rush, Nevada County had an influx of vineyards and winemaking, mainly sold to fortune seekers in the area. Prohibition brought both into sharp decline, and it remains one of the smallest planted areas for vines today, with 632 acres (256 hectares) planted. The highest point of Nevada County reaches 7,000 feet (2,134 meters), while the few vine plantings remaining reach around 2,000 ft (610 meters).

Soils in the area are highly diverse, derived from granitic and volcanic parent material, as well as some metamorphic stones. Summers are dry and warm, but not as hot as further south in the foothills. Winters are chilly but warmer than more extreme areas in the foothills, protecting most vines from frost or winter freeze. Cabernet Sauvignon is the most planted variety. Nevada County has no recognized AVAs. Wines can be bottled with the county name.

Vines were first planted in neighboring Placer County during the gold rush but went through a swift decline that was never repaired after Prohibition. The highest elevation peak reaches 9,044 feet (2,757 meters), with the few vineyards that remain sitting below 1,500 feet (457 meters). Soils in the area are from decomposed granite, including course granitic sand.

EL DORADO

The gold vein that triggered the gold rush of the mid-1800s has its northern terminus in El Dorado. The county served as an important destination for the period, with fortune seekers camping on hillsides exposed to the elements in the hope of finding gold. A series of villages emerged in the hills to serve the new population arriving from around the world. Zinfandel plays an important role in the county's wine history, but El Dorado has offered a series of firsts for the state as well.

In the 1970s, winemakers in the coastal parts of California believed they'd planted Gamay and used it to make a Beaujolais Nouveau inspired wine called Napa Gamay. In 1980, ampelographer Pierre Galet visited California and recognized the vines as an obscure cultivar from southwest France, called Valdiguié. The variety is relatively unknown in its country of origin. Valdiguié had mistakenly been planted as Gamay from Napa and Sonoma all the way to Paso Robles. The variety was pulled in many locations, while others kept it to use in blends rather than afford the expense of replanting.

It wasn't until 2000 that Steve Edmonds of Edmonds St. John talked Ron Mansfield into planting true Gamay Noir in a vineyard at 2,800 feet (853 meters) elevation. In 2002, California's first Gamay was released, made by Edmonds St. John. It would be years before any more Gamay was planted but it's experienced a small but steady expansion. Jolie Laide makes Gamay from some of Mansfield fruit. Arnot Roberts did previously. The variety has expanded into coastal areas of California as well. Megan Zobeck focuses entirely on the variety for her wine brand, M. Zobeck. As for Valdiguié, it is now celebrated as its own variety (at modest scale) in California, which has even managed to spark attention (and questions) for the variety in France.

Today, El Dorado is an exciting locale for Rhône varieties, which deliver some of the most interesting wines of the region. As in so much of the world, Cabernet Sauvignon has the most plantings in the area. But Iberian, Italian, and German varieties can be found there as well. And, like the rest of the foothills, also Barbera.

El Dorado's planted area sits at 2,219 acres (898 hectares). Elevations in the area reach 10,886 feet (3,318 meters), but vineyards sit between 1,200 and 3,500 feet (366–1,067 meters). The rugged slopes and shallow soils of the region limit planting. Soils are derived from a mix of granite, shale, and volcanic parent material.

El Dorado County AVAs

El Dorado (1983): the northern end of the gold vein found in the mid-1800s
Main varieties: Zinfandel, Cabernet Sauvignon, Syrah

Fair Play (2001): higher-elevation areas fully within El Dorado County
Main varieties: Syrah, Grenache, Mourvèdre, Tempranillo, Barbera

AMADOR

The oldest vineyards in North America are found in Amador County. Deaver Ranch in the Shenandoah Valley of Amador grows the oldest surviving *vinifera*, the Mission grape planted in 1854. As the first wine grapes to arrive in the region, residents started thinking of it as the *first* wine in the sense of the first to ever settle the area. Long-standing Amador families continue to call the grape "Native" as a result. There were, after all, never any Spanish missions in the Sierra Foothills so Amador vines didn't have a mission-era association. Outside settlers didn't push into the region until California gained independence from Mexico.

Tegan Passalacqua began working with Deaver Ranch for a vineyard-specific bottling of Zinfandel for Turley. When he started his own Sandlands in 2010, Passalacqua looked to some of his favorite vineyards in the state. When the old Native vines became available, he began bottling a dry, red table wine with the variety. Sandlands certainly wasn't the only winery working with the Mission grape, but Passalacqua's respect in the wine community brought unique attention to it. The wine helped revitalize interest in the grape, with a small boom of winemakers working with it and others even establishing plantings. More recently, Sandlands has also released an Angelica wine, fortified in the fashion made during the mission era.

Some of the oldest Zinfandel in California is also found in the Shenandoah Valley: Grandpère is one of them. Continuously making wine for more than 150 years, Grandpère is dry-farmed, head-trained, and organically grown.

The historic importance of Amador has led to its recognition through three AVAs: Amador, Shenandoah Valley, and Fiddletown, all made official in 1983. Together, the area has a uniquely high concentration of vineyards of at least 65 years old. The preservation of such vineyards has depended on one of the jokes of the wine industry. White Zinfandel has been disparaged for decades as cloying, too sweet, too pink. But regardless, it was a crowd-pleasing, market-leading powerhouse for decades. To get the required volume for the wine's high demand, more vineyards were needed to provide more Zinfandel. Producers of white Zinfandel scooped up fruit from older vineyards throughout the Central Valley and Sierra Foothills. As red Zinfandel sales waned, its pink counterpart surged, ensuring historic vineyards stayed in production. More recently, Barbera has become a dominant variety throughout the foothills. Even so, the region is still planted primarily to Chardonnay and Cabernet Sauvignon, closely followed by Syrah and Zinfandel.

One of the best-known vineyards and growers in Amador County is Shake Ridge Ranch from Ann Kraemer. Kraemer developed her career in Napa Valley, then headed to the foothills so she could afford to lead her own project alongside her family. Shake Ridge offers an impressive range of varieties, including Tempranillo, Mourvèdre, Graciano, and Barbera among others, grown at different aspects with a variety of training methods. Kraemer has successfully used the site to give newer projects and producers a chance to make wine grown to their needs in an organically farmed vineyard. Producers such as Tegan Passalacqua for Turley, John Lockwood for Enfield, Hardy Wallace, first for Dirty & Rowdy and now Extradimensional Wine Co. Yeah!, Terah Bajjalieh for Terah Wine Co., Morgan Twain-Peterson's Bedrock, Ferdinand, Keplinger, Favia, A Tribute to Grace, Newfound, and Emme have all made standout wines and built their brand recognition in partnership with Kraemer and Shake Ridge.

Soils in Amador include more volcanic material than some other parts of the foothills as well as granite. Total planted vineyard area is limited by the rugged terrain and steep slopes of the area. Vineyards that remain are found along plateaus and upper elevation valleys. Vineyards total 3,644 acres (1,475 hectares).

Amador County AVAs

California Shenandoah Valley (1983): continues into both El Dorado and Amador Counties
Main varieties: Zinfandel, Cabernet Sauvignon, Barbera, Petite Sirah

Amador County (1983): contains some of the oldest vineyards in the country
Main varieties: Zinfandel, Barbera, Petite Sirah

Fiddletown (1983): slightly lighter-bodied wines than neighboring Shenandoah Valley
Main varieties: Zinfandel, Barbera, Petite Sirah

CALAVERAS

Calaveras joined the modern wine industry in the 1960s, when Barden Stevenot took a property on steep slopes in the forest just east of the town of Murphy and began planting vines. Some of his original plantings of Wente clone Chardonnay still grow on their own roots on site. Since his tenure, the vineyard has expanded to include a rugged mix of Iberian and Italian varieties, generally delivering tension and length. The surface soils come from eroded schist but a little bit deeper vines grow into limestone. Chalky texture makes even varieties you don't expect to find, like Touriga Nacional and Albariño, inexplicably compelling. Today, Stevenot's vineyard is owned by Matthew Rorick, founder of Forlorn Hope. He farms the site, making wine of his own while also selling fruit to a collection of talented winemakers: Newfound, Etxea, Ferdinand, and Enfield among them.

Schist rock can be found throughout Calaveras County, often overlying calcareous rock and soils below

The area experiences hot summer temperatures, without the cooling influence of the ocean. Even so, the area is tucked against the enormous Sierra Nevada mountains. Cool air descends from the mountains, pouring into the foothills along Calaveras County, helping wines to retain natural acidity, most especially when harvested with that in mind.

The total planted area in Calaveras County is around 725 acres (293 hectares), with no certified AVAs. The highest point of the area stands at 8,174 feet (2,491 meters), but vineyards grow at between 1,000 and 3,000 feet (305–914 meters).

TUOLUMNE AND MARIPOSA

Tuolumne and Mariposa Counties are the furthest south in the Sierra Foothills AVA. They're also furthest from the ocean winds and breezes enjoyed by El Dorado and Amador. Few vineyards grow in the hills of this area. Even so, interest has been slowly increasing with a small number of producers anchored there.

Sierra Foothills wineries

Clos Saron

Oregon House
clossaron.com

Gideon Beinstock made an impression with some of the most terroir-expressive and compelling red wines in the state during the heydays of the Renaissance winery. Then, in 1999, he and Saron Rice founded Clos Saron, making rustic wines from the northern portions of the Sierra Foothills. His Syrah drew a new audience of wine lovers to the state and changed the perception of what's possible in the region's wines. His wines are made with minimal intervention, so little sulfur dioxide added that the final effect is essentially no sulfur, and without other additions. Clos Saron also makes what they consider their flagship wine, a Pinot Noir, as well as Zinfandel and a delicious range of white and red blends.

Forlorn Hope

Murphy
www.forlornhopewines.com

Matthew Rorick founded Forlorn Hope with the aim of devoting the business to vineyards and varieties easily considered hard luck cases. The wines were made from regions less familiar to the consumer (and even many winemakers), from varieties few had heard of. In some cases, the grape might be familiar, but Rorick made it in a style almost lost to history that he chose to revive. Today, the wines revolve primarily around his Rorick Heritage Vineyard tucked into the steep slopes of Calaveras County near the town of Murphy. The vineyard includes some of the first vines planted in the region, including own-rooted Chardonnay from the 1970s. It also grows almost feral, enticing Syrah, Tempranillo, some of the best Albariño I've ever tasted, Zinfandel, Touriga Nacional, and one of the Sierra Foothills' signature grapes, Barbera. The queens of the Sierra wines are made from unusual blends in a white, rosé, and red cuvée for affordable gulpability.

Frenchtown Farms

North Yuba
www.frenchtownfarms.com

Renaissance vineyard still provides fruit. Sections of its vineyard are farmed and maintained by Aaron and Cara Mockrish, who admire its older, dry-farmed vines. The couple live nearby, where they are also planting their own site. They're farming without pesticides to instead rely on organic methods and an attentive relationship with the vines. In 2015, they started Frenchtown Farms and make Syrah, Malbec, and red and white blends that reflect their own distinctive character with a commitment to light-touch winemaking and vibrancy.

La Clarine Farm

Somerset

www.laclarinefarm.com

La Clarine Farm by Hank Beckmeyer and Caroline Hoël makes unfiltered, minimal-intervention wines from organic vineyards. The wines are smaller in production, rolling with natural acidity, and highly drinkable, a fresh and delightful approach to wine. La Clarine has been a star in the US natural wine community, while delivering wines with a sense of purity and brightness. Their wines tend to be gulpable blends, but they also make varietal Gamay Noir, Mourvèdre, and seductive Syrah.

Renaissance

Oregon House

www.renaissancewinery.com

When it started at the end of the 1970s, Renaissance winery was a game changer in California for multiple reasons. It was founded as a community during the back-to-the-land movement of the time and members flocked to the remote Sierra Foothills location to live and work there. The vineyard and winery stood near Oregon House, in a remote northern part of the Sierra Foothills, on a site that had originally been the home of a religious commune some considered to be a cult. In the 1990s, Gideon Beinstock took the helm as winemaker. During his tenure, the winery made some of the most ageworthy, distinctive, terroir-expressive reds in California from its own organically farmed mountain vineyards. Wines are no longer being made under the Renaissance label, but the winery still offers old vintages of the wines that are worth seeking out. It is worth tracking down and enjoying the 1994 to 2006 vintages.

Turley

Amador, Paso Robles, St. Helena
www.turleywinecellars.com

Larry Turley started as a motorcycle-riding emergency room surgeon, before launching his way into wine. In 1993, he founded his own winery, Turley, committed to older vineyards that carry the heritage of California, farming them organically. Turley works with some of the oldest vines in the northern half of the state, mostly field blends, though the Petite Sirah is also gorgeous. Today, Turley operates three wineries to be closer to the vineyards, in Paso Robles, Amador, and St. Helena. Only the Paso and Amador locations have tasting rooms. For those wishing to experience the dynamic, site-expressive character of Zinfandel, Turley is the ideal winery. Turley has also made a dry, white Zinfandel, a tongue-in-cheek nod to the variety's history while elevating the idea of rosé from Zinfandel. They also now make a single bottling of Cabernet.

12

THE FAR NORTH AND SOUTHERN CALIFORNIA

THE FAR NORTH

Common knowledge still tells us Mendocino is the northern limit for grape growing in California. From the perspective of reasonable yields (you know, large enough for the grapes to pay for the farming) and adequate ripening, that idea is probably right. Viticulture is marginal in the northern parts of California. Most of the area is filled through with redwoods, and parts of the coast are part of a rainforest that stretches all the way north to southeastern Alaska.

Logging began in northern California in the 1800s and dominated as an industry until mere decades ago. In the 1990s, an activist, Julia Butterfly Hill, helped change things. While attempting to stop the clearcutting of an old growth forest, Hill took a surprising direction in protest. She climbed into the canopy of an enormous tree she called Luna and lived there for a little over two years. Other activists brought her food and supplies. The logging company tried to force her down with helicopters and fire hoses. But Hill's efforts brought national attention to the value of redwoods, and the timber company agreed to reserve a section of old growth in the forest. As forests have regrown, independent thinkers have too. California stands for nothing if not for its creative, independent, and innovative spirit. Vineyards depend on it. So vines have modestly found their way into pockets of forest further north.

Known for: Mount Shasta, Shasta Lake, mountain beauty
Hidden strengths: glacial formations, the second highest peak in the volcanic ring of fire
Leader in: growing some of the tallest trees in the world, biodiversity
First peoples: Kahosadi, Ku'wil, Oohl, Natinnoh-hoi, Wint^{h}u, Chimariko, Achomawi, Atsugewi, Nomlāqa Bōda
First vines: late 1800s
Elevation: up to 14,179 feet (4,322 meters)
Number of AVAs: 5
First vines: 1882
First AVA: Manton Valley, 2014
Primary grapes: Cabernet Sauvignon, Syrah, Barbera, Zinfandel

The US AVA system allows anyone with the wherewithal and clarity of mind to complete an adequate AVA application to go through the process. AVA approval doesn't depend on the vines of an area being established, or wines having being made. Instead, AVAs are recognized by history, growing conditions, and distinction of place. The northern part of California doesn't grow very much wine, but it still has five official AVAs.

Humboldt and Trinity County share the Willow Creek AVA. In Trinity there is also Trinity Lakes. Willow Creek reaches up to 1,000 feet (305 meters) in elevation. Cooling ocean air reaches the AVA, though it is warmer than the coastal parts of Humboldt, and cools significantly at night. Soils are rocky and in places have calcareous rock. In Trinity Lake, vines can grow in the valleys tucked close to the lakes of the region. They provide a mitigating influence and proximity to water needed in the otherwise dry region. Elevations here start at around 2,000 feet (610 meters), keeping temperatures cool and nights cold. Cooler climate varieties are grown in both.

Near the slopes of Mount Shasta are two more AVAs: Inwood Valley and Manton Valley. Mount Shasta rises to 14,179 feet (4,322 meters) elevation, but vineyards in Manton grow at around 2,000 to 3,500 feet (610–1,067 meters). In Inwood they're closer to 1,500 feet (457 meters). Protected from wind, Inwood Valley grows Malbec and Cabernet Sauvignon. In Manton Valley there is Chardonnay.

Nearby, Siskiyou County includes the Seiad Valley AVA, a short distance from the Oregon border. Seiad grows Riesling. More recently, vineyards in Siskiyou have also started growing French Alpine varieties, showing off the genuine alpine nature of the area.

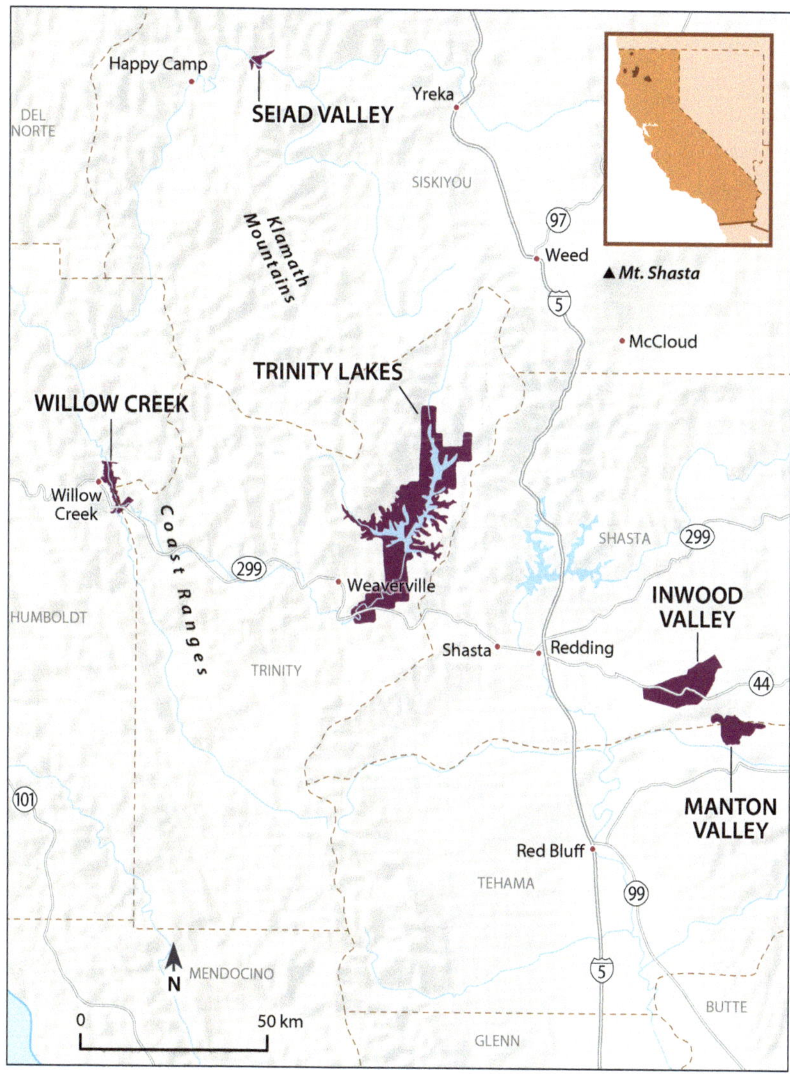

Map 6: The Far North

Humboldt County AVA

Willow Creek (1983): overlaps Humboldt and Trinity Counties

Main varieties: Syrah, Cabernet Sauvignon

Trinity County AVA

Trinity Lakes (2005): remote, mountainous and surrounded by lakes

Main varieties: Syrah, Cabernet Sauvignon

Shasta County AVAs

Manton Valley (2014): overlaps Tehama and Shasta Counties

Main varieties: Syrah, Cabernet Sauvignon

Inwood Valley (2012): in the foothills of Mount Lassen and Mount Shasta

Main varieties: Syrah, Cabernet Sauvignon

Siskiyou County AVA

Seiad Valley (1994): a few miles below the Oregon border

Main varieties: Syrah, Cabernet Sauvignon

Wineries

Iruai

Etna

iruaiwine.com

Chad and Michelle Westbrook Hinds started making wine while they were based in Berkeley. In 2019, they uprooted and moved north to the Shasta area, working with vineyards in the Trinity and Siskiyou mountains. The Westbrook Hindses call the area the California Alps. There were clear contenders for planting the steep sloped regions. Syrah loves a view, as the saying goes, meaning a mountain slope overlooking an expanse or valley. But they've also shown what California can do with other mountain varieties. They've developed Mondeuse, Gamay, Nebbiolo, Trousseau, and Poulsard. Their wines are fresh, mouthwatering, and minimal intervention but stable. The success of Iruai has brought a new generation of interest to the remote areas of California and a new host of varieties.

New Clairvaux Vineyard

Vina

www.newclairvauxvineyard.com

Founded by Trappist-Cistercian monks, New Clairvaux vineyard has recently experienced a revitalization. Today, winemaker Aimée Sunseri works with the monks who still lead the property to deliver wines unique to Tehama County. The winery grows Trebbiano, Tempranillo, Syrah, and Assyrtiko. With the Tempranillo they even make a nouveau style. The Trebbiano goes back to an even older history of the site and was first planted by the founder of Stanford University, Leland Stanford, in the 1800s.

SOUTHERN CALIFORNIA

Known for: bikinis on roller skates, palm trees, Hollywood, Disneyland
Hidden strengths: t-shirts and shorts on roller skates, psychics
Leader in: high-intensity traffic, forgotten vineyards being revitalized
First peoples: Tongva, Taaqtam, Ivilyuqaletem, Ipai, and Tipai
Elevation: up to 5,905 ft (1,800 meters)
Number of AVAs: 14
First vines: 1789
First AVA: Saddle Rock Malibu, 2006
Primary grapes: Zinfandel, Palomino, Grenache, Cabernet Sauvignon, mixed field blends

The Sacred Mission of Spain into Alta California established itself initially in what is now San Diego. There, they encountered the Ipai and Tipai people and slowly converted them to Catholicism both to save their souls for Christianity and so they would farm the vineyards and food crops needed to survive within the new settlements.

In the 1770s, *vinifera* was planted, and the variety grown came to be known as the Mission grape. San Diego was the first site to grow wine grapes in what is now California, but soon after, San Juan Capistrano and San Gabriel became the most planted vineyards in California. When the mission period ended in the 1830s, new settlers moved into the region, welcomed by Mexico with free trade and citizenship not offered by Spain. The Indigenous neophytes remained the indentured viticulturists of the region but wine growing became a commercial venture. Vines expanded especially through the greater Los Angeles area.

The LA Basin and the Inland Empire

Before Napa Valley and Sonoma County became synonymous with wine country, Los Angeles was the center of California's wine trade. In the late 1800s, vineyards lined the banks of the Los Angeles River and dotted the area, supplying grapes for the finest wines in the state. By the early 1930s, the impact of the gold rush, an outbreak of Pierce's disease, and the scourge of Prohibition had nearly wiped out LA's vineyards. Soon, they were converted to citrus groves or cleared for urban development. Winegrowing moved to the north coast of California.

Viña Madre, the mother vine

The first *Vitis vinifera* vines in California were established along the southern coast at mission San Juan Capistrano in the late 1770s. Within a few years, cuttings had been brought north to Los Angeles and established at mission San Gabriel. There, wine took hold in Alta California. San Gabriel became the largest and most celebrated mission winery in the region, known for making both dry table wine, and Angelica sweet wine. The original vineyard at San Gabriel became known as Viña Madre. Cuttings from the vineyard supplied the first commercial grape plantings in the territory after Mexico gained independence from Spain. Thus, San Gabriel was the mother vineyard, the source for all others.

In the 1800s, a series of floods and then fires destroyed agriculture throughout southern California. The vineyards of San Gabriel were badly damaged, with only a few vines surviving. Those that kept growing were assumed to be the original *vinifera* brought to California during Spain's mission era. Other plantings of this variety were found in downtown Los Angeles, and at other historic vineyards and pueblos in the extended region. One such cutting was established at a private home near San Gabriel in 1861 and continues to grow today, an enormous vine much like a tree, providing canopy and shade for the surrounding area. More recently, genetic testing has been done on the vine and the other plantings found in the extended area and it has been discovered to be an interspecific hybrid between the original mission *vinifera* grape and a native California species, *Vitis girdiana*. All indications are that the crossing occurred naturally from vine proximity, and that its hybrid characteristics are what has allowed it to survive through disaster for centuries. Today, cuttings of Viña Madre have proliferated in other vineyards of southern California as well as Sonoma County. Winemakers in Los Angeles are using the original vine to make Angelica sweet wine.

In honor of the original vineyard, this variety came to be called Viña Madre.

Today the history of wine in Los Angeles and other parts of the southern coast is being revived, and a new generation of winemakers is making wines in the region again. In 2019, three wineries, Angeleno Wine Company, Cavaletti, and Byron Blatty, banded together to form the Los Angeles Vintners Association (LAVA). Since then, a surprising number of others have joined – Acri Wine Co., Golden Star Vineyards, Cielo, Flowers & Cheese, Friendly Noise, Aja, Herrmann York, Moraga, Nabu, and Warson – all making wine in greater Los Angeles.

They've also revitalized wine history. Angeleno, Cavaletti, and Byron Blatty helped uncover the history of Viña Madre, a vine from the mission era that still grows at San Gabriel, with a growing habit more like that of a tree than a grapevine (see box, opposite). Local investigation has revealed Viña Madre grows in multiple parts of the Los Angeles basin, including the center of downtown LA. Today, winemakers are working to revitalize the grape. Cuttings have been sent to the north coast of California where it is now growing. And together, Caveletti, Angeleno, and Byron Blatty are using it to recreate California's first wine, Angelica, a fortified wine made with fresh pressed juice or sweet wine.

Growing conditions

The extended LA area includes growing regions as far west as the Santa Monica Mountains, while the eastern limit is Cucamonga. It extends north to the desert of Antelope Valley, and south to Temecula, approaching San Diego.

Malibu includes three AVAs: Saddle Rock Malibu, Malibu Coast, and Malibu-Newton Canyon. Almost as soon as the appellations were approved, a rush of opposition emerged from wealthy residents blocking the planting of vineyards, which would encourage wine tourism and mark the natural hillsides. Still, some plantings can be found. Byron Blatty makes wine from the area. Elevations of Malibu vineyards vary significantly, stretching from sea level to around 3,100 feet (945 meters). They face the Pacific Ocean with persistent air flow and fog at lower elevations, yet also sun exposure that brings ripening. Soils are largely derived from shale and sandstone.

To the northeast of LA, three AVAs grow in desert conditions. The Sierra Pelona mountains abut the Mojave Desert, creating night-time cold plus daytime heat, with a whisper of ocean breezes. Elevations reach 5,905 feet (1,800 meters). Soils in the area are derived from volcanic and granitic parent material and are relatively loose with minimal structure.

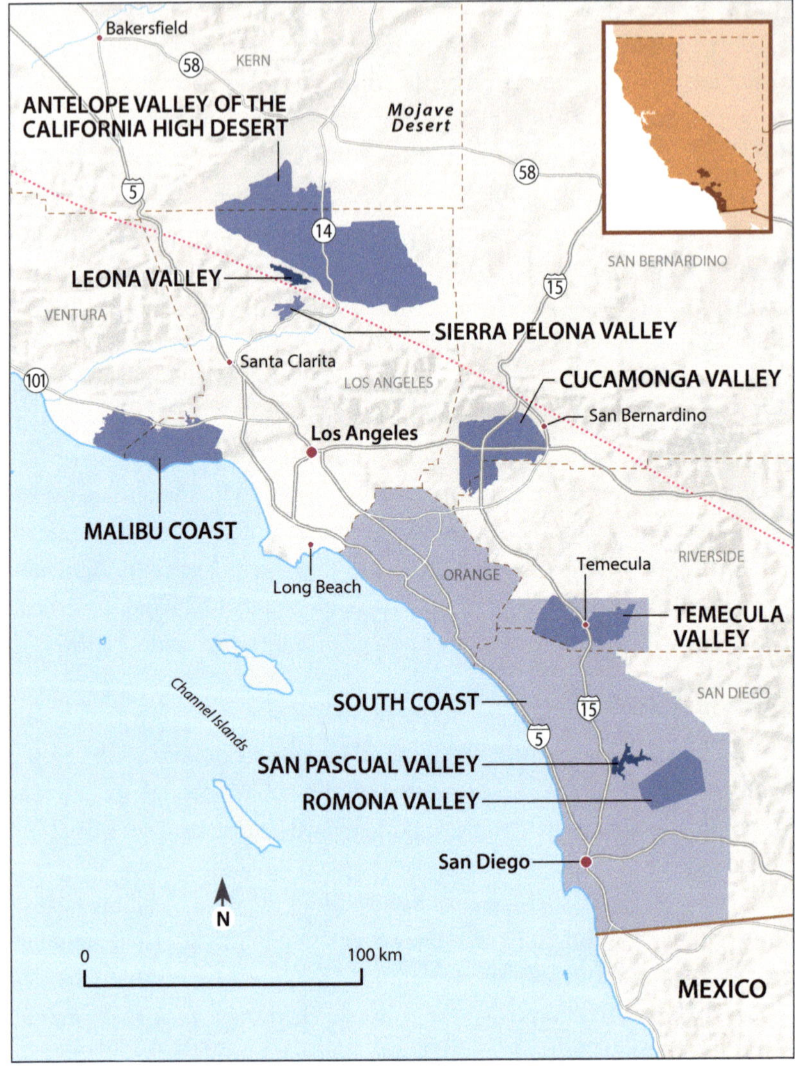

Map 7: Southern California

AVAs here include Sierra Pelona Valley, Leona Valley, and Antelope Valley of the California High Desert as well as Palos Verdes Peninsula. Vineyards outside these AVAs can use Los Angeles County as their designation.

To the east, Cucamonga and Temecula are both part of the Inland Empire, though some locals in Temecula refuse the designation. It captures the sense of urban development that has stretched into San

Bernardino and Riverside Counties. Vines reaching back to the 1800s can be found in both regions. In Temecula a rush of plantings from the late 1900s to today can also be found, most established to serve the influx of wine tourism from Los Angeles and San Diego. Both Cucamonga and Temecula are recognized AVAs.

Cucamonga grows further from the ocean with comparatively warmer temperatures. Historically, vineyards in the area have provided grapes for fortified, sweet wines. But winemakers such as Carol Shelton have been making dry Zinfandel from the area, particularly Lopez vineyard, since she founded her winery in Sonoma County. The AVA grows at around 500 feet (152 meters). Soils here are sandy with pebbles and hold very little water. Temecula includes vineyards from the 1800s but the vast majority have been planted since the 1990s. Its location between Los Angeles and San Diego has made it a prime destination for wine tourism. Most wines remain local, sold directly to visitors through the cellar door. Vineyards grow at between 1,000 and 1,600 feet (305–488 meters). Surrounding mountains bring a cooling down-draft. The decomposed granite soils are well draining.

San Diego County

The Indigenous peoples of what is today known as San Diego County, especially those closest to the border with Mexico, were the first to face the incursion of outside explorers, followed by the Franciscan monks and soldiers of the Sacred Expedition that brought Spain into Alta California. The mission San Diego de Alcalá, also the origin of the city's name, was established. Eventually, San Juan de Capistrano, further north though still south of Los Angeles, became one of the region's most valuable mission outposts, and the first to successfully establish *vinifera*.

The number of wineries and the planted acreage of San Diego County remains modest. Avocados and citrus are both larger crops for the region. Oranges unique to the area enjoy sweetness with still good acidity, making them a coveted crop from the area that is sold overseas as well as locally, most especially in Asia.

The vineyards of the region primarily grow Cabernet Sauvignon and Syrah, with selections of other well-known grapes like Chardonnay, Zinfandel, and Sangiovese too. Vineyards in San Diego County tend to grow in inland valleys and are thus protected from direct ocean exposure.

Southern California AVAs

South Coast (1985): encompasses five counties in Southern California
Main varieties: a great diversity of varietal plantings

Los Angeles County AVAs

Malibu Coast (2014): overlaps Ventura and Los Angeles Counties
Major varieties: Cabernet Sauvignon, Syrah

Saddle Rock Malibu (2006): along the coast of the Pacific Ocean
Main varieties: Cabernet Sauvignon, Syrah, Sauvignon Blanc

Malibu-Newton Canyon (1987): in the high elevations of the Malibu mountains
Main varieties: Cabernet Sauvignon, Cabernet Franc, Petit Verdot

Palos Verdes Peninsula (2021): hugs the coastline west of Los Angeles
Main varieties: Pinot Noir, Chardonnay, Cabernet Sauvignon, Merlot

Leona Valley (2008): sits within the Sierra Pelona Valley AVA
Main varieties: Chardonnay, Syrah

Sierra Pelona Valley (2010): in the high desert region near Antelope Valley and the Mojave
Main varieties: Zinfandel, Syrah, Tempranillo, Muscat

Kern County AVA

Antelope Valley of the California High Desert (2010): overlaps Kern and Los Angeles Counties
Main varieties: Tempranillo, Zinfandel, Symphony

San Bernardino County AVAs

Cucamonga Valley (1985): one of the first commercial wine regions of California; overlaps Riverside and San Bernardino Counties
Main varieties: Zinfandel

Temecula Valley (1984): originally named Temecula, renamed 2004
Main varieties: Cabernet Sauvignon, Syrah, Petite Sirah

Yucaipa Valley (2024): high elevation vineyards in the San Bernardino mountains
Main varieties: Cabernet Sauvignon, Merlot, Zinfandel, Syrah

San Diego County AVAs

San Pasqual Valley (1981): showcases well-drained, granitic soils in northern San Diego County

Main varieties: Cabernet Sauvignon, Syrah, Sangiovese

Ramona Valley (2006): a significant diurnal shift helps retain natural acidity

Main varieties: Cabernet Sauvignon, Viognier, Muscat

San Luis Rey (2024): from coastal town Oceanside to the Merriman Mountains

Main varieties: Cabernet Sauvignon, Merlot, Cabernet Franc

Wineries

Angeleno Wine Company

Los Angeles

www.angelenowine.com

Tannat, Graciano, Godello, Loureiro, and Treixadura are the varieties of Angeleno Wine Company. Jasper Dickson and Amy Luftig are making some of the most dynamic, mouth-watering, palate-friendly and flavorful wines in the area. Exact cuvées and varieties are flexible to reflect each vintage, but the wines are consistently delicious with fresh-tasting fruits and vibrant acidity. They're the kind of wines that just taste happy.

Byron Blatty

Los Angeles

byronblatty.com

Byron Blatty opened to make smaller production, luxury wines showcasing the mountain and ocean character of the Malibu Coast. They started with a focus on Cabernet and have evolved into an exploration of varieties and vineyards throughout the region. Mark and Jenny Blatty work with winemaker Joey Perry bringing together a love for fine wine and the desire to spark the potential of Malibu's character in wine.

Herrmann York

Redlands

www.herrmannyork.wine

Three families, including two brothers, joined forces to form Herrmann York, making wines from the inland valley of Southern California. They partner with some of the oldest sites in the extended area, including historic vineyards in Cucamonga. Thanks to the region's vineyard history, they make wines from Palomino, Zinfandel, Grenache, and newer plantings of Cabernet Sauvignon. Uniquely, they bring a significant portion of whole cluster to their Cabernet. California includes a lesser-known history of Palomino and Herrmann York are working with some of its oldest plantings in the state.

PART 3

WHAT WE'RE FACING

13

THE CULTURAL SIGNIFICANCE OF WINE

Warren Winiarski's earliest memory was pressing his ear against the side of barrels to listen to the fermentation inside. In Chicago, the wines his father made from honey and dandelions sparked a fascination for Warren. Winiarski arrived in Napa Valley in the 1960s, and wine growing became a way for him to live an agrarian life as he started Stag's Leap Wine Cellars. It also symbolically connected him to the community where he grew up in Chicago, and his own family history. Wine became a way to link the future to the past.[1]

In the early 1970s, winemaker Richard Sanford searched for property and a way to connect with the land. After returning from fighting in Vietnam, with the sense of detachment that followed, planting a vineyard became his solution. He found a parcel in a remote part of southwestern Santa Barbara County. It was exposed to cold ocean winds and fog and without electricity. Sanford sold his car to buy vine cuttings, lived on site for seven years without power, and established the first vineyard in what would become the Sta. Rita Hills AVA. His efforts demonstrated the winegrowing potential of the region. The isolation, sweat equity, and connection to place gave Sanford physical labor as well as what he describes as a spiritual exercise. Wine became a way for him to recover from the traumas of war.[2] His success led an entire industry to be built around him.

For those impassioned by wine, it extends beyond an agricultural product to a transformative beverage. In Ancient Greece it served as medicine. In Rome it was a symbol of empire. For the Christian church, wine became a union with the holy and divine, as well as a way to convert more people to a new way of life. In the United States, in the second half of the last century, wine became a symbol of cultural sophistication. For many, wine has also been a way to endure insurmountable circumstance.

Serge Hochar survived both the Lebanese war of the 1970s and the Syrian–Israeli war in the 1980s. Although Lebanon was unstable, Hochar made wine throughout fifteen years of violence. For Hochar, taking his winemaking seriously was a way to get through the war. During more isolated vintages, he continued to make wine even without knowing where it might sell. At the height of violence in the early 1980s, his wife and children left the country for safety. Hochar stayed to tend his vineyards and continue to make wine. He described his 1982 vintage as "a pure wine of war." The following year, he traveled by hovercraft for more than six hours from Cyprus back to Lebanon to arrive home and immediately harvest. "If you ask me on which side I fought during these 15 years," Hochar said, "I can only tell you that I was fighting for my wines."[3]

WINE'S TRIVIALITY

Wine can seem trivial compared to global events. Yet, Philokalia vineyards near Bethlehem continues to make wine during the Israeli–Palestinian conflict.[4] During fighting between Azerbaijan and Armenia in 2020, the Keushguerian family donned bullet-proof vests and harvested their vineyards anyway. They also traveled into Iran to source grapes and make more wine.[5] Amid peril, wine may seem meaningless to those unable to help. Yet, for Philokalia, the Keushguerian family, and Hochar, making wine during war became a way to revolt against the political situation, and retain a sense of personal identity through cultural tradition. The need to preserve culture also emerges in less violent circumstances.

Anti-alcohol lobbying has increased in the decade since 2015. But in 2024, the wine community banded together to protect fine wine. The group Vitævino created a declaration and online petition "to defend the role of wine in society, and its cultural heritage."[6] It's an attempt to build a pro-wine lobby amid international pressures against it. The same year in California, the writer Karen MacNeil, author of the world's most sold book on wine, *The Wine Bible*, joined forces with other wine professionals in the state. Together they launched what became an international campaign, Come Over October. The initiative focused on moderation rather than abstinence and emphasized how wine can create community and shared experience. Wineries, wine businesses, professionals, and wine lovers from around the world buoyed the effort.[7] Media coverage, retail displays, and social media support reached more than a billion people. Wineries reported an associated sales increase.[8]

Wine can also provoke cultural change. In San Francisco, urban farmer Christopher Renfro decided to work against the exclusion of Black farmers in America, while also increasing access for other people of color. He created the Two Eighty Project,[9] partnering with vintner Steve Matthiasson of the North Coast and UC Davis researcher Beth Forrestel to develop a paid apprenticeship, training, and winemaking program for Black, Indigenous, and people of color seeking access to the inner workings of fine wine in California. Winiarski and Sanford sought connection to the land to establish a new way of life. For Renfro, that same connection is a way to build new community possibilities.

Part of wine's transformative power lies in the transformation it makes itself. Through fermentation, grapes become an entirely new elixir, wine. Its complexities, nuances, texture, detailed aroma and flavor, and, yes, alcohol compel its followers to fill it with personal meaning. Its significance impregnates entire cultures. Even those that fight to remove it do so in recognition of its greater influence. The social allure of wine exceeds that of other agricultural commodities. Its ability to reach people brings powerful edification as well. Wine has informed consumers about conservation, sustainability, farmworker safety, and climate change more than almost any other farm product. Wine also transforms economies.

ECONOMIC IMPACT

For vintners like Sanford, Hochar, or Renfro, the value of wine is personal, as well as community based. It also depends on time and capital. The annual harvest hangs on a year's worth of weather, farming, and preparation. Picking the grapes and making their wine means capturing at least some of the investment of time and resources already given to the vines. Miss harvest, lose an entire year of work. The financial impact can be severe. It can threaten the survival of the winery and affect its surrounding economy.

In 2022, global vineyard acreage was estimated at 18.04 million acres (7.3 million hectares) across all grape types (wine, table, and dried). Grape harvest throughout the world that year reached 81.1 million tons, half of which was for wine.[10]

The global wine industry contributes to the economy of countries around the world. It provides jobs, taxes, and exports, it spawns wine tourism and supports ancillary industries such as restaurants, bars, hotels, tank and barrel makers, cork and oak forest farmers, testing laboratories, yeast and nutrient manufacturers, and research investments.

In the United States, all 50 states produce wine from locally grown grapes. Across the country, in 2022, the wine industry employed an estimated 1,007,459 people, plus 364,234 jobs in associated wine supporting industries. The total wages generated that year amounted to $96.5 billion,[11] while wine tourism reached $16.69 billion in tourist payments the same year.[12]

In 2022, California wine contributed $88.12 billion to the country's economy. The benefit for businesses that support wine was additional. The state industry employed 255,734 documented workers as well as 116,192 more in ancillary industries. The number of undocumented workers was also significant. Total wages generated the same year for employees were around $32.05 billion. California wine tourism created 25.22 million tourist visits, and $8.56 billion in tourist spending. Associated taxes were additional.[13] Tax gains include undocumented workers. In the United States, taxes are paid before wages, so even undocumented workers in an ongoing job contribute to the system.

IMPACTS ON WINE

The economy of wine helps build surrounding communities and support other businesses served by wine. Farming traditions and the vines themselves also act as a symbol of cultural heritage, and our ongoing connection to nature, its preservation and enjoyment.[14] Wine's cultural meaning as well as its economic force make it important to consider impacts on wine from outside. The loss of wine would harm more than simply the vineyards or their wineries.

Climate change has already changed farming. Regions are moving where they grow as well as what, and when they harvest. It is changing rain cycles and regional aridity, as well as the chemistry of grapes themselves.

The volatility of the market has made wine sales unpredictable. Producers successfully reaching newer consumers are thriving while those relying on sales channels that worked twenty, or even five or ten years ago are struggling.

The beverage industry has diversified. Ready-to-drink options like canned spritzes and bottles of pre-mixed margaritas are popular. Even the easy-to-blend gin and tonic can now be purchased ready-made and canned. Cocktail culture is re-emerging. Bartenders have turned to foraging and seasonal produce to make ever-changing drinks lists, even as the classics like the martini, the Hemingway, and whiskey on the rocks are also growing in popularity again. Younger drinkers are more willing to experiment. The Boomer generation tended toward brand loyalty, and specific wines or cocktails as a statement of identity, wealth, and values. Since the COVID-19 pandemic, consumers are more willing to taste across categories rather than sticking with one type of drink per night. The change has made it harder for wineries to build a stable sales base.

And, just as the so-called dry party and puritans of the early 1900s helped create the Eighteenth Amendment of 1920 that made alcohol sales illegal, and the 1980s with its raised concerns of physical health also had an impact on wine, the alcohol industry is now grappling with neo-prohibitionism. The combination of health questions and ideas about "demon liquor" are impacting the industry.

The United States' fear of immigration also creates challenges. As states make visas for migrant workers more difficult to get, or legally becoming a working resident or working toward citizenship more challenging, the pool of service workers and vineyard labor has shrunk. The effectiveness of such laws remains questionable. For many industries, including wine and especially agriculture in general, the difficulty of legal visas has simply made workers without documentation more desirable. Employees are needed regardless. When legal options diminish, farmers turn elsewhere.

The challenges wine is facing are both social and economic.

WINE'S IMPACTS

But wine's significance is not merely social or economic. The farming choices of individual producers affect soil health, erosion, and aridity of an area. The addition of chemicals like pesticides and fertilizers create pollution in soils, water supplies, and the air. It also threatens the health of workers and community members near such farming. Water use for irrigation or the winery can affect both surface water and an area's aquifer. Reliance on community electrical systems taxes energy supplies. Depending on the source of power, it can increase local temperatures or air pollution, further tax the water supply, or risk starting wildfires.

It's also been shown that farming choices significantly contribute to global warming. Exposed soils become drier and create dust particles in the air that can change the regional water cycle. Diesel engines for frost fans, tractors, and other equipment contribute to greenhouse gases, which further contribute to climate change.

The remaining two chapters of *The wines of California* consider what the wine industry is facing. That is, what will need to be addressed in order for wine to innovate and succeed yet again? Chapter 14 looks at social considerations: neo-prohibitionism, market volatility, and the evolving consumer and generational differences. Chapter 15 explores climate change and climate action.

California wine has a long history of facing what feel like insurmountable challenges. It also has a long history of people coming together to problem solve, innovate, and succeed once again. Honest examination of pressures on the industry can reveal ways the people of California wine might plan its future.

14

SOCIAL CONSIDERATIONS

THE CHANGING MARKET

Wine came of age in the 1990s. Lured by its sophistication and guided by the rising wine media of the United States, the American consumer invested. The success of the middle class grew the audience for wine, and the California industry increased. Pushing into the new millennium, the industry experienced steady growth in both volume and sales value. Global events like Y2K, 9/11, and the global financial crisis caused by the housing crash created market volatility in the century's first ten years but amid generally upward growth. More recently, COVID-19 marked another change.

The crisis of the pandemic initiated lockdowns, closed borders, and caused population losses worldwide, in some countries reaching into early 2022. In its first two years, US wine consumption increased. Consumers gripped by uncertainty opened their wine cellars and spent savings on more home consumption. Many in the wine industry assumed that the world reopening post-COVID would return us to a pre-COVID marketplace. Instead, sales channels and buying practices have both changed.

Stay-at-home policies transformed work life across the United States. A lucky core of the population moved to work from home options, holding onto continued income. Those in essential industries such as deliveries, food services, or public utilities kept paychecks by accepting an increased risk of exposure. However, many others in public-facing industries, especially hospitality, simply moved to unemployment or life in a gig economy, working when work appeared. Those with the means stepped away from work entirely.

Restaurant closures caused by the pandemic pushed wine sales from on-premises wine lists to retail and direct-to-consumer initiatives. Ongoing hospitality challenges post-lockdowns meant restaurant sales were slow to return. In 2020–25, restaurants closed, while others are still struggling. Wine sales declined. The distribution system was slow to adjust, and stock became comparatively static. An oversupply of wine in the marketplace developed. The expansion of vineyard acreage in the last twenty years created a grape glut, making some grapes impossible to sell in 2024.

The economic impacts of COVID-19 fostered market volatility, labor shortages, supply chain issues, a cost of goods increase, and inflation. With rising inflation, the spending power of the dollar decreased, as the cost of living increased. Moving out of pandemic lockdowns into a post-COVID economy spending plateaued, eventually decreasing in the second half of 2023. From the first airing of "The French Paradox" on the CBS show *60 Minutes* in 1991, and into the first twenty years of the new millennium, sales of California wine experienced steady growth. The industry was in an economic upswing. By 2018, consumer spending had shifted. People were drinking less wine by volume but buying better wine by value.

The economic impacts of COVID-19 changed wine buying further. With increased inflation hitting alongside a higher cost of living, consumers developed new buying strategies. Less expendable income overall has led to a new stratification in the marketplace. Studies show what has emerged is a two-tier wine market.

Two-tier spending

In the post-COVID economy, tighter budgets are guiding buying decisions. Consumers are still spending but with new buying patterns. Today's consumer can access more information, more quickly than ever before. The result is greater sophistication and insight into the quality and cost of goods purchased. The result, researchers say, is a segmented marketplace built by two primary, differing values.

In one market segment, consumers devoted to quality continue to buy wines and other products that have already proven their worth. They are buying from fewer brands with greater loyalty. They are also comparatively more conservative in spending, but willing to buy what they know is good rather than seek a discount. Here, consumers are sticking with wineries but tending to buy more bottles of the more affordable wine. That tends to be wine from multiple vineyards or regions, rather than the single-vineyard cuvée.

Tablas Creek has capitalized on this change. In 2010, in response to the global financial crisis and associated consumer spending decreases, the winery launched their Patelin de Tablas tier. The wines were made by combining estate fruit with that from vineyards farmed nearby in blends resembling their estate-only (and more expensive) Esprit de Tablas. Today, the Patelin tier continues to be a market driver. In 2023, they launched a similar concept, Lignée de Tablas. Lignée wines are also made incorporating fruit farmed outside the estate but all are single vineyards growing clonal material imported from France by Tablas Creek. Both the Patelin and Lignée brands allow the winery to release wines that are more affordable yet pay tribute to the quality for which the winery is known.

Ridge made a similar move in 2001 by launching its Three Valleys cuvée. Primarily Zinfandel, Three Valleys is the only wine Ridge makes that brings together vineyards from multiple regions. The blending means making a more affordable wine. More recently, Caymus has taken the approach farther. They now sell a Caymus Cabernet labeled with the broad appellation California, rather than using North Coast or Napa Valley. It's made in the style beloved by Caymus drinkers, but broader fruit sourcing means they can make less expensive wine. In all three cases – Caymus, Tablas Creek, and Ridge – wineries are delivering wine that respects their established consumer base by delivering quality at greater value, rather than relying on discounting the brand.

For the quality segment, purchasing decisions also connect to brand values. The quality of the goods is essential, but so is provenance, and the environmental and social commitments the brand supports. Here, economists believe fine wine producers will be able to maintain a loyal following, but overall fine wine will likely occur at smaller volumes. Brand loyalty tends towards already known wineries, making it harder for newer producers to burst into the quality marketplace.

For younger Gen-X, Millennial, and Gen-Z populations, there is also an expectation from quality brands for transparency, established relationships, and compelling stories about their founders, the wines, and how they're produced. Newer brands in the fine wine space are helping to speed the growth of support by aligning with wine buyers who reach an audience in line with brand values and type.

In the other segment, cost differentiation proves essential. These consumers are price sensitive, moving from brand to brand or store to store for greater affordability. Loyalty is not the goal, saving money is. At the same time, they aren't looking for "cheap." Wines under $10 per bottle are disappearing. Even those taking price-sensitive decisions want a decent wine.

Alternative packaging, creative marketing, and atypical styles also do well in this price tier. Any can give one wine the edge over another of the same price. O'Neill Vintners has created several brands that hit this segment. Allegro combines affordability with brightly colored labels and a spritzy, slightly sweet style. FitVine makes a plea to the health-conscious consumer by marketing with athletic images and the promise its wine is "low sugar for healthy people." Harken captures something different. The wines rely on recognizable varieties, like Chardonnay, even made in something like a typical style (oak flavor, creamy fruits), but with playful labels, understandable language to describe the wines, and the promise they pair well with food. Harken wines are affordable while delivering the notion that classic styles can be fun and easy.

When it comes to price savings, consumers are more willing to purchase creatively and try experimental products if the price delivers a saving. Buying decisions can be motivated by an added benefit as well as a discount. Unique packaging, a distinctive message, and more affordability combine to create a compelling sale. It is easier for larger wineries and higher-volume brands to succeed in this market share. To reduce margins and stay economically sustainable, wineries must increase efficiency and reduce costs.

Brick & Mortar, located in Sonoma County, is one of the few wineries succeeding in both tiers. Their bottled wines offer classic Pinot Noir and Chardonnay from admired vineyards and relatively small production. They also make an impressive traditional method sparkling wine. Here, wines offer the opportunity to explore the nuances of Sonoma and Mendocino Counties. But they also make highly affordable canned wines of five types: blanc, rouge, rosé, plus a sparkling rosé and sparkling blanc. The wines gained a significant distribution contract for the national grocery chain Whole Foods and reaches markets throughout the United States. The two-tier crossover has worked for Brick & Mortar partially because they've been transparent about both efforts since they started. But the canned wines also don't tend to sell in the same venues the fine wine does, and they no longer sell canned wine direct as they do the fine wine.

In the new two-tier marketplace, medium-sized wineries have the greatest work to do. The challenge resides in failing to make a significant distinction either in quality or lower costs. An already established customer base can help. Incorporating more discounts to attract the price-sensitive buyer can backfire. Such practices tend to push away the quality purchaser and while discounts might create immediate sales, they don't mean return sales.

The fine wine industry is also facing a reckoning. The history of success from the 1990s well into the 2000s made it possible for wineries to deliver pricey wine – sometimes even with lower comparative quality. As an example, the name Napa Valley allowed the formation of new (essentially unknown) brands that were reliant on lower quality fruit still from within the region; the label Napa Valley meant they could upsell the wine for a higher price. But, arguably, the approach has negatively impacted the prestige of Napa Valley. There are innumerable wines in Napa worthy of the soaring reputation of the region, but if you don't know their quality already, how do you tell them apart from the overpriced wines?

Wine investors

One of the most significant market developments since the 1990s is the emergence of a serious and persistent wine investment market. The growth in collecting has simultaneously spurred new wine trading businesses and a rise in wine brokers. Collectors are not simply buying for later drinking. The investment rests in buying to sell for higher prices later.

While other investment types such as stocks tend to decline in market uncertainty, wine investing is less responsive to market volatility. Wine is also more accessible compared to other specialties like fine art, and more stable than precious metals. As the number of bottles of a particular wine diminishes, the value of the wine increases. It's an inverse correlation of supply and demand. Even amidst post-COVID volatility, wine investments have continued.

Brokers report wine investments do experience some changes. Buying during market variability shifts from fewer multi-thousand-dollar bottles when the economy is surging, to more bottles around the $1,000 mark as it declines or remains uncertain.

As top French and Italian regions have shown more variable resale values, Liv-EX,[1] the world's top wine trading marketplace, reports California wines are more stable and on the rise. Brands like Screaming

Eagle, Harlan, Scarecrow, Dominus, Ridge, and Diamond Creek tend to top the list.

Auctions have returned since the pandemic but in new ways. Top buyers, auctioneers report, were rarely in the room during live auctions, but since auctions have begun utilizing a Zoom marketplace, even more collectors buy by phone rather than in person.

Changing categories

Economic changes are not the only cause of shifts in the wine market. Global trends have been moving from the hearty reds of the 1990s and early 2000s to fresher styles in the mid-2020s. Older generations still prefer red wines, but as they age, their spending power declines. Facing retirement and health changes naturally decreases drinking and reduces long-term investments.

Increased sales instead have built in sparkling wines. Champagne proves an important part, but sparkling wines in general have strengthened. For the first half of the 2020s, sparkling wine has been the top purchasing category for consumers. In 2023, revenues grew 0.8 percent, reaching 8.9 billion euros in value, even with decreasing volumes.[2] The same year also marked a significant decline in the sales of still, fortified, and semi-sparkling wines. Together, they saw an overall decrease in value of 5.3 percent, or 1.4 billion euros, and a volume decline of 7.6 percent.[3]

The overall trend in buying has also moved towards premiumization. Consumers have been buying fewer bulk wines since the early 2000s. The economic downturns of 2009 and 2013 paused the shift slightly, but the move to less but better wine quickly returned. Post COVID, the increase in value and decline in volume continues. Throughout 2024, market declines have caused panic in the wine marketplace, yet the move towards premiumization continues.

The market's future

Will the ongoing market bring more stability? The high stakes of the 2024 US presidential election increased spending declines. Presidential changes in the United States consistently mark periods of market volatility. As candidates propose different ways to manage the nation's economy, both investors and consumers tend to pause spending. But just as the post-pandemic boost did not continue, the pre-election decline won't either.

The second Trump presidency includes numerous changes. His promised return to international tariffs means a further increase in cost of imported goods for American consumers. In his first presidency, the US dollar went up in value, making exports difficult, leading to an impact on domestic businesses. Trump's attachment to deportation also creates challenges. President Barack Obama initiated deportation practices that Trump then increased with more virulent scare tactics and imprisonment. The combination led to not only increased deportation numbers but also more voluntary departures by people leaving the atmosphere of increased hostility. The effect was a decline in the workforce.

The US food and service industries rely on immigrant labor. The United States has consistently experienced a national recession alongside labor shortages. For agriculture, public utilities, construction, food, and wine, labor shortages put even more at stake. At some point, harvest becomes impossible. Mechanization is only a solution in some cases. Most produce needs hand farming and fine wine relies on hand labor. In these cases, as access to labor decreases, costs to consumers go up.

Research from Hochschule Geisenheim University in Germany examines global industry changes in viticulture, wine, and its marketplace. Their studies show further changes in public interest for wine. For the first time since the 1980s, interest in wine-based drinks has returned. New blends and co-ferments akin to more sophisticated versions of the wine coolers of the 1980s have taken hold. The category offers a refreshing, creative opportunity for producers needing to utilize oddball lots of wine, or wine that doesn't work for traditional sales. Consumer interest in new wine products like co-ferments and wine cocktails demonstrates evolving consumer interests and a move out of traditional categories into experimentation and curiosity.

Geisenheim researchers point out that international policies will need to change to support such efforts in the industry and make such sales affordable for producers. In the United States specifically, the three-tier system (see appendix V) and its associated policies significantly slow sales. Interestingly, American-owned cruise lines have become a powerful workaround. Cruise lines can purchase outside the three-tier system, and thereby directly from wineries, improving margins on both sides.

For the industry to succeed, it needs to align production more readily with consumer interests. California wine is known to be traditional. As the market has changed, most wineries have maintained the same model aimed at Boomers and older members of Gen-X that built their original success. But California wine also has a history of adaptation and innovation that leads to problem solving and strengthening the industry. Amid so many simultaneous market changes, that creativity is even more essential.

New trends in the market have differing origins. Some, like the turn to sparkling wine, are driven by a shifting consumer palate. Others are caused by the latest generational changes in the marketplace.

GENERATIONAL CHANGES

The spending power of the Boomer generation built the California wine industry. The wealth of the middle class was spurred by their parents following World War II, then furthered by the Boomers, who supported growth in wine from the 1990s well into the twenty-first century. By the new millennium, California was one of the top wine-producing regions in the world. The United States became the world's largest consumer wine market, and California both its most substantial producer and biggest consumer.

Simultaneously, generational changes coincided with economic declines. Younger generations were raised exposed to a different world from that of their parents. The combination created a new approach to spending as well as a new range of interests to enjoy. The Millennial and Gen-Z generations represent a new way of valuing money. Spending is not only for consuming goods. It's also a means to support businesses that act in ways aligned with consumer values. Buying power channels moral values.

Gen-Z also represents a significant demographic change that has been building since the Boomer generation. Young people today are more racially, ethnically,[4] and religiously diverse, as well as more openly gay, lesbian, bi, trans, and asexual.[5] As younger consumers become more invested in buying by values, they are also more diverse. The combination influences their spending habits in ways that differ from older generations. The generation's size makes their spending power impossible to ignore. They're also beginning to reach drinking age.

New generations

Generational spending studies show that Boomers (born roughly between 1946 and 1964) rely on brand permanence. If they believe Chevy trucks are the best trucks, they are more likely to spend accordingly over decades. Brand loyalty is not merely a familiarity choice. It is an assertion of identity and moral commitments as much as desirability.

For Millennials (born between 1981 and 1996), brand loyalty exists but for differing reasons and has less longevity. A brand that demonstrates environmental and social commitments aligned with the consumer's values receives repeat purchasing. If those brand practices change, or someone else demonstrates them more strongly, buying will likely shift. For Millennials, the brand is less an attachment of identity than the values the brand expresses. Boomers prove less willing to pay more for additional values. Environmental and social considerations are not worth spending more on for the older generation. Millennials also demonstrate greater willingness to experiment through tasting and changing products.

In the middle, sits Generation-X. Gen-X (born roughly 1965 to 1980) has a smaller population than its older and younger counterparts. And it also has less buying power. The older half shows buying habits most akin to Boomers, while the younger half more closely resemble Millennials. The rise of the California wine industry coincided with and was spurred by the Boomer generation. Alongside it came a growth in wine media. As wine grew, American beer came to be dominated by a few high-volume brands that were comparatively bland and less inspiring. Beer imports occurred but were harder to find. Excitement existed in wine. Gen-X came of age with the emergence of craft beers. The beer industry's expansion to a greater range of styles, quality, and producers overlapped with this generation's young adulthood. Into the twenty-first century, specialty beers began taking up significantly more supermarket space. By 2020, there were large refrigerator aisles that included only beer. Wine continued alongside a new energy in fermented and malted grains.

Millennials were raised with the greatest expansion of beverage options ever seen. They also enjoyed the greatest access to quick information via the internet and, later, social media. The combination means Millennials developed greater discernment to navigate a crowded market at younger ages than their older counterparts. Purchasing based on provenance, quality, and business practices are more important even in young adulthood.

When asked in surveys what beverage they would bring to a party, answers differ according to generation.[6] Boomers overwhelmingly answered a bottle of wine. Gen-X evenly selected between wine and beer.

For Millennials, what to bring to a party evenly included wine, beer, malt beverages, and hard seltzer. As of 2025, Gen-Z (1997–2012) only the first part of the generation has reached drinking age. There is not yet enough data to identify their alcohol buying trends.

Spending habits

Generation-X, Millennials, and Gen-Z have all enjoyed unprecedented access to information and advice thanks to the internet and social media. Though Gen-X was the last generation to experience a time before the internet, their lifetime coincided with its emergence, making its adoption easier.

Such an increase in access and speed of information has created new spending habits. At the same time, each generation has grown in a different economy. Boomers have had more overall expendable income as well as comparatively greater financial stability, more long-term savings, and generally more investments such as property and stocks, as well as wine. Millennials, by comparison, have far less disposable income. With inflation the money they have holds less spending power. They are also less likely to own a home, tend to have less savings, and are more likely to carry both student loan and credit card debt.

Younger generations are also more health conscious. Their activities are guided by their health commitments. But the desire for health is also a cost trend. It has created a move to moderate drinking driven not only by health awareness but also economic reality. With less expendable income comes reduced buying of optional goods. Alcohol is not a necessity, so as budgets tighten, it's easier to let go.

For those willing to drink alcohol, wine also suffers a cost challenge. A full 750 milliliter bottle of wine costs more money and is often harder to drink alone, for people more likely to live on their own, than beer or liquor. Studies also show people tend to drink more wine when married than single. Yet, millennials are less likely to marry than previous generations.

Yet these challenges also present unique opportunity. Knowing a standard bottle of wine presents higher comparative cost and so lower value, as well as a greater likelihood of waste (and so then even higher cost) means the wine industry can explore alternative packaging and sizing to capture a market segment searching for options. Lower perceived cost might not come from changing the actual price per volume of a bottle of wine, and instead come from offering smaller sizes or formats such as bag-in-box that deliver less wasted beverage. Until recently, it has been easier to invest in a can of beer or ready-to-drink (RTD) beverage, or even a cocktail, simply because the smaller volume creates less immediate cost. A move to canned wine, as well as smaller format glass bottles, has been building in the last fifteen years. Both are good options for outdoor activities such as picnics, hikes, and stadium seating (although glass is often banned in areas where it breaking could pose a safety risk). These tend to work for easy, quick drinking wines. But more premium wines need a different option.

Premium producers have been turning to bag-in-box packaging for many of their lower and mid-tier wines. The format gives the opportunity for consumers to buy more volume at a lower price, while selecting an option that provides greater freshness for longer and less waste. Producers have also found the box delivers more room to convey a message or use playful graphics. Rather than relying on a small bottle label, each side of a box can be used to inform or attract a customer.

Broadening the base

Another aspect of generational change prioritizes social diversity. More than merely a demographic curiosity, recognition of the unique backgrounds and cultural distinctions of the American consumer supports business success and buying choices. Studies have shown that companies with greater diversity at the decision-making level consistently enjoy greater long-term success as well as stability. But integrating a broader base of cultural backgrounds, abilities, races and ethnicities, as well as gender expressions into a business depends on creating a welcoming workplace.

Businesses that move to incorporate diversity in response to social pressures often do so too quickly. Welcoming a more diverse workforce depends on shifting company culture. A single diversity hire, or spending on diversity, equity, and inclusion (DEI) training for employees will not on its own change employee and leadership behavior. DEI must be approached not as line-item thinking, but as part of a holistic plan.

Though initial steps to shift into greater diversity practices can be challenging, studies show that it supports greater long-term success as well as financial sustainability. From the simplest perspective: greater employee well-being increases employee satisfaction; greater employee satisfaction means reduced staff turnover. One of the highest business expenses for a company is employee turnover. Businesses that can minimize such changes increase efficiency and reduce spending. Additionally, greater diversity support fosters more diversity success in a positive cycle of growth. As more employees feel welcome, they attract consumers from a broader demographic as well as employees who can add to the wherewithal of the business. Interviewing wine distribution salespeople, a simple truth emerges. A salesperson who understands the needs and interests of their customer is more likely to increase sales with them. This occurs in relation to entrepreneurship as well. During the Barack Obama presidency, the creation and success of Black-owned businesses increased throughout the United States.

Greater company diversity also leads to increased company sales. As more employees can reach more market segments, the company broadens its consumer base and sales stability. It's an advantage during market volatility. Companies connect to new customers by understanding their taste preferences, spending habits, and food-and-drinks culture. They find new audiences by making sense to them. Greater company diversity means greater reach.

Harnessing purchasing trends

Age, brand responsiveness to environmental or social issues, wine knowledge, place of purchase, and marital status all influence purchasing decisions. For wine to succeed, businesses need to respond to consumer interests.

The increased interest in environmental and social issues can be fostered by certifications that denote such practices. Brand messaging, even on the wine label, can also be effective. But with greater discernment, younger generations are also more skeptical. Certifications alone readily appear performative. Brands succeed by finding additional means to demonstrate commitments. Clear communication demonstrates values, but for younger generations, so does supporting events and causes that matter to them.

International research suggests the wine industry can maintain quality wine with loyal customers but likely at a smaller scale. Return purchasing occurs when the winery or business continues to demonstrate its commitments over time and through multiple means. Brand loyalty must be continually earned. For cost savings to both the business and consumer, exploring creative alternatives in packaging formats, as well as fresher styles of wine, and even wine-based products or atypical blends can support consumer curiosity as well as a producer's need to find new ways to sell what a grape provides. Younger generations are more open to non-traditional packaging, especially when it increases the value of the wine. Bag-in-box and cans also mean lower packaging weight, and thus lower carbon footprint from transportation. New formats can demonstrate environmental commitments.

The wine industry tends to be traditional, slow to change its approach to making or selling wine. But the traditional practices of the California wine industry are largely built in response to the buying habits and interests of the Boomer and older Gen-X generations. There is room for wine to adapt to the habits and interests of consumers today. To move forward, the industry must innovate in response to Millennial and Gen-Z interests as well.

NEO-PROHIBITIONISM AND HEALTH

In 2024, calls against the demon alcohol, rebounded. The year before, the World Health Organization (WHO) published its severest claims against alcohol to date, announcing, "No level of alcohol consumption is safe for our health."[7] Anti-alcohol lobbying increased internationally. Canada re-wrote its health guidelines encouraging no more than two drinks of alcohol per week (down from per day) in 2023.[8] In 2024, Japan produced alcohol guidelines for the first time in its history.[9] In Alaska, the State Legislature passed a bill requiring businesses to display cancer warning signs.[10] South Korea already requires them.[11] In 2025, the US Surgeon General recommended cancer warning on alcohol labels.[12] Then the Department of Health & Human Services (HHS) released a draft report saying alcohol increases mortality rates.[13] The results are questionable.

Drinking discouragement isn't new. US Prohibition of the 1920s was founded on it.[14] The 1980s created MADD (Mothers Against Drunk Driving).[15] In 1989, US alcohol regulations required warning labels on bottles discouraging drinking during pregnancy, or when operating machinery.[16] It also said alcohol may cause health problems, but didn't say more. National sales of alcohol declined in the United States, and wine sales specifically decreased each year of the 1980s. In the same era, health consciousness was increasing, and consumers were avoiding drinking in the expectation that such abstinence would benefit their health. The trend only turned around in 1991, when *60 Minutes* aired its special on the benefits of red wine with dinner.[17]

The 1980s also ushered in international efforts to encourage abstinence. The WHO's fight against alcohol began when it declared it a carcinogen in 1988.[18] Warnings and statements asserting the cancer-causing effects of alcohol have increased since, but scientific understanding of their possible links remains limited.[19] Researchers do see correlations between drinking and increased breast or throat cancer. But likelihood of other forms of cancer and health ailments seem to decrease with drinking. And a significant study published in 2024 found that detrimental health effects and increased mortality rates associated with drinking only occurred in certain groups. Those already dealing with a pre-existing condition, or poor socio-economic circumstances, did show higher mortality rates. Healthy adults and those with ample financial resources did not.[20] WHO even admits there is no adequate evidence to declare alcohol unsafe.[21] Regardless of this, global beliefs around health and drinking are changing.

Health claims

In the United States, health guidelines, which include alcohol recommendations, are reviewed on a regular cycle and released generally around every five years. Referred to as the Dietary Guidelines for Americans, the 2020–2025 edition states that up to two drinks per day is a safe limit for adult men, and one is safe for women.

"The French Paradox," or recommendations to follow a more Mediterranean-style diet with its inclusion of wine at dinner, has remained a standard in the United States since its 1991 announcement on *60 Minutes*. Studies affirming the Mediterranean diet show that enjoying red wine with a meal reduces the effects of cholesterol.[22] Decreased mortality rates are associated with healthy low-to-moderate drinkers. It was also understood that binge drinking has a different impact, significantly increasing mortality rates.[23] Heart-health benefits in drinking red wine continue to be affirmed. Research shows a decreased risk of rheumatoid arthritis symptoms from low-to-moderate drinking.[24] Increased risk of breast cancer does correlate with alcohol. But other types of cancer do not. Studies show that 4 per cent of all cancers worldwide link to alcohol consumption.[25] Other products that are consumed carry much higher risk. Tobacco causes 20 per cent of all cancers. Diet with low consumption of whole grains but high consumption of processed meats is implicated in 30–40 percent of all cancers.[26]

Experts looking at the recommendations associated with Canada's recent changes have argued that the data relied on by the Canadian Centre for Substance Use and Abuse (CSSA) doesn't match the conclusions the organization made. The study showed drinking decreases risk of the most extreme causes of stroke, coronary artery disease, and brain hemorrhage. The CSSA recommendations don't mention those results.[27] Similarly, the HHS report doesn't align with its evidence.[28] There are also concerns that pushing abstinence can increase problems for people with disordered drinking. Until recently, abstinence was seen as the most effective path to avoiding extremes for people with drinking problems. Evidence now suggests that for people with unstable drinking, harm reduction strategies that don't require abstinence can sometimes be more effective. Some individuals do need abstinence, but it can be a mistake for others.[29]

Market impact

COVID-19 significantly changed drinking. Lockdowns in the first year of the pandemic led to increased drinking for people stuck at home. A rebound effect was expected when people were let outside again, but instead, drinking slowed. The pandemic acted as a sort of wake-up call for consumers to consider health and wellness more deeply than they had previously. An increased awareness of and care for mental health, digestive well-being, mood, altruism, and health of the planet all increased post COVID-19.

Supply chain delays, and lower production rates of goods overall created increased costs for all businesses. It's increased the cost of wine and liquor. Inflation rates are also increasing. The effect is that the cost of living keeps going up while the value of income goes down. People simply have less money to drink wine. Studies also show that Gen-Z and Millennial generation consumers are more concerned with the ethical values of products they buy. Health considerations and environmental impact drive purchasing more than they have done for Gen-X or Boomers. Decision making with those concerns in mind has only increased since COVID-19. These market changes have increased alongside the rise in anti-alcohol messaging and health-impact claims. Consumer studies seem to show the approach is working.[30] More than 30 percent of American consumers believe alcohol increases health risks. For most, the belief is based more on marketing than research.

Piggybacking on increased cost-of-living concerns, anti-alcohol groups are also working to increase taxes on alcohol, while other groups also seek to increase tariffs. The combination will reduce exports of California wine as it becomes more expensive to move wine between countries.[31] Tariffs on international wines are also more likely to negatively impact California wine than increase its sales. Wine lovers do not readily switch from one beloved wine to a lesser known one, so the idea that tariffs mean consumers will end up replacing wine from Europe, for example, with wine from California is not guaranteed. The entire three-tier distribution system works in unison to sell wine. An increase in cost of wine from other countries will hamper sales through the distribution system, ultimately also impacting California wine.

As wine becomes more expensive due to outside causes (like taxes and tariffs), numerous associated businesses will suffer. It will become harder for producers to sell wine, especially to international markets. Distributors and importers will either lose revenue through smaller margins or decrease the likelihood of sales due to higher prices. Restaurant margins on alcohol will also change. For anti-alcohol lobbyists the economic impact is likely part of the strategy.

The big picture

But market impact is not the only consideration when it comes to alcohol or anti-alcohol lobbying. Past experiments in pushing abstinence, the most vivid of which was the United States' own Prohibition,[32] have shown how such policies can covertly increase drinking as well as lower understanding of its actual effects. When policies simply exclude alcohol, studies on the reality of harms or health benefits caused by alcohol also tend to disappear. That makes it harder to know the actual impact of alcohol on human health. Multiple health studies that could give greater insight into the potential health impacts of alcohol have been canceled due to fear of the anti-alcohol lobby.

A comprehensive study, and one of the first of its kind, investigating the link between drinking, diabetes, heart attack, and death was canceled in 2018 when anti-alcohol lobbyists and an article in *The New York Times* described the study as sponsored by the alcohol industry and recommending moderate drinking.[33] The study had not at that point started so couldn't yet have made any conclusions. Loss of research has also occurred due to pressure from the other side of the argument. In Canada, the Yukon territory instigated a study with researchers in 2017 to determine the impact of warning labels on sales.

The government canceled the study when representatives of the alcohol industry complained.[34]

Studies continue to demonstrate that the Mediterranean diet, with its glass of wine at dinner, is the healthiest approach.[35] But history has shown that prohibitionist efforts have power as well as patience. Attempts to make drinking in the United States illegal began in the 1850s and persisted until their success in 1920. Recent research has demonstrated that such efforts at the international, national, and state levels also operate with ample funding.[36]

Accepting reduction over abstinence

In the prohibitionist efforts of the early 1900s and 1980s, those against alcohol delivered absolutes to change public minds and sway policy makers: demon liquor was deadly, full stop. It's an approach that is seeing a revival.

The wine industry has historically responded to such efforts with nuance. Wine *is* okay to drink. Even good for you. It's not bad for your health. That's a less persuasive message simply for being less direct. History shows these public debates tend to be more marketing than fact. Persuasive change rises more from truthful insight delivered in digestible bites than big messaging. And both prohibitionist and safe-wine lobbying depend on group effort.

In recent years, the wine industry has created new avenues to build potential responses. In 2023, experts from around the world specializing in health and wine joined forces for the Lifestyle, Diet, Wine & Health congress in Spain. They analyzed the WHO's recent public health recommendations. Then they presented research from additional studies on the known health benefits and detriments of drinking. They were joined by heads of wine businesses worldwide.[37] The challenge, as described by Felicity Carter, is that it takes significantly more effort to prove false information wrong than it does to spread it. How the industry will use the insights gained at the congress in Spain remain unclear.

The Come Over October effort originated in California in 2024 and quickly went global. Wineries, trade partners, and ancillary businesses joined forces to spread the message of convivial moderation. It was a reminder that wine is about community and sharing. Neither drunkenness nor abstinence were part of the campaign. Wineries reported that, at least in the short term, it worked. Those that held events and special offers connected to the idea for consumers saw increased sales.

Millennial, Gen-X, and Boomer consumers enjoyed wine together at home. Its impacts seemed to reach these generations more than those newest to drinking. Gen-Z consumers, after all, largely do not yet have (or own) homes. At the same time, encouraging convivial moderation at home could threaten the hospitality industry on which wine depends.

Historically, rhetoric around the acceptability of alcohol (or not) has seemed to trade in absolutes. Doing so misses the reality that most people who drink alcohol have it occasionally. Binge drinking and total abstinence are each far sides of a scale rather than its center point. Drinking treatment programs have turned from mandating absolute avoidance of alcohol to instead suggesting harm reduction and reduced consumption. Similarly, to address the neo-temperance movement, wine must shift from fighting the rhetoric directly and instead offer a realistic, persuasive response. When the industry responds to the statements fully against alcohol, the anti-alcohol lobby simply wins. Getting involved in that debate means questions of the safety of wine are repeated again and again.

Trying to overtly increase wine sales while marketing the benefits of wine also inadvertently aids the neo-prohibitionist effort. The appearance of such promotion strengthens the claim that alcohol is pushing an agenda for the sake of sales. In other words, it weakens the argument for wine by playing into the dynamic those against drinking have supplied. Borrowing from sociology, a practice of harm reduction means accepting one's tendencies as they are, then working with them to broaden the feeling of well-being and success. For wine businesses, that means accepting the tendencies of today's consumers, then working with those tendencies.

Consumers generally have stepped into a period of reduced consumption. Inasmuch as that is true, individual consumers will have fewer overall purchases. The alternative is not to persuade them to again increase wine buying, but instead to work in support of such moderation in new ways. Make the consumer feel supported in their efforts. Now, what does it look like for the wine industry to amplify the consumer's joy of convivial moderation, and thereby broaden its audience and means of success?

15

CLIMATE CHANGE

UNDERSTANDING CLIMATE CHANGE

Climate change occurs naturally. Volcanic activity, dust storms, the earth's angle and proximity to the sun, increases in solar flares, sunspots, solar wind, and coronal mass ejections all contribute. Scientists have confirmed multiple climate change cycles in earth's history. An ice age followed by warming periods, then glacial expansion repeats over eons. There are also shorter-term changes. Every few years weather over the Pacific Ocean goes through natural cycles of either warming or cooling known as El Niño and La Niña respectively.

Large volcanic eruptions also contribute to climate disruption, shielding the sun's warmth and cooling the planet.[1] Ilopango in El Salvador in the year 540, Samalas on the Indonesian island of Lombok in 1257, and Mount Pinatubo in the Philippines in 1991 each released millions of tons of sulfur into the upper atmosphere and created planetary cooling for years after. The evidence can be observed in tree rings. In the late 1700s, the Laki fissure in Iceland erupted. The sulfur released into the upper atmosphere led to crop failures, the death of farm animals, and a layer of haze across Europe that cooled the region significantly for more than eight months.[2] It instigated drought and food shortages, not only in Iceland but elsewhere in Europe, and may even have been implicated in the outbreak of the French Revolution.[3]

It was the Age of Enlightenment, when philosophers were rethinking the value of humankind. It produced a cultural shift prioritizing science, the use of reason, equality, progress, and liberty. In France, it triggered a revolt against royalty and the upper class. In Britain, it was the driving force in an effort to provide greater resources to the population (while retaining the royals) known as the Industrial Revolution. The drive to increase efficiency in manufacturing and agriculture led to new farming methods. They transformed agriculture, and increased food supply. Improved access to food accelerated population growth, and the greater ease of farming meant fewer jobs for rural workers and a shift to urban centers looking for work. The expanded urban workforce increased productivity and factories boomed. An increase in manufacturing, machinery, and mechanized transportation took off across Britain, slowly making its way to other parts of Europe, and eventually to other areas of the world.

Medicine and access to healthcare improved during the nineteenth century, as did access to affordable food. As a result, death rates declined. Another population boom triggered the demand for more food, goods, and medicines. As the population surged, it supported more factories. Chemical production emerged, and fertilizers and pesticides were developed that strengthened crop yields.

Factories, transportation vehicles, and machinery all needed power and coal became a primary source. Burning coal was used to create steam that would then spin generators and produce electricity. New power enhanced agricultural efficiency, human comfort, and the productivity of factories. At the same time, the burning of coal created pollution in the air, water, and land. Ever-increasing amounts of water were required to run generators. The mercury, cadmium, and arsenic in coal ash that filled the environment led to new diseases in humans and wildlife. And, over time, it created a new form of climate change.

Greenhouse gases

Climate change caused by human populations occurs from the massive release of greenhouse gases (GHG) into the atmosphere at rates that exceed what would happen naturally. As atmospheric greenhouse gases increase they wrap the planet, creating a barrier. Short wave radiation (ultraviolet light or UV) from the sun is able to penetrate the barrier and reach the earth's surface. However, warming long-wave radiation (infrared) emitted from the planet remains inside the barrier, warming the atmosphere instead of escaping back into space. As heat is unable to escape, the planet warms.

Carbon dioxide (CO_2), nitrous oxide (N_2O), methane (CH_4), and synthetic fluorinated gases such as hydrofluorocarbon fill the atmosphere as greenhouse gases. Naturally occurring CO_2, N_2O, and CH_4 are essential for life on the planet, but industrialization has led to them being released into the atmosphere at rates higher than those at which they would naturally occur, strengthening the atmospheric barrier. However, the most abundant greenhouse gas is not man-made.

Water vapor (H_2O) dramatically increases the warming effect of greenhouse gases. Humans do not directly generate the water vapor in the earth's atmosphere. But the increase of other greenhouse gases heats the planet, and as the planet warms the atmosphere holds more water vapor, increasing the barrier surrounding the planet. As the barrier strengthens, the planet warms further, and more water vapor enters the atmosphere in a perpetual loop.[4] While humans do not directly cause water vapor to increase, human activity that increases the other greenhouse gases does. It is estimated that as CO_2 leads to an increase in atmospheric water vapor, the presence of that water vapor doubles the warming impact of that level of CO_2.[5]

The accumulation of greenhouse gases in the atmosphere increases global warming and thus climate change.

Weather changes

Planetary temperature increases make weather systems more variable. Both heat events and precipitation are intensified. Heat events increase in temperature and last longer. Rainstorms happen less frequently yet more powerfully. The combination increases flooding, erosion, and also the frequency of cyclones.

But increased rain is counterbalanced by increased aridity. As rain comes faster, the ground is less able to absorb it. It floods over the surface and back into rivers, seas, and the ocean instead. More intensive rain events don't necessarily support increased groundwater. And while rain intensity goes up in wetter parts of the world, drier areas become drier. The switch from massive rain to increased dryness is more severe in Mediterranean climates, like that of California, where a year's worth of rain arrives only in winter. By late autumn, the region's dryness has reached peak levels. Dried plants and vineyard canopies wait as kindling, fuel to spawn wildfires.

The bad news is that farming practices contribute to the cycle. The good news is farming practices can mitigate the cycle. As growers have become more focused on improving quality by making farming changes in response to climate change, these changes have also shown support for decreasing GHG and lessening water vapor impact. As the wine industry reduces its GHG emissions and improve its environmental and carbon sequestration practices it becomes part of the climate solution.

WILDFIRES[6]

On September 10, 2020, California woke to a dark orange sky. By midday it had grown so dark that residents of San Francisco had to use automobile headlights to navigate the city, and house lights to see inside. The apocalyptic scene was the result of dense smoke caught in the lower atmosphere filtering out the sun. It stretched along the California coast from Mendocino in the north all the way down to the southern side of Paso Robles, a distance of more than 350 miles.

Just a few weeks before, a rare summer storm in California's driest season had caused thousands of lightning strikes to ignite hundreds of wildfires throughout northern California. Then conditions worsened, and new fires erupted, covering California and Oregon in blazes.

The governor of Oregon called the fires "a once in a lifetime event." But in California burns had occurred in each of the preceding four years, each time forcing mass evacuations and firefighting efforts of such a large scale that first responders from around the world flew in to assist. In wine country the fires also brought fears of winery loss, vineyard damage, and smoke taint. In 2020, in California alone, more than 4 million acres[7] burned in 9,639 wildfires. It was more than double the state's previous record. In 2025, wildfires burned parts of the city of Los Angeles. They burned fewer acres (and outside wine country) but are projected to be the costliest fires in U.S. history.

Climate's contribution[8]

Climate change research shows not just a global increase in temperatures, but with them also an increase in the air's vapor pressure deficit (VPD). As the VPD goes up, plants tend to respire more quickly, making it easier for them to dehydrate. To put that another way, as the planet warms, plants and soils are effectively getting drier. In California, the combination has lengthened the state's natural fire season.

For California, wildfires are an expected phenomenon as rain only falls during half of the year. By late summer the state includes a bevy of dried plants that serve as kindling. Historically, residents have understood the typical heat spells and windstorms of late summer or early autumn bring with them the risk of wildfire. But climate change has increased the frequency of hot weather events and it's also made the landscape drier. The number of days with that perfect combination of low vapor pressure and high temperatures has increased at the same time that forests and hillsides are gaining kindling.

According to climate change researchers, in the US west coast between the years 2000 and 2015, there was a 75 percent increase in the proportion of forested area experiencing the optimum conditions for forest fires. At the same time, the west coast has also gained nine additional days *per year* of wildfire potential.[9] Thanks to climate change, wildfire season in California has literally been getting longer while potential wildfire areas have been getting larger.

Impact on grapes

As the fire season extends, it also begins earlier. For wine growers and winemakers, the change means fires are more likely to overlap the harvest season, or veraison, the period before harvest when grapes are ripening.

Studies show grapes are most vulnerable to smoke impact from veraison onwards. The grapes' skins soften as they also change color. The softening makes it possible for the volatile phenols that give smoke its smell to penetrate the skins and bond with natural sugars in the fruit. For the most heavily affected fruit, smoke taint is perceptible immediately upon pressing. In other cases, it might not appear until after fermentation. In still others, only after aging will it become apparent. The bond between phenols and sugars can "hide" the impact of smoke initially. So fresh juice and even fermented wine might not reveal the aromas or flavors of smoke until later. Even worse, current tests are unable to detect the presence of some of these phenol–sugar bonds. Over time, the wine's own acidity severs the bond between phenols and sugars. The smoke becomes a recognizable aroma and flavor within the wine. But how long this process might take is unpredictable, making it difficult to assess the wine quality.

Since smoke impact is primarily within the grape skins, in most cases juice inside the berry remains untouched. (Extreme conditions can "push" the smoke further into the fruit, affecting the juice as well.) With the proper technique, white wines can be made successfully from grapes impacted by smoke. While wine's acidity cleaves the bonds between phenols and sugars, fermentation does to some extent too. In making red wines, the skins serve as an important component as the structure of the wine is largely extracted from the anthocyanins and tannins of the skins. Soaking these in the fermenting juice provides both color and form to the resulting wine. In fruit impacted by smoke, including the skins in fermentation or aging ensures the wine will carry smoke character.

But what about rosé? Rosé is made by allowing a short period of skin contact with the pressed juice of the grape.[10] Researchers have demonstrated that smoke character is fully extracted from grape skins in two to three days.[11] Extraction of the anthocyanins and tannins takes longer. In other words, shortening maceration time for red wine, or making rosé instead, is unlikely to eliminate or reduce smoke taint in the final wine though it will affect both its structure and color. It does seem that, since it is possible to make white wine without smoke impact, a still *blanc de noirs*, a white wine from red grapes, could be possible.[12] Several producers in California have successfully done this in recent years.

Studies are also being done on ways to remove smoke impact from wines. Filtering, carbon fining,[13] reverse osmosis, and spinning cone technologies have each proven unsuccessful. Each seems to succeed at temporarily removing some smoke impact while also stripping desirable aspects of the wine. During aging, volatile smoke compounds release into the wine even after earlier attempts to correct them.[14] In 2024, new techniques for removing smoke impact emerged. Synthetic polymers have been developed that bond with the volatile phenols of smoke. Specific filtration technology then pulls both the polymers and the smoke phenols from the wine.[15] While promising, the approach has not yet been used by the industry. Until we find solutions, quick release and early enjoyment can allow such wines to be enjoyed before smoke appears.

The financial impact of smoke taint can be high and less predictable than the immediate impact of vineyard, building, or equipment losses. For producers, releasing wines quickly, or blending potentially impacted lots into higher-volume wines, thus diluting smoke's presence, offer possibilities for salvaging income.

Vintners across California are collaborating with researchers, as well as with producers in other parts of the world, to seek solutions. New studies and techniques look promising. Research institutions in

Australia and the United States are both exploring further potential methods for mitigating smoke impact with promising results.

Researching smoke

The 2020 fires included five of the ten largest fires in state history, though 2025 exceeded them in cost.[16] Knowledge gained from the 2017 Wine Country Fires, and to a lesser extent those in 2008, helped establish ongoing research.

In 2008, some wines in northern Napa and Sonoma Counties were hit by smoke pushing south from fires in Mendocino. Little was known in California about fire impact on fruit at the time. So, wines that tasted fine on release became examples of smoke taint with age.

As in California, bushfires are a normal part of the Australian landscape. The problem for both regions has come as fires have pushed closer to agricultural and urban centers. In viticulture, Australia's experience with bushfires was initiated when they became prevalent in the 1990s. In response, fire and smoke research began there in 2005, and a smoke taint research group was founded in 2006.[17] Since impact from the 2008 fires in California was comparatively isolated, few studies were done but cooperative research between the state and Australia did begin in 2008 and was expanded in 2017.

In 2017, 85 to 90 percent of grapes within the affected region of the North Coast AVA had been harvested already when fires started.[18] Damage was significant. Fires led to loss of life, buildings, infrastructure, and some vineyards. Impact on wines, however, was relatively minimal. Later ripening sites or varieties, such as Petit Verdot, had not yet been harvested, but most of the industry remained intact. Almost immediately, collaboration with researchers at home and abroad began. Wineries with smoke-affected fermentations sent samples to researchers at UC Davis. Fruit left unharvested due to surrounding fires was also studied. New testing protocols to identify a broader range of volatile smoke compounds were created. Winemakers sent grapes, fermentations, and finished wines to labs in California, Canada, and Australia for the most comprehensive data possible. Scientists in Australia shared findings with those in California. The unusual nature of the 2017 fires added to the knowledge base for Australian studies.[19] The efforts prepared the industry to gain even more insight in further fire events.

The resulting wines from California's 2020 fires suggest that the way smoke is absorbed into grape skins is profoundly varied not only within a region but also across a vineyard. It appears that air currents, temperature, and vapor pressure have a role in how smoke settles into and is absorbed by fruit in different areas. The way that the topography of a site changes airflow can shift which vines suffer the most. Significant differences in smoke compound markers can even be found in neighboring rows. The age of the smoke itself affects the impact on an exposed vineyard. The volatile phenols[20] responsible for creating smoky character in wine seem to dissipate quickly. As smoke ages, its components change, reducing the likelihood of impact on surrounding vineyards. As an example, the 2020 fires that started in Sonoma County had significant impact in neighboring vineyards. The same fires sent smoke clouds to the Sierra Foothills on the eastern side of the state. Wines from that area were not affected by smoke taint. The smoky aromas, it seems, had dissipated as the smoke traveled.

The length of exposure and concentration of smoke change the intensity of absorption into the fruit. The different types of material burned, including different forest or plant types as well as more urban materials such as those from vehicles or buildings, change what volatile phenols are present and how they are absorbed. Humidity and rainfall within a region or site also seem to play a role in smoke absorption. Studies have already shown that water on the fruit or available in the soil changes the permeability of grape skins,[21] and so it makes sense that it would also change the absorbability of smoke through those skins.

Smoke is also absorbed differently by different grape types and at different parts of the season.[22] Further research is being developed in both California and Australia to expand the varieties studied. As of 2024, there is no quantitative data comparing vulnerabilities of cultivars.[23] Eventually, planting grapes less susceptible to smoke absorption might be one key to evolving wine in response to wildfires. More recent work also suggests farming techniques might influence how grapes absorb smoke compounds. Early observations suggest vineyards practicing regenerative farming techniques focused on soil health seem to produce grapes with less apparent smoke absorption. It is believed that increased soil health might correlate with denser cuticles in the resulting vine and fruit. Vintners hypothesize this might make the fruit less susceptible to smoke penetration. We do know denser grape cuticles reduce the potential for mildew or bunch rot,[24] so it is possible they play a similar role in smoke absorption. At the University of Oregon, researchers are experimenting with a protective coating that can be sprayed onto grapes to completely or partially prevent smoke compounds from bonding with sugars in grape skins. As of 2024, their research demonstrated at least partial effectiveness on three wildfire smoke compounds.[25]

Alleviating vineyard damage

In 2017, few vineyards burned in California wildfires. Vineyards that did burn were generally those planted alongside forests in the hottest part of a wildfire. Individual vine damage seemed to result from secondary causes, such as irrigation tubes melting, rather than from the wildfire itself. The belief was that the moisture within vines during the growing period protected them from fire.

The fires of 2020 demonstrated the limitations of that view. The most intense fires, fueled by substantial forests, burned through vineyards completely as they advanced. In Napa Valley, Cain, at the top of the Spring Mountain District, lost its vineyard as fire pushed through forest on the Sonoma side of the Mayacamas and then rushed over the slopes into Napa Valley. Several vineyards in the Fountaingrove District on the Sonoma side were also severely damaged. The heat of the fire in question appears to play the most important role in vineyard impact, but the dryness of the vine also varies at different times in the year and can be a factor.

Regulations around land management to reduce fire potential have increased. Property owners are now responsible for clearing and managing areas that could provide fuel for fire. The use of goats and other ruminants to clear under forests and areas of weeds is more common. What they eat reduces fuel for the fires.

Fire suppression versus fire management is also being reconsidered. Fire scientists have found that over the last hundred years, communities have tended to focus on fire suppression. But in areas with a natural wildfire season, it seems that fire suppression has increased our susceptibility to fires. In California, the increase in fire damage has come partially from population expansion. More people have built homes in areas of the state traditionally prone to fire. To protect these residents the state reduced prescribed burns and forest clearing efforts, but leaving smaller trees and undergrowth in forests untouched increases effective kindling.

Carefully managed prescribed burns in forests and grassland areas are now recommended. For California, that means low intensity, controlled fires when it's naturally cooler and wetter, and there's no wind. The goal is to use managed fires to lessen the potential impact of naturally occurring wildfires. The practice lessens the accumulation of fallen branches, pine needles, and dried leaves on the forest floor. It also removes the smaller trees and brush that catch fire easily and act as a sort of fire ladder lifting flames into taller canopy. Controlled burns won't stop wildfires, but they do reduce fire fuel, lowering the severity of naturally occurring fires. And they can also create fuel breaks in a fire-prone landscape, helping to slow or stop the advance of wildfires.

It is important to rethink where we build and farm, how we manage the surrounding landscape to create fire breaks and how we handle the power grid to reduce it triggering wildfires.

Adjusting approach

As dire as wildfires feel to the wine world, their increase alongside climate change will not result in a complete loss of wine. Nor will it lead to a total loss of a particular region's wine growing ability, or a widespread loss of grape varieties. But it will demand a rethinking of how we grow grapes, and what we plant where.

As wildfire season becomes longer, that means it also arrives earlier. Shifting vineyards to earlier ripening varieties will help increase the likelihood of ripening and harvest before possible threats of smoke or evacuation. This depends too on wineries, regions, and their marketing strategies being willing to change.

For already established vineyards, some farming decisions can shift vine development to encourage slightly earlier ripening. Those that have been able to shift to dry farming, for example, harvest earlier than when they were still using irrigation. But, ripening earlier also means fruit is picked during warmer temperatures. In some regions, this could create other challenges.

Fine wine is intimately connected to ideas of terroir and site expression. As regions get warmer, more intervention is required in vineyards to grow the same varieties farmed there before the effects of climate change became apparent. Where possible, shifting to year-round cover crops keeps soil temperatures cooler. Vine training to grow broader canopies (rather than simply vertical ones) also cools the soils and shades fruit. These are comparatively simple changes. Others are more involved. Some vineyards increase irrigation or overhead sprinklers before or after heat events. Sunnier or warmer regions have begun using sunshade cloth around the fruit zone, and in some parts of the world as a canopy above vineyards.

There comes a point where such interventions also begin to change flavor and structure characteristics, altering the overall expression of a wine. The effect would seem to be one of expressing farming choices more than innate site character. And here is where bigger changes need to be made. It is unlikely that a region like Napa Valley, for example, will have to stop growing Cabernet Sauvignon *in toto*. But it could become more difficult to grow vineyards close to forests for fear of wildfire, or to grow Cabernet in the warmest parts of the region without significant intervention.

Newer wineries might be better placed away from forested areas to avoid risk of fire loss. This alters some of the romance of wine and yet also acts as insurance. At the same time, those already in fire zones can increase their land management to mitigate the likelihood of fire. They can also partner with neighbors to create fire breaks. Wineries built in potential evacuation zones have begun incorporating remote control sensors into fermentation tanks, and external generators for use during power outages. The newest tank sensors allow winemakers to read the wine's chemistry and instigate punchdowns or pump overs through their phone from anywhere in the world. In many regions, power companies are shutting down parts of the power grid to prevent electrical events from triggering fires. On-site generators help ensure presses, pumps, and cooling systems stay operational during the peak of harvest.

The frequency of wildfires combined with the drive to study their impact means our understanding of how to combat wildfires, what causes them, how they affect vineyards and how to alleviate their impact in wine have increased exponentially. Researchers are evolving how they study climate, collaborating with wine regions and wineries to utilize already established weather and vineyard data to evolve both our understanding of wine growing and our changing planet. Remote sensing technologies and vine sensing equipment are making that easier. Most of all, wine's ongoing success will depend on the wine community's willingness to evolve.

SUSTAINABILITY, TECHNOLOGY, AND RESOURCES

Sustainability has become a new mantra for the wine industry. Winery websites and bottles mention it. Wine shops have sections devoted to it. Importers in other countries demand it. Certifications have proliferated.

The conversation originated as an ecological plea. It asked for a particular goal: improve farming practices to conserve the environment. As discussions expanded, the goal did too. To improve farming practices, a business must be able to afford the upgrade. Economic sustainability is essential. To foster a healthy environment, the winery must employ better practices too. It's not just farming but the making that must be sustainable. For practices to improve, employees, managers, and business owners must all participate. It's the people that make it possible.

Questions of sustainability include the entire supply chain: barrel makers, glass manufacturers, label and foil producers, cork or closure suppliers, vineyard products, tasting room needs, transportation of employees. As businesses have become more willing to seek conversion from their suppliers, sustainability improvements have percolated out to ancillary industries as well.

Farming choices directly affect soil health, plant life, and the surrounding ecosystem. Farming plays a role in wine quality, as well as either contributing to or helping mitigate climate change. But sustainability reaches beyond the vineyard into the entire production chain: wine growing, winemaking, distribution in tasting rooms and beyond, as well as waste management. Each is a link in the success of a wine business. The water used, the power supply, air quality, resource efficiency, worker well-being, they all help determine the health of the industry.

Wine originates in the vineyard, but its epicenter is the winery. Fruit processing, fermentation, pressing, aging all occur in the production facility. Laboratory testing, bottling, labeling, packing, and shipping often do too. In some, tasting rooms, tours, and broader sales for distribution are also happening. The winery functions through complexity. Production interacts with hospitality. Water use combines with weather considerations and waste management. The lighting, sound, air flow, and size of the building impact employee well-being.

Achieving sustainability is an ongoing process. It depends on improvement over time. As conditions change, businesses must too. The goal is met in finding solutions. Bigger businesses with the capital to invest in research and experimentation hold the key to problem solving. Their willingness to share solutions can speed the process of overall improvement for the industry. At the same time, smaller wineries have found solutions by utilizing collaborative capital. Together, their resources can have more impact.

Collaborative groups

Regional bodies have created local certifications to provide guidelines plus a nod of approval to producers within their area. California helped build some of the early examples.

In 1992, the Lodi Winegrape Commission launched a program to help farmers reduce pesticide use and improve farming practices. It started as an integrative pest management program and evolved to one of the more rigorous sustainability standards in the country, known as Lodi Rules. Its multi-pronged protocols consider all aspects of a vineyard and winery, and guide a state version, California Rules, and outside the state, Certified Green. Lodi Rules have also become a model for similar initiatives in other countries and wine regions. Lodi Rules served as a starting model for the California Code of Sustainable Winegrowing, a collaboration between the California Association of Winegrape Growers and the Wine Institute. The California Code provides a workbook to growers and wineries that helps them examine more than 200 practices across all aspects of the production chain and assess how to improve them.[26]

Sustainability protocols and collaboration extend beyond production into sales. Importers and distributors have begun setting sustainability requirements for wines they represent. Sweden remains the biggest wine market in Scandinavia, buying from regions around the world. Its liquor sales are managed almost entirely through their state monopoly, the Systembolaget.[27] Wine proves a popular drink in Sweden, accounting for around 45 percent of the monopoly's sales. And it's an important export destination for California wine. Sweden was also one of the first countries to sign the Kyoto Protocol, and then the Paris Agreement, with their commitment to reducing GHG emissions federally. The Systembolaget purchasing and sales program is guided by these international treaties. Monopoly guidelines include restrictions for producers sold within the Systembolaget around GHG emissions, recycling or reusability of packaging, use of packaging with lower climate footprint, reliance on renewable energy, water management, and human rights commitments based on those written by the United Nations to reduce farmworker exploitation and improve safety.[28] A sustainability committee reviews the ability of wine regions and producers to meet requirements and adjusts purchasing plans accordingly.

The challenge for a region like California comes from the difference between state and federal regulations. State employee standards, for example, exceed those of the rest of the country. Legal protections for farmworkers started in California. If the Systembolaget considers human rights and employee protections at the state level it will get a different result than it would by examining the national guidelines. The complexity of the relationship between state and federal governance within the United States makes sustainability requirements more complicated. Whether a sustainability initiative like that at the Systembolaget determines fulfillment at the producer, regional, state, or national level can dramatically change which wines are included.

By using its purchasing power towards climate action and sustainability goals, the Swedish monopoly is pushing change in wine regions internationally. But how do wineries meet such goals? Discussion groups designed to foster conversation and spur ideas have formed worldwide. The Porto Protocol describes itself as "a global wine community dedicated to sharing solutions for tackling climate change."[29] It includes around 250 members from 20 countries. The Sustainable Wine Roundtable in the UK brings together around 70 members worldwide to advance sustainability initiatives, from production to sales.[30]

As awareness of climate change has increased, so has the focus on greenhouse gas (GHG) emissions. Among them, carbon dioxide (CO_2) proves one of the easiest to measure and seemingly easiest to control. It's also wine's largest contribution to the GHG problem. Without addressing climate change, sustainability proves almost irrelevant. Mitigating CO_2 output has become a primary goal.

The Jackson family in California joined forces with the Torres family of Spain to invite other wineries around the world to form an industry program for mitigating CO_2 emissions. Together, they created International Wineries for Climate Action (IWCA). IWCA brings together businesses of varying sizes. California wineries include Ridge of the Santa Cruz Mountains, Spottswoode, Opus One, and Cakebread Cellars in Napa Valley, Medlock Ames of Sonoma County, Silver Oak, Twomey, and St. Supéry. All are acting alongside international producers including Yealands and Felton Road of New Zealand, the Okanagan Crush Pad of Canada, Symington in Portugal, Champagne Lanson from France, and bigger companies such as Ste. Michelle in Washington and the Crimson Wine Group and Constellation.

What makes IWCA stand out is that it provides scientifically supported guidelines for businesses as well as practices for meeting them. Reducing CO_2 emissions depends on more than changing tractors.

Diesel running frost machines, generators, and other farm machinery all play a role. The weight of packaging and how it is transported are crucial considerations. And the means for CO_2 reductions extend through the entire production and supply chain. To work out how to address so many levels of consideration, IWCA partners with consultants and experts outside wine. Together, they build tools to measure changes, such as carbon calculators and measurement standards calibrated for wineries. They also explore new solutions based on the local conditions in a participating country or region.

Climate action proves to be an urgent need. At the same time, climate action must be supported by sustainable practice along other axes of production. It is about more than farming or mitigating CO_2 emissions, and factors in the well-being of employees, the water supply, power needs, and more. The ongoing practice of sustainability requires shifting a system that extends beyond the winery into the surrounding ecosystem and community.

Water

In California, repeating drought cycles have forced wineries to innovate on water use and availability. The Mediterranean climate means rainfall occurs almost entirely in winter. Conditions become drier as the growing season and harvest progress. This is a normal process in a Mediterranean climate, yet the regularity of drought has been increasing, making water considerations more crucial. This is true in regions around the world. With less predictability in the rain cycle, the pressure on wineries and vineyards has changed. According to UC Davis associate professor, Ron Runnebaum, "it takes approximately 5 gallons of water to make 1 gallon of wine".[31] Wineries are looking at how to alter that ratio, decrease water needs, and become more efficient at both water use and reuse.

Dr. Roger Boulton, professor emeritus of UC Davis, helped design the first Leadership in Energy and Environmental Design (LEED) Platinum certified winery in the world, opened at the campus in 2011. The US Green Building Council developed the LEED program to offer guidelines and ratings for environmental design and constructions on buildings worldwide. Platinum is its highest standard. Boulton sought to make the most sustainable and efficient winery strategy possible to educate current students while also serving as a working model for other production facilities. As part of the initiative, Boulton designed a capture and storage system for rainwater that was planned for three years of water needs.

The roof of the building combines solar panels with a rainwater catchment system to harness two resources at once. The solar panels provide all power needed during harvest. Rain is funneled from the roof panels into tanks through a series of pipes.[32] The captured rainwater then goes through a reverse osmosis filtration system to remove impurities. Such filtration can create potable water good for drinking or dishwashing. The university uses it to ensure water for winery and landscaping needs.[33] The reliance of rainwater capture helps preserve local groundwater without tapping the regional aquifer. Allowing long-term water storage helps alleviate pressure during drought years by extending the rain's benefit over a longer period. Even outside extended drought cycles, ensuring wetter winters can support drier years increases efficiency.

In Sonoma County, wineries are expanding efforts to create similar measures. Governor Jerry Brown approved the Rainwater Capture Act in 2012, allowing businesses and residents to collect and store rainwater for landscaping and industrial needs. La Crema winery in Windsor has integrated rainwater capture into its winery tanks. Rain funnels into tanks before being filtered and stored for use in the winery cooling tower, which accounts for 20 percent of the winery's water need.[34] (Regional schools have also been increasing rainwater systems. In 2017, Sonoma County expanded the initiative, approving appropriately filtered rainwater for drinking and cooking.[35])

Other wineries are rethinking rain for groundwater storage as well. Increasing the number of ponds on vineyard properties increases their capacity to store winter rain. It also allows for release during dry periods to support local habitat and improve groundwater conditions. The open nature of pond storage means evaporation in summer can be a problem. To help reduce water loss, some wineries are installing solar arrays atop the ponds. This approach captures renewable solar power while also reducing water loss.

The UC Davis winery facility has also worked to increase water efficiency. In 2024, they completed installation of a new clean-in-place system for barrels and tanks. It relies on less water during the cleaning cycle, while also capturing the water used for reuse on other tanks.[36] The university has been working on developing water reuse systems that can safely and effectively use water up to ten times. Wineries have also begun water-capture systems to improve water reuse. In Sonoma County,

Kendall-Jackson has improved the barrel-washing process and increased its water reuse abilities from single use to three or four cycles of reuse. Silver Oak in Alexander Valley has reached similar reuse efficiency with the final cycle of water recapture being used for flushing toilets in the facility. The approach conserves fresh water and alleviates the need for pumping groundwater or reducing the local aquifer.

More efficient equipment can also help. High pressure water hoses, for example, provide more impact with less water. Ultraviolet (UV) cleaning machines have been developed that use shortwave light to kill bacteria and other microorganisms. In the winery the technology can be used on tanks and barrels as well as other equipment.[37] UV is also being used to purify winery air systems and protect grapes from powdery mildew.[38] Ozone disinfectant systems also reduce water need. Ozone levels must be monitored for safety, but when done correctly it is safer than chlorination with less chemical impact.[39] While water conservation is the goal, improving a business's water efficiency also helps avoid it becoming a waste product.

Waste management

Waste management extends beyond refuse, toilets, and recycling. Thoroughgoing approaches consider the overall efficiency and output reduction of the facility. This includes water systems, energy sources, power needs, and greenhouse gas emissions.

While CO_2 is not generally described as a waste product, such framing emphasizes the value of reducing carbon production from the wine industry. Agriculture in general contributes 22 percent of global GHG emissions.[40] CO_2 proves to be wine's biggest contribution to climate change. Packaging and transportation account for 68 percent of the industry's total emissions. Farming and winemaking account for around 16 percent.[41]

Traditionally, wine relies on glass bottles for distribution. But glassmaking historically depends on the combustion of natural gas, which releases high levels of CO_2 into the atmosphere. Innovations are being made to transform how glass is produced and alleviate the problem. Collaboration between glass producers in Europe has led to the creation of production designs for hybrid furnaces that significantly reduce the reliance on natural gas and thus also CO_2 emissions. Gallo Glass announced in 2024 its plans to install a hybrid electric furnace to remove CO_2 from its glass production. Gallo's plant in Modesto is the largest manufacturer of glass in the country, producing around 2.5 million bottles for wine and spirits per day. Gallo plans to demonstrate the success of such furnaces at its plant to hopefully expand its use elsewhere.[42] In Europe, the world's first fully electric furnace is being tested.[43]

Reducing the weight of glass, or other possible containers, reduces transportation emissions as well. Additionally, the shape or style of the container can change how much space it takes up, thus also lowering the total weight per container being transported.

Packaging

Packaging proves to be one of the biggest contributors to potential GHG emissions. Glass production as well as the weight associated with its distribution are among the most significant aspects of a winery's carbon footprint. But packaging also creates potential material waste.

Studies show generational differences in purchasing habits associated with packaging changes: the amount of influence packaging will have on buying decisions plays out differently by generation. Boomers are less likely to change from glass bottles to newer alternatives. Millennials, on the other hand, are both more curious and more environmentally conscious of the container impact of their spending. Across generations, more people are demanding the elimination of synthetic polystyrene in shipping packaging.

More packaging has shifted to recyclable materials, but its effectiveness depends on the recycling systems in a particular market. For some areas where recycling is difficult or impossible, shifting to packaging reuse or compostable packaging proves to be the best alternative. Alternative packaging is taking hold. The inert nature of glass, its impermeability, and its greater ability to protect liquids from outside factors like temperature changes or oxygenation, mean it will likely remain the ideal vessel for fine wine. But wines designed to be enjoyed sooner can rely on other types of packaging. Canned wine weighs significantly less than glass, and aluminum has a higher recycling rate than glass in the United States.[44] Flat plastic bottles,[45] paper bottles,[46] and bag-in-box for premium wine[47] are also becoming options. Each lowers the weight to volume ratio, and thus also emissions during transport.

Smaller formats also reduce immediate costs to consumers, and in some cases by reducing container volume there is also less waste of the wine purchased. Larger formats, on the other hand, such as 3 gallon bag-in-box versions of premium wines and higher quality options allow consumers to buy more at better value while also retaining the freshness of the wine being enjoyed over months.

Rethinking standard practices

Although its impact is smaller than that of packaging and transportation, winemaking itself also produces CO_2 emissions. Newer technologies have made it possible to even do carbon capture within a winery. Carbon release from fermentation may seem modest on a global scale, but its effects in the cellar are strong enough that wineries must ensure it is released to avoid asphyxiation or damage to mucous membranes. Collectively, CO_2 emissions during fermentation do make an impact. In Burgundy, winemaker Diana Snowden-Seysses of Domain Dujac, as well as Snowden in Napa Valley, has adopted carbon capturing in the Dujac cellar, and is working to bring it to other wineries as well. Rather than being released into the atmosphere, the CO_2 from fermenting wine is held and converted into organic bicarbonate.[48] Currently, there is no market for selling the bicarbonate. Researchers are investigating how to change that.

The UC Davis research and teaching winery was one of the first in the world to develop and practice carbon capture from all fermentation vessels. The system pulls CO_2 off fermentation and removes it from the cellar environment before it is ever released, thus increasing safety for people present. The approach reduces energy attached to other methods of managing air quality and safety within the building.[49] The UC system has also been designed to convert the captured CO_2 into chalk, thus sequestering it while modifying it for other uses.

The impact from any one winery seems minuscule, but Snowden-Seysses emphasizes the impact of regions collaborating on such efforts. As she describes it, if California implemented winery carbon capture it could harness 300,000 tons collectively and prevent it from entering the atmosphere.[50] Similarly, other regions working together can collectively mitigate their GHG emissions as well. The broader adoption of such an approach could even support the wine industry becoming carbon negative. Zero emissions are one of the conditions of the Paris Agreement.[51]

Emissions standards as well as recycling protocols can also be examined through connected businesses in the supply chain. Barrel manufacturers produce GHG emissions during the process of drying and charring the barrels. The wood used depends on forest management, which can have an impact on emissions as well as carbon sequestration. Cork producers similarly determine forest health. As both wine barrel and cork industries depend on the longevity of their forests, most tend to practice sustainable forest management. Cork producers retain the necessary trees, which alleviates desertification from exposed soils, and allows the forest to continue sequestering carbon. Cork suppliers today argue that use of cork has lower environmental impact or GHG emissions than other closures. Aluminum is infinitely recyclable (when collected), but the recycling process creates significant emissions, and mining for aluminum creates other forms of environmental harm. Cork has issues for wine quality, but those have improved significantly in recent years. Additionally, cork harvesting supports local economies without the land extraction associated with aluminum.

Groups like IWCA are making it easier to calculate carbon contributions from outside the winery, and more businesses have started measuring their own emissions. Collaborative work with suppliers can make it easier for both to measure, share, and mitigate environmental impacts. Businesses have started utilizing purchasing commitments in relation to such changes, which is making it more cost effective for more companies to join the effort.

Winery waste and its secondary impacts through GHG emissions, land and water contamination, or damage to habitats can be reduced by increasing resource use efficiency. It simply means using fewer resources, and thus reducing both requirements and waste. Using compostable dishes and utensils in kitchens, tasting rooms, and elsewhere can shift waste to reusable material. Avoiding single-use products, while recycling carboard, and reusing glass or other materials can help. Still usable barrels can be sold or donated to other wineries. Decommissioned barrels can be reused as a component of water filtration systems, or as part of landscaping.

Rainwater catchment, as well as water capture and reuse within the winery, helps mitigate wastewater production. Switching to UV and ozone cleaning technologies, where possible, helps limit chemical waste.

Composting grape pomace produced during winemaking, cuttings from the vineyard, or cover crop is also an option. Or these can be fed to farm animals. Some wineries are also extracting by-products from grape pomace and seeds for beauty products, creating a new form of reusability.[52] Researchers have also identified ways to convert compost biomass into renewable energy. This creates an important solution for dealing with the planet-wide increase in organic waste. For wineries, it offers a cost-effective approach to energy production available every year.

Renewable resources

Wineries rely on power. Coal- and natural gas-fueled powerplants produce significant GHG emissions. Wineries are therefore turning to reliance on solar arrays and windmills to generate renewable power.

In Paso Robles, J. Lohr winery unveiled a 3-acre solar-tracking array in 2009. It was the largest of its type in the country, and as of 2020 continues to be the largest of any winery. The winery pulled 3 acres of grapes to plant solar panels instead. The array's ability to track sunlight over the course of the day and with changing seasons increases its ability to capture solar power by 15 percent. The J. Lohr array in Paso Robles accounts for around two-thirds of the winery's energy needs. In 2020, they added another solar array for their Monterey winery, which accounts for almost 100 percent of its power needs. Also in Monterey, Ryder Estate Winery has since 2017 produced 100 percent of the energy it needs via wind turbine. The turbine also provides power to the surrounding grid, supporting around 125 homes in the region. In areas where wind is persistent, turbines are more effective at energy capture than solar.[53]

Reducing power needs increases energy efficiency as well. More efficient machinery and lighting systems are becoming available. Producers can reduce power demands by well thought out placement and construction of winery buildings. Largely underground wineries and caves need less insulation and less cooling or heating power; likewise, buildings beneath forest canopy are also cooler. More recently, wineries have reduced the heat absorption through roofs by creating what is called a living roof. Small plants suited to the region grow on the rooftop, decreasing heat absorption and keeping the building inside cooler.

Ridge Lytton Springs, in Sonoma's Dry Creek Valley, opened in 2003. The area was not suitable for caves or underground buildings so Ridge took a different direction. The Lytton Springs facility relied on haybale construction. It became the first winery in the country, and at the time the largest production facility, built using haybales.[54] The density of the straw provides natural insulation, maintaining even temperatures inside and thus reducing energy needs.

Rethinking winery design can also help. The Living Building initiative and certification program offers guidelines for creating more sustainable buildings that rely on less power, have a lower environmental impact, and offer improved air flow and a wealth of natural light, also making them more physically and psychologically supportive spaces for employees.[55] In 2020, Silver Oak Alexander Valley became the largest Living Building certified structure in the world. The Living Building Challenge is considered one of the most rigorous sustainability certifications for buildings in the world. Silver Oak is also LEED-Platinum certified.[56]

Providing incentives to help employees change practices to become more energy efficient or reduce emissions is an effective tool. Jackson family wines provides subsidies to help employees purchase electric vehicles and has installed charging stations (billed to the winery) at their offices.[57] Supporting employee well-being can reduce employee turnover, which creates some of the highest people-related operating costs for a winery. Retaining employees also improves winery efficiency. It takes time to learn and practice sustainability efforts in the vineyard and winery. Long-term employees alleviate that shift in protocols. Fair wages, healthcare, mental health resources, family leave, and emergency funding are all beginning to appear in larger wineries throughout California.

Beyond production

Developments in environmental research and engineering have demonstrated how wine businesses can also act for habitat restoration. In Lodi, LangeTwins have pulled vineyard plantings away from riverbanks and allowed natural waterflow to return to previous pathways, while at the same time helping to restore native plants. The restoration of streambeds and an increase in wild land surrounding them has increased the biodiversity and wildlife of the area. This is also the basis of certifications like Salmon Safe, which demands wineries lessen the impact of their production on local rivers. Wetland and lake restoration has also occurred naturally at the LangeTwins property. They now use the restoration project as part of an education program for regional schools. Classes take fieldtrips to LangeTwins farms and restored habitats to study the water cycle and water use, as well as sustainable farming through inclusion owls and other birds. Such programs support the knowledge of local communities, which can influence their sustainability and voting commitments as well.

Wineries have also begun partnering with other groups such as the National Oceanic and Atmospheric Administration (NOAA) and Sonoma Water to create positive impact rain, weather, and water management on site, which also supports habitat restoration. During some of the worst drought years, wineries with more plentiful containment ponds have released some of their storage water to help refill creeks and tributaries, supporting fish and other wildlife.

Since 2010, landowners, especially vineyards and wineries, along Dry Creek River in the northern portion of Sonoma County have worked together to create river restoration to reduce erosion while also improving fish habitats. Quivira, Seghesio, Rued, Amista, Dry Creek Vineyard, Yellow Dog, Truett Hurst, Meyer Family, and Lone Star vineyards have all been part of the effort. It marks a significant, ongoing collaboration between wineries as well as scientists and regional experts to improve conditions in the surrounding area.

Ongoing consumer education through improved communication and more trade education can help improve consumer understanding, explaining not only why corporate social responsibility matters but also how it is being practiced. Today, most consumers perceive such claims as a mere marketing tool and so corporations genuinely practicing social responsibility must be able to demonstrate how.

Acting for more than climate

Sustainability means more than improving efficiency and reducing impact. Its goal is to use resources responsibly. To be effective, business practices and community commitments must shift from extracting resources merely to increase ease and profits to instead recognizing that the earth is a limited resource. For sustainability to be effective, it must occur in relation to effective climate action, but it also needs to focus on more than just carbon emissions. Water, energy, animal welfare for those with farm animals, employee well-being, and waste management are all part of robust sustainability protocols.

Ultimately, such changes must also be economically sustainable. Businesses aren't sustainable unless they are able to afford their own means of production for their environmental and community commitments.

The problem solving of collaborative groups and businesses with the capital to research and experiment is providing the tools needed. Encouraging consumer support helps it continue.

CLIMATE ACTION AND FARMING

The 1920s welcomed the introduction of tractors and farm implements across the United States. Tractors were invented late in the 1800s, but just before the advent of Prohibition they became practical. Farming could be done with machinery instead of by hand. Increased yields across a greater area became possible.

The population of the United States had increased. Immigration from Europe brought a steady flow for decades. The eastern seaboard was becoming crowded and the country needed more room for growing food. So the US government worked to increase western expansion. The Homestead Act had been passed in 1862. People willing to settle the west were granted 162 acres (66 hectares) for minimal expense. As tractors became available, the promise of making money through agriculture drew even more settlers. By 1900, the Indigenous of the region had already been confined to reservations. Their land was up for farming.

At the same time, agricultural consultants instructed that, in drought-ridden areas, like the Great Plains, it was crucial to expose soils by clearing land to increase water absorption through the area. They were wrong.

As farming expanded, the native grasslands of the region were cleared. Land was exposed throughout the expanse. Then rains stopped. Topsoils became so dry they were reduced to powder. When winds arrived, the area was subsumed in dust storms, making it practically uninhabitable, and certainly not suited to agriculture. Western expansion to the region had failed. While the Great Plains experience natural drought cycles, the Dust Bowl exacerbated them. The great influx of population that had settled in the center of the United States was forced further west where, until the advent of World War II, many became farmers and farm hands in California.

The Dust Bowl crisis of the 1930s reveals some of the ways farming can impact climate. Prior to clearing the region, the native grasslands that covered the area had adapted over millennia to the drought conditions of the plains. They created a habitat that provided soil structure. It also retained moisture throughout the year even through drought. The soil humidity and grasses in turn provided food for wildlife. We now also know the grasses of the Great Plains are one of the most significant carbon sinks in North America. They sequester more carbon than almost any other ecosystem due to the depth of their roots and their adaptation to the region. They also provide more reliable carbon storage as they are highly adaptable to fire and climate change. Clearing the land stopped their carbon sequestration and released carbon dioxode into the atmosphere. But it also created a Dust Bowl that expedited aridity not only in the region but across more of the continent.

How land is managed plays a significant role in climate change. Plowing the grasslands emitted carbon dioxide into the atmosphere as it dried the soils and produced the dust that dominated for almost a decade. Dust is a side effect of farming that exposes bare ground. Exposed earth, especially darker soils, quickly heats from sun exposure and becomes drier. As the soil heats, plant ripening and plant dehydration accelerates too. For vines, the combination significantly changes fruit character, can reduce acidity and speeds not ripening so much as concentration through dehydration. Dust production also impacts rain cycles.

As dust particles become airborne, they draw water vapor in the air, forming a mass of dust and water droplets. The resulting water droplets are smaller, making them more likely to evaporate before they can produce larger droplets essential to rain. The effect impacts the water cycle in the region at the same time as the exposed soil increases surrounding temperatures. And as these dust clouds become airborne, they quickly spread beyond the region where they were formed by wind, impacting the water cycle more broadly.

Though it can seem counterintuitive, dust reduces the likelihood of rain at the same time as it increases water vapor in the atmosphere. Water vapor has a greater impact on the warming impacts of climate change than other greenhouse gases. This is partially due to the way water vapor increases the impacts of carbon dioxide in the atmosphere.

The push for no-till farming not only reduces soil loss (a critical consideration if we are to maintain the ability to farm) but it also helps lower local temperatures and encourage rain. An individual farm might not have significant impact, but an entire region uniting around such practices could influence the weather cycle of its area by reducing temperatures without obscuring rain. This kind of change could have a strong positive impact on the ripening cycle of wine grapes in the area, while also mitigating warming.

While regenerative farming has become a powerful buzzword, investment in genuine regenerative farming practices can play a powerful role in climate action. But what is regenerative agriculture?

The Rodale Institute has been a leader in understanding and developing techniques for regenerative farming, just as it was a leader in the advent of organic farming. It has also done extensive studies comparing the results of conventional farming to those of organic as well as regenerative organic approaches across multiple considerations. The institute describes regenerative agriculture as "a system of farming principles that rehabilitates the entire ecosystem and enhances natural resources, rather than depleting them." It depends on recognizing the natural interactions of the specific environment and engaging them in a way that fosters greater resilience. For example, observing how an ecosystem responds when disturbed gives insight into those natural processes, which can then show how to support them.

Regenerative farming practices include keeping ground covered, in other words avoiding tilling, eliminating chemical pesticides, including fertilizers, fostering biodiversity in the environment as well as the soils, and as such, integrating considerations of wildlife as well as livestock and the people involved in farming. Each contributes to the holistic well-being of the farming environment.

In understanding regenerative practices, the use of tilling (or not) is one example. Increasing soil coverage reduces soil temperatures, which reduces surrounding air temperatures, which slows the ripening cycle of vines. This can support more ideal grape chemistry for those growing wine. Cooling soil temperatures also encourages greater microbial diversity in the soils. Too hot, and the microbes suffer significant die off.

Microbial life in the soil improves the soil's ability to sequester carbon. Parent plants, like vines in vineyards, or grasses in the plains, depend on microbes to convert nutrients and minerals in the soil to forms plants can absorb. The plants, in turn, feed the microbes with liquid carbon released through their roots. As the symbiotic relationship between microbial life and plant roots improves, more carbon is pulled into the soil, biodiversity of the soil increases, plant roots deepen, and the plant transports even more carbon into the soil. The plant also becomes more resilient to surface changes, such as increased temperatures and reduced rainfall. As the plant becomes more resilient, it more successfully sequesters carbon. It's an increasing virtuous cycle.

At the same time, it must be recognized that cover crops have different levels of success in different areas. Cover crops adapted to a region's growing conditions have greater likelihood of success. In transition, some parts of an established vineyard might need a longer process of adjustment before tilling is fully stopped. In other regions, growing conditions simply do not include much, if any, soil cover. Growers in these areas will need other strategies to mitigate erosion, soil temperature, and aridity.

The western United States has an additional consideration. Burrowing rodents, such as the vole, populate gardens, farms, and vineyards west of the Rockies, demanding a different form of pest control than the eastern United States and elsewhere. Underground pests such as voles or gophers often eat the roots of vines, or the bark at their base. Their activity tends to increase during droughts, when plants become a needed water source. Cover crops or grasses, especially around the base of a vine, provide protective covering for such animals, increasing their number and impact in vineyards. Viticulturists in California need other forms of problem solving to take advantage of the benefits of cover crops while also avoiding voles. Ivo Jimenez of Grgich Hills in Napa Valley has had success by establishing irrigated insectary (flowering plant) areas between vineyard blocks. With the ready water around the flowering plants, voles have a preferred water source instead of the vines themselves.

The practice of farming itself releases only a modest amount of direct carbon dioxide, relatively speaking. Significantly more is released by converting the land, as in the example of the Dust Bowl. But the presence of water vapor in the atmosphere doubles the warming impact of the level of carbon dioxide present. So, increasing the amount of soil covered in farmlands can play an important role in alleviating global warming.

However, carbon dioxide is not the only greenhouse gas. In agriculture, the biggest GHG impact comes from the cattle and dairy industry where the animals release methane as part of their digestive process. Livestock create around 14 percent of global GHG emissions, primarily through cows.[58] It's a practice that barely relates to wine. The more significant impact of farming related to wine is the contribution of nitrous oxide (NO). Farming with chemical fertilizers plays the greatest role in GHG emissions for agriculture, more than methane or carbon dioxide.

The use of chemical fertilizers to feed plants readily causes an excess of nitrogen in the soils that plants cannot absorb. So, while initial applications of fertilizer can give plants a boost, too much fertilizer creates a challenge for soil microbes. When overloaded, they can't convert the fertilizer into a usable form for the plant. Excess nitrogen is changed from its nitrate form, which plants can absorb, into nitrous oxide, which they can't. This process occurs for microbes in oxygen-free environments. Rain events deplete oxygen in soils, thus speeding the process of N_2O release from soils into the atmosphere.

Fertilizers are often applied to vineyards at the end of the season, after harvest, when plants are most depleted. For California's vineyards that generally means fall. Due to the state's Mediterranean environment, rains typically begin in late fall.

The move from chemical farming to regenerative practices improves climate action by reducing GHG emissions, but it also improves biodiversity by counteracting species loss. Greater biodiversity in the environment increases microbial biodiversity and overall habitat restoration. Over time, the effect is improved plant resilience and comparatively lower impact from climate change.

Improved soil management also reduces erosion. Covered soils are protected from wind as well as erosion due to water. They have greater ability to absorb water rather than lose it to runoff, and the combination improves fertility of the soil to sustain agriculture.

Regenerative farming won't stop climate change. But it can play a substantial role in mitigating it – without increasing farming costs. The move to regenerative organic viticulture can increase carbon sequestration while also reducing carbon release. While common knowledge says organic farming is more expensive, guidance through the process of conversion to the unique conditions of a site can reduce such costs and even make the long-term cost of farming organically lower than that of chemical farming. Without the use of chemical fertilizers, the impact of nitrous oxide is also significantly reduced. A series of recent studies estimates that a move to regenerative organic viticulture can sequester more carbon dioxide than that same farm currently emits.

Further, the farming practices needed are already available. While carbon dioxide is not the only consideration in climate action, reducing its emissions has a significant impact. Farming is an important aspect of climate action alongside reducing emissions through packaging, transportation, and production, which the wine industry has already identified how to do. In other words, we have the tools we need to make positive change. We need to implement them.

EPILOGUE: REGENERATION AND INNOVATION

California wine is again facing the need to regenerate. Consumer interest has moved into other beverages while inflation has decreased consumer spending. The promise of tariffs means supply-chain cost increases are anticipated. The likelihood of a strengthening dollar means loss of a chunk of export sales.

Climate change has altered how growers farm. Too much wine and too many grapes, a wine glut globally, have led to a buyers' market. People with fruit or wine to sell face uncertainty.

California has faced much worse, however. Consistently, the wine industry has found solutions. Phylloxera in the second half of the 1800s almost collapsed production. A hundred years later, it appeared again and radically transformed Napa Valley. It forced replanting that became an opportunity to redesign the region. Prohibition stopped winemaking almost completely, yet it expanded the vineyard area. After repeal, grower cooperatives were formed to rebuild winemaking and guarantee that fruit was purchased. Post-Prohibition, the market cared little for table wine. The state established programs for at-home wine education and certification; consumers responded and steadily became wine buyers. The state's wine expansion started in the 1970s. Yet, in the 1980s, the American public in their penchant for VHS-tapes of *Jane Fonda's Original Workout* and *Buns of Steel* stopped drinking wine.

The double miracle of Morey Safer with his *60 Minutes* segment "The French Paradox" at the start of the 1990s, and Robert Parker shifting from a love of Bordeaux in the 1980s to fanfare for California in the

1990s helped change things. But it was the dogged determination of California winemakers to create a new relationship with the media that built a road for Parker and helped create an interest in wine columns across the United States. Without the heritage or prestige of the classic wines of Europe, California wine generated a new story. The young guns of the west were besting the top wines of history. California wine helped make wine marketing via wine media a global phenomenon.

LESSONS FROM HISTORY

The painful moments of the state's history in wine cannot be denied. Wineries have gone out of business or been sold for economic reasons. Succession planning from a founder generation to their kids creates problems that can also lead to the need to sell up. The central focus on Zinfandel field blends for at least a century gave way to fighting varietals, single varieties in their own bottle of wine, in the 1980s and 1990s. Pierce's disease and a new financial boom in northern California moved vineyards from the southern coast to the north coast in the late nineteenth century. Los Angeles' original vineyards were all but forgotten as grape growers in the south instead turned to citrus. Today, Napa Valley has become a benchmark region for the world's Cabernet. The decimation of phylloxera in the 1980s thanks to the wrong rootstock, AXR1, made that possible. The region's demand to replant but with greater insight into what works helped make Cabernet the king grape of California and supported its global elevation to grape royalty. Forty years later, as consumer tastes change again, perhaps Napa faces another reinvention. Each time California wine has renovated itself success has depended on a staunch group of vintners working together to innovate and fix things.

Amid a national recession in 1894, the state's leading wineries combined capital to create California's first wine corporation, the California Wine Association. The effort meant they dominated the market until the advent of Prohibition. Shortly after Prohibition ended, emerging small wineries across the state joined forces and created the Wine Institute, which continues to problem solve and support the state's industry today. The Wine Advisory Board operated for decades in the mid-twentieth century, through a portion of proceeds from every producer in California, to build market interest. In the 1930s and 1940s, UC

Davis collaborated with the Napa Valley technical group of winemakers to improve quality and stability in wine. UC Davis has planted research vineyards in the Sierra Foothills, Lodi, Napa Valley, and in Davis to study the longevity of varieties as much as the impact of climate change.

Following World War II, it was the farmworkers who first understood the harmful impacts of pesticides. They made the broader American public aware and farming regulations were then made. The government, it turned out, had to learn from the differing experience of agricultural laborers to understand what regulation was needed. In the 1960s, Alice Waters' trip to France made her realize how bland the mass-produced food of the United States was at the time, triggering her determination to find locally grown produce in California to fix it. It took a new experience to make her realize what had been missing and what she wanted to change. When Waters returned, she found area farmers and local winemakers to help build the menu for her restaurant, Chez Panisse. Her success came from the broader community of food. As that community expanded, so did Waters' sphere of influence.

And this is how it happens in wine. Innovation depends on problem solving to create new solutions to service changing needs. As people enter wine from differing educational, business, and life backgrounds, they bring new perspective with them, the key to wine gaining new insight. The history of California wine has repeatedly demonstrated it was never a single leader that created significant, productive change, not Junipero Serra, not General Vallejo, not Agoston Haraszthy, not Robert Mondavi, not Cesar Chavez, not Jess Jackson. None of them transformed California wine alone. This is not to deny their importance or influence but to point out that they were given credit that depended not only on others who did enormous portions of their work, but also on those who provided enormous portions of their insight and inspiration.

California wine has faced incredible market volatility since 2020 (though one might easily argue as far back as 2008). At the same time, it is producing the greatest range of styles, varieties, and regions in wine it has ever had, as well as arguably the best quality. The challenges of the 2020s will undoubtedly transform the industry, but they will not eliminate California wine. In the face of climate change, viticulture and varietal types grown will change, but wine will not lose any specific variety completely. The resilience of California wine depends on its ability to harmonize its wine growing with a changing environment as much as a changing market. That depends on expanding who we include in decision-making and leadership as well as learning from mistakes of the past.

LESSONS FROM FARMING

Farming done for a singular outcome is the view that created the Dust Bowl. Reveal bare ground, remove competing plants, focus on crop alone. During the 1930s, the Great Plains in the central United States were consumed by dust blizzards and sandstorms. In 1935, a storm like a tsunami on land consumed Oklahoma and parts of Texas. Winds of over 40 miles per hour (64 kilometers per hour) blew, carrying walls of dirt so dense the sun did not penetrate, and temperatures dropped severely. Livestock died across the region. People were lost in the countryside unable to see to find their ways back home. Such dirt-weather events lasted for almost a decade.

Through the 1920s, wheat was one of the nation's most important crops, farmed throughout the Great Plains. The area was known as the breadbasket of the nation. The success of wheat farming on the plains was regarded as the gold rush of grain. Families filled the plains, gaining from government incentives offered to increase the wheat crop. For a time, the region also built the nation's economy. However, by the early 1930s, an oversupply of wheat had collapsed prices; farmers suffered in a wheat glut. Then the winds came. A mass migration from the region to more stable growing areas was necessary.

The Dust Bowl is considered the worst, human-made ecological disaster in the history of the United States. More than 100 million acres were impacted. Around 2.5 million people became migrants, forced to leave the area. It changed how America managed land and how farming was done.

Escaping the Dust Bowl depended partially on the United States entering World War II, which transformed the US economy and gave farming time to move outside the Great Plains into other parts of the United States (one being California). But the success of World War II was paid for with new consequences. The rise of chemical agriculture following World War II transformed farming yet again: DDT eradicated pests; chemical fertilizer fed plants. It was named the golden age of pesticides. Crops were again farmed with a singular focus: grow single crops to increase yields. Chemical agriculture is now regarded as yet another ecological disaster, speeding the extinction of innumerable species, introducing toxins to food, building pollution, and, as in the farming of the Great Plains, increasing the loss of topsoil.

The most important lessons of the Dust Bowl are numerous. Speedy expansion creates an oversupply in the market, which then injures the industry. The impacts of farming today can create damage that persists decades from now: consider the long-term impact. Agricultural choices must be done with respect for a region's growing conditions, in balance with its surroundings. Learn from history; understand a problem to solve it. In other words, slow down, farm with patience, grow crops in relation to their environment.

We are still learning the lessons of chemical agriculture, but some are already clear. If humans can damage the environment, they can also work to restore it. If bare land causes soil loss, covered land can minimize it. Chemistry causes harm when we forget the limits of biology. If pesticides cause pollution and introduce toxins to food, find other farming solutions.

Since the 1980s, regenerative agriculture has been percolating into vineyards. The Rodale Institute coined the phrase to describe cultivation that sees humans, their crops, and how those crops are grown as small parts of a much larger whole. Humans can damage the environment. They can also work to conserve and restore it.

Investment in regenerative farming has increased since the 1980s. Patagonia, a multi-billion-dollar company focused on outdoor clothing, joined with the Rodale Institute and Dr. Bronner's, a family-owned company making organic personal care products, to create the Regenerative Organic Certification (ROC) and spawned a movement. In 2020, the first businesses received ROC, after a pilot program in which more modest sized businesses from multiple industries helped translate ROC standards to industry needs. Tablas Creek in Paso Robles served as the pilot vineyard and winery business. It was also the first to receive ROC.

Studies have shown that agriculture plays a significant role in climate change. Around 20 percent of human-generated greenhouse gas emissions are caused by agriculture. This includes nitrous oxide (N_2O) released from the use of excess fertilizer, methane created from beef and dairy farming, and carbon dioxide (CO_2) emissions from the use of diesel machinery. Added to that is the loss of carbon sequestration potential through land-management choices.

The good news is that farming can also play an important role in climate action. Research has shown that the removal of chemical fertilizer can reduce direct global greenhouse emissions in agriculture by around 20 percent.[1] The Rodale Institute has done long-term studies over decades on the impact of such change on yields. According to their research, after a five-year transition period, crop yields return to similar levels as before the transition and, depending on an area's other growing conditions, can even exceed pre-transition levels. In organic farming, chemical fertilizers are not allowed. As a result, the shift to organics alone can play a significant role in reducing the greenhouse gases from farming.

For those not ready to change their farming methods completely, mitigation strategies can have an important positive impact. The type of fertilizer used plays a serious role. Several fumigant pesticides significantly increase N_2O emissions, over other types. Other mitigation strategies like the development of sensors to determine appropriate fertilizer applications as well as timing of application can do even more to reduce N_2O emissions. This will depend on innovation.

Regenerative organic farming and biodynamic farming (which is also organic) both enrich soil health. Multiple studies demonstrate that pesticide applications (not allowed in biodynamics or regenerative organics) significantly contribute to the loss of micro-flora and fauna in soils, which are essential to carbon sequestration. The switch to either regenerative or biodynamic practices removes such pesticide use and, over time, increases soil life, which improves carbon capture. Actively farming to improve soil life, rather than merely eliminating harmful applications, does even more.

Farming alone will not stop climate change. Even so, Rodale Institute studies show that farming changes can (over time) increase the ability of soils to sequester carbon to the point of absorbing significant levels of CO_2, indicating that farming can play a positive role. In some cases, carbon sequestration in combination with overall mitigation practices can not only bring a farm to carbon neutral but even shift the site to carbon negative (absorbing more CO_2 than it produces).[2] Regenerative practices include maintained cover crops, which significantly reduce soil erosion, and thus also the production of dust in the atmosphere. This helps support the rain cycle, increases the ability of soils to absorb rainfall, and over time plays a part in reducing the potential for drought. Imagine multiple farms or an entire region implementing these practices together.

LAND BACK

The regenerative farming community recognizes that to some degree what they have developed is not entirely new. Instead, it depends on recapturing older farming techniques and a more holistic perspective on our engagement with the environment, then translating those into a modern approach. This includes technology. Others recognize how doing this capitalizes on the worldview, food production, and land management approaches native to innumerable Indigenous communities. In California and elsewhere, an increasing practice of returning land back to tribal nations has emerged. Indigenous communities being gifted or purchasing their ancestral homelands allows them to manage areas according to their own views of regenerating biodiversity, while also overseeing their own economy. Such practices have already provided a new perspective to government programs, environmental organizations and more.

In southern Oregon and northern California, a dam removal project along the Klamath River was completed in 2024 and marks the largest dam removal in US history so far. Chinook salmon returned to the river within mere weeks. Tribes along the Klamath are receiving land back within the region. They are working with other groups to restore and replant the dam removal area consisting of thousands of acres to native habitats. In Washington state, a similar project was completed and the habitat restored by the area's tribal people, who gathered hundreds of thousands of seeds from native plants before spreading the seeds on the newly exposed ground. It is now one of the most beautiful areas in the state.

Also in California, Governor Newsom spoke in 2024 about the increased need to return ancestral land to Indigenous groups. The Shasta nation has been striving in northern California to receive a little under 3,000 acres of ancestral homelands back to restore the area and rebuild the salmon population. Newsom announced his support. California is also working on building tribal partnerships with other communities to restore around a third of the state's coastal line.

The land back movement has also helped improve housing needs and given tribal nations the ability to determine their own economic well-being. Restoration goes beyond wildlands. It must include restoring and supporting well-being in the human community as well as its economy.

In the 1980s, the Yocha Dehe Wintun Nation was given land from their ancestral homeland in Yolo County, California. Today they manage around 22,000 acres (8,903 hectares) and operate the Séka Hills agricultural business, which is now one of the largest farming ventures in the county. Included, is a 9-acre (3.6-hectare) vineyard, producing 10 estate-grown wines for their winery, Séka Hills. The site is farmed organically with sections of the broader property farmed sustainably. It serves as a model for combining the practices of restorative farming in harmony with community needs and economic sustainability through wine. Such approaches can be an example beyond tribal communities as well.

CHANGING DEMOGRAPHICS

Economic downturns and changes in consumer spending are having an impact on the wine industry. How will the industry solve the problem? Will producers collaborate? Will they work with consumer interests to build something new? Will groups like the Wine Institute, regional associations, the government of California, or the US department of food and agriculture incentivize productive industry solutions? How is any of this done?

Understanding the buying habits of Americans is one place to start. So, what do we know about consumers today, in the mid-2020s? The Boomer generation's purchasing power is declining. As they retire or face health issues, they are buying less wine. But marketing approaches that worked for Boomers are less effective for younger generations. The older side of Gen-X has resources to buy wine, and does, but perhaps to lesser degree than their older counterparts. They are also often drinking beer, splitting spending across two categories of alcoholic beverages. The younger side of Gen-X faces economic challenges like those experienced by the Millennial generation. Both generations are more beverage curious. Wine is merely one option among many. Beer, cocktails, ready-to-drink beverages, sake, and to a smaller extent cider can all be considered on any given evening. Spending is dispersed even further across alcohol categories, decreasing the attention on wine.

Looking at wine specifically, challenges occur due to sheer container size. A standard 750-milliliter bottle is often too big for a single person to finish in a day, increasing the amount of waste, which increases the perception of cost. While dedicated wine lovers might use Coravin, or other wine-saving techniques, average consumers often say they're intimidated by the size of the wine bottle. Those new to wine are simply unclear how long it can last in bottle. So, flexible drinkers, willing to turn to other beverage types, drink elsewhere. Studies show people are more likely to buy and drink wine when married. Whether this is due to the increased stress of marriage or it is simply easier to finish a bottle with another person merits further study. Both are likely true. Millennials are less likely to marry than previous generations as well as being more curious across a range of beverages. This means their spending habits differ significantly from those of Boomers. It also means there is an opportunity in finding new packaging sizes for wine that are either easier to finish by a single person or mean less waste over time.

Millennials and Gen-Z are both more determined to make buying decisions based on company practices that foster positive environmental and social change than previous generations. Studies show Boomers are generally unwilling to spend more money because of such practices. Younger generations will. We are still learning what Gen-Z tends to buy. We do know they are the most ethnically and racially diverse generation so far in US history. Businesses reaching their communities depend on respecting that diversity and welcoming it. They support a greater range of gender expression and sexual orientation perspectives than any generation before. They are also more empathetic of neurodiversity. While Boomers used decisions like Ford trucks or Chevy, Republican or Democrat, McDonald's or Burger King to assert their identity, Gen-Z already has theirs. They want it respected and expect companies to help build a better world. Wineries are unlikely to gain the depth of loyalty from Gen-Z consumers like they did from Boomers, but a winery's social and environmental commitments do create return customers. Businesses that sponsor and appear at Pride events, for example, report increased return spending from attendees that found them through such events.

REAL WORLD EXPERIENCE

Regenerative viticulture understands the vine is never alone. It pushes into the soil to grow roots. The leaves of the vine canopy pull carbon dioxide from the atmosphere and push it through root hairs into the soil as liquid carbon. Microbes in the soil consume the liquid and deliver the nutrients the vine needs to survive. Cover crops help cool the soil, fix nitrogen, avoid erosion, and improve the ground's ability to absorb water. They also attract beneficial insects that can act as nature's own pesticide, protecting the plants from harmful bugs. Over time, cover crops left untilled adapt to the site itself. They combine with native plants to create their own holistic ecosystem, an aspect of the overall farm.

And regeneration depends on the farmer, the farmworkers, and the surrounding community. A single site farming organically fears chemical inputs from neighboring farms. But protective corridors, now mandated in many parts of the world, lessen the impact. The organic area is effectively expanded. That farmer becomes a teaching model for others nearby, demonstrating success can be possible. The farmworkers carry their knowledge of such techniques to other sites in the area. The sphere of influence for organic practices slowly expands.

It's a principle that can build resilience. A human community in harmony with its plant community increases its own elasticity through market changes, drought, heatwaves, flooding, and the erratic weather of climate change.

Valuing such diversity also spawns innovation. In the community of plants, ecological diversity creates unique interactions between species that can prevent disease as well as pests. Vineyards near conifer or redwood forests attract wild mushrooms and fungi when left untilled. Studies have shown that the fungi and vines foster connection. The vines provide food for the fungi, which in turn increase water availability to the vine.

In human communities, welcoming diverse perspectives invites a broader range of first-hand knowledge and experience. That expansion of perspective, when given a platform, helps create collective problem solving and can anticipate challenges before they arise. Diversity integrated into the decision level makes organizations and businesses more successful for exactly these reasons. It is easier to benefit or reach a community when the decision-makers of an organization have connection to the experiences and needs of that community. In wine, this means not only creating a product outsiders *think* the community would like, but understanding a community's eating habits, food and beverage preferences or limitations, how people celebrate, what they avoid, and more. Market analysis can help this but greater employee diversity in sales and product development gives more insight. This is also why the Regenerative Organic Certification (ROC) ran a pilot program testing the ROC standards across industry types. They needed a range of real world experiences, to find out if they work.

Innovators understand invention comes from understanding challenges honestly. The solution arises in response to a community's or the market's need. Charles Krug recognized a cider press could give better quality than other options available in California in the 1800s. He'd essentially created one of the state's first basket presses for wine. Late in the nineteenth century, H. W. Crabb planted his To Kalon vineyard in Napa Valley to an enormous range of varieties beyond those typically seen in the state. From the experiment he identified eventually that Cabernet Sauvignon should be one of the great grapes for the region. Ernest Gallo created new ways of selling wine by studying what was missing in the marketplace following Prohibition and providing it. Jess Jackson did something similar in the early 1980s when he collaborated with Jed Steele to create Kendall-Jackson Vintner's Reserve Chardonnay and a new tier in the wine market. ROC founders responded to the disruption of climate change by working to create real world approaches for businesses to join in climate action. Today, it is one of the most admired certification programs in the wine industry.

California has demonstrated again and again the determination of its people to fight for positive change, even as its history includes innumerable obstacles and harms. Civil rights, marriage equality, protections for LGBTQIA communities, disability accommodations, free speech, have all been fought for and won in California. California farmworkers made California the first state to pass farmworker protections, and helped the state pass DDT and organophosphate restrictions before the rest of the country did. Government incentives encourage reliance on renewable energy. The state's wineries have partnered with technology companies and created water reuse and management solutions, rainwater capture and filtering systems, renewably powered generators, electric vehicles including trucks and tractors. The wine industry and UC Davis are now advancing climate change solutions and insight in viticulture. The important difference in the struggles of history is not whether hardships and harms have happened – they consistently do almost anywhere – but how the region's communities work to foster positive change. California wine is in yet another era of reinvention. It has the lessons of history, the resources, and the know-how for its people to innovate and regenerate yet again.

APPENDIX I: WHAT IS AN AVA?

In the United States, the appellation of origin system for wine is overseen by the Alcohol and Tobacco Tax and Trade Bureau (TTB), a bureau within the United States Department of the Treasury. The TTB was created in 2003 in response to government changes made because of the September 11 attacks in 2001. Prior to the change, the Bureau of Alcohol, Tobacco and Firearms (ATF) oversaw the creation and management of wine's appellations of origin.

In the United States, legally recognized appellations are known as American Viticultural Areas, or AVAs. An AVA is specific to wine made from grapes. Other systems are used to denote place of origin for other food products. It is possible to label a wine without an AVA designation. However, if the wine label includes any of the following, a place of origin or AVA is required: vintage date, varietal designation, a claim of "estate bottling," or a brand name. These are clarified further by requirements in TTB regulations. The place of origin and these other requirements must appear on what is called the brand label of the wine. The brand label should appear on the side of the bottle and must include the wine's brand name, wine type, and origin, as well as alcohol content, volume, sulfite declaration, health warning label, and the name and address of the bottler of wine (or importer in the case of an imported wine).

There are two types of designation of origin for wine labels in the United States: a political specification and a viticultural place of origin. A legally recognized political boundary, such as a country, state, or county, can be named on a wine bottle as the wine's location of origin. A viticultural

AVA has gone through an extensive process to be legally recognized by the TTB for use on wine labels. For example, the California county of Contra Costa had no viticultural AVAs until 2024, but wines grown there could use Contra Costa as the designation on the label because a county is considered a legally recognized place of origin. The nearby Lodi AVA, on the other hand, has gone through the TTB process of being recognized as a viticultural AVA to be used on wine labels.

A viticultural AVA references a specific grape-growing region with clearly described boundaries. It has been recognized by the TTB as having distinctive climatic or geographical conditions that distinguish it from surrounding areas and affect the way the grapes are grown. Viticultural AVAs are recognized based on an extended approval process that includes assessment by the TTB as well as commentary from the public. Applications for the AVA must first fulfill specific requirements to be submitted to the review process.

The requirements for a viticultural AVA include three main elements:

1. A proposed name, and evidence that it is a known and used name in the area of the proposed AVA.
2. A description of the climatic and/or geographical conditions that distinguish the proposed AVA from its surroundings and affect how grapes in the proposed AVA grow. Accompanying this must be evidence that supports these claims.
3. A clear, point-by-point description of the boundaries for the proposed AVA, that correspond to the outline drawn on a United States Geological Survey (USGS) topographical map.

These same conditions are required to modify an already existing AVA.

Importantly, an AVA only denotes place of origin for the wine. Unlike appellation systems in some other countries, a US AVA does not include requirements or restrictions for grape type, farming methods, harvest times, or irrigation controls. It also does not include quality levels like a *cru* system would. The AVA is only used to designate where the grapes were grown. The United States government, and individual states within the United States, do not mandate or differentiate farming, variety, or quality conditions.

The addition of quality designations (or a *cru* system) or other types of farming, variety, or harvest conditions as a legally recognized system is unlikely in the United States. The government wishes to appear neutral in these assessments and so only regulates place of origin.

APPENDIX II: WINE LABEL REQUIREMENTS

Wines from the United States made to be sold commercially must include specific label requirements.

The AVA

The label must denote an appellation. This can be a legally recognized AVA that has gone through the TTB approval process. It can also be a legally recognized political boundary. These include the country, a specific state, a county. Multistate and multicounty appellations are also allowed under certain conditions. A multistate AVA is made of two to three states that are contiguous. A multicounty appellation includes two to three counties in the same state.

The AVA on the label denotes where the grapes for the wine are grown. For country, state, and county designations, at least 75 percent of the fruit must be grown within the relevant boundary. For multistate or multicounty designations, all the fruit must be grown in the states or counties indicated. For multicounty designations, the percentage of fruit grown in each county must also be stated. For example, the wine label might state: "55% Sonoma County 45% Napa County."

A wine denoting a TTB-approved viticultural AVA must be made with at least 85 percent of its grapes grown in that AVA. It also must be fully finished within the state that includes the AVA.

Alcohol content

The percentage of alcohol by volume must be listed as a numerical amount on the wine label. It can be on any label on the container. The percentage number must be written with either the word "alcohol" or the abbreviation "alc." either before or after the number. Alcohol labeling requirements fall into two main categories: wines 14 percent abv or higher; wines 14 percent abv or lower.

For wines that are 14 percent and higher, alcohol labeling must fall within an allowance of plus or minus 1 percent of that listed. For example, a wine can be labeled 15 percent if the actual alcohol content does not exceed 16 percent or fall below 14 percent.

For wines with less than 14 percent alcohol, the alcohol stated can fall between 1.5 percentage points more or less than that listed. For example, if a wine is labeled as 12.5 percent, the actual content can be no more than 14 percent and no less than 11 percent.

Additionally, alcohol can also be listed as a range of alcohol. Wines over 14 percent can name a 2 percentage point range on the label. For example, 14–16 percent is acceptable. Wines under 14 percent can list a range of up to 3 percentage points. For example, the wine might be listed as including 11–14 percent alcohol.

Additional requirements

The label must be clearly legible without need of extraordinary assistance. The text must appear on a contrasting background. That is, white text on a white background is not allowed, as the text must be readable under normal conditions. The legally required elements must be more noticeable than any merely descriptive information. That is, an AVA, for example, must be named separately rather than within a paragraph describing the wine. Text sizing requirements vary by container volume. The appellation text must be as clear as the text for the wine type.

Wine labels for wines sold across state boundaries or internationally must go through an approval process with the TTB known as a Certificate of Label Approval or COLA. Some wines instead qualify for a Certificate of Exemption from Label Approval. Either the COLA or Certificate of Exemption must be obtained before wine may be sold or shipped across state or international lines for commercial purposes.

All requirements are explained further on the TTB website.[1]

APPENDIX III: SUSTAINABILITY CERTIFICATIONS

Winery certifications connected to the question of sustainability have radically increased over recent decades. Certification programs center around different aspects of such goals. In farming, vineyards can be certified sustainable, organic, biodynamic, or regenerative organic. Specific environmental concerns can also be certified such as Salmon Safe, focused on restoring and maintaining fish-friendly habitats near vineyards. Broader efforts are also certified. The movement 1% for the Planet helps businesses donate to environmentally effective organizations. For business certification, there is B Corp. For building design and efficiency there are LEED and Living Building. There are also now certification programs for fighting climate change specifically, including IWCA and a Zero-Emission certification from the California Air Resources Board. Below is an accounting of certifications relevant in California since 2024.

Climate action certifications

Living Future includes third-party certification that ensures certified buildings have maximum energy efficiency, are powered with renewable resources, and are either combustion free or showing progress towards phasing out combustion. It relies on decarbonization rather than carbon offsets, and considers both operational carbon and stored, or embodied, carbon.

The California Air Resource Board (CARB) offers voluntary certifications for vehicles. CARB programs seek to reduce smog-causing pollutants and greenhouse gas emissions in transportation vehicles. The Zero-Emission Vehicle Program (ZEV) certifies passenger vehicles. Specific years and models of vehicles are recommended when they meet the program's clean vehicle standards. The Zero-Emission Powertrain Certification applies to medium- and heavy-load vehicles.

International Wineries for Climate Action (IWCA) uses science-based and international standards to create zero-emissions guidelines for wineries and wine-related businesses. It is based on GHG Protocol, which is the most commonly implemented GHG inventory standard in the world. IWCA also partners with the Race to Zero campaign to help persistently reduce wine industry GHG emissions. Member businesses can be certified at Gold and Silver standards, and also join as applicant members to access resources and work towards certification.

Buildings efficiency

LEED delivers a framework for achieving health, efficiency, and cost-saving green-building standards in relation to measures of sustainability for the environment, as well as social benefits. Guidelines address GHG emissions, the surrounding biodiversity, water efficiency and waste management, employee health and safety, and energy efficiency. These are made in relation to United Nations Sustainable Development Goals. LEED provides standards for new building designs as well as guidance in upgrading existing structures and offers four levels of certification. Its highest is Platinum, followed by Gold, Silver, and Certified.

Living Building offers guidelines, and an overall philosophy for achieving self-sufficient, regenerative building design that under-utilizes surrounding resources of the building site, and creates a healthy and positive internal environment for employees, guests and other building users. Building design includes elements of the natural environment, especially the use of natural lighting over electric, water reuse, renewable energies and energy efficiency. It certifies individual buildings and infrastructure projects, and offers insights on renovation. Certification includes examination of seven areas of design: interrelationship with nature, materials safe for all species, reliance on solar energy, water balance in the building's location, equitable practices, health and happiness of people in the physical space, and uplifting beauty of the space.

Broader environment

Salmon Safe is a peer-reviewed ecolabel utilized throughout the United States that certifies farms, vineyards, parks, urban centers, and other types of land management or development. The certification is practice-based, utilizing expert insights to create salmon-positive practices throughout the relevant habitat. Salmon have spawning needs that differ from those of other species and are directly impacted by the form of a creek or riverbed. The Salmon Safe program guides land owners on how to protect and restore waterways to elevate salmon habitats. Certification can be earned by public buildings, corporations, universities and schools, urban centers, public parks and natural spaces, golf courses, farms, and vineyards.

Fish Friendly Farming certifies farms, vineyards and other agricultural properties that have fulfilled requirements to foster a healthier environment for fish and other wildlife and improve the quality of naturally occurring water systems on and near the property. The certification reviews farm progress on a five-year cycle, seeking improvement over time. The California Land Stewardship Institute, a non-profit based in Napa Valley, administers the program. It includes certified properties in nine counties in the North Coast, the Sierra Foothills, and the northern portions of the Central Coast.

Broader framework

B Corp offers certification for a business's social and environmental practices and performance. It is run by a non-profit organization, B Lab. B Corp certified businesses operate in consideration of their impact on the surrounding community, their employees, the planet, and their own profitability. Certified businesses are required to operate with transparency, accountability, and continuous improvement.

1% for the Planet certifies businesses that donate 1 percent of their annual sales to environmental organizations vetted by the program. The organization helps businesses and corporations partner with non-profit groups working on environmental initiatives.

Farming frameworks: sustainable, organic, biodynamic, regenerative organic

Sustainable

Lodi Rules originated as a comprehensive pest-management program and evolved into a rigorous and multi-pronged third-party certification for vineyards and wine grapes. It offers guidelines for sustainable management of soil and water, relationship to the surrounding ecosystem, community, and employees, and pest management. Lodi Rules also offers the California Rules and Certified Green seals, which are based in the same framework but available for vineyards outside the Lodi region.

California Certified Sustainable offers certification and protocols meant to develop farming that is good for the planet, the community, the grapes, and the wine. The system includes certification for wineries, vineyards, and/or the resulting wine. The program was developed by growers, wineries, and sustainability experts through a partnership of the California Wine Institute and the California Association of Winegrape Growers. Certification is audited by a third party and standards continue to be updated through peer-review.

Napa Green certifies vineyards and/or wineries based around six pillars of sustainability: water efficiency, soil health, climate action, energy efficiency, waste management, and social justice. It approaches sustainability through a whole systems leadership model and provides toolkits for members to make improvements towards sustainability requirements. Certification standards are third-party certified.

Organic

California Certified Organic Farmers (CCOF) provides certification as well as support for reaching organic standards. It is accredited by the United States Department of Agriculture to certify farms, livestock management, restaurants, private labelers, and food processors. CCOF also provides grants and support to members, promotes consumer understanding of the importance of organic practices, and advocates for organic standards.

Biodynamic

Demeter Biodynamic Certification is the only organization to certify biodynamic farms and products in the United States. It is part of the global program Demeter International. Organic requirements must be met along with standards for soil fertility, farm-based disease solutions, pest management and disease control, water conservation, and biodiversity. Demeter also works to increase consumer awareness and understanding of the value of biodynamic principles.

Regenerative organic

Regenerative Organic Certification (ROC) offers agricultural guidelines for soil health, animal welfare, and fairness for farmers and workers. The goal of ROC is to regenerate soils and environmental health, respect animal welfare, and improve the lives of farmworkers and farmers. By focusing on soil health, guidelines help strengthen carbon sequestration in order to improve soil and plant nutrition, improve yields, and support the broader community. Regenerative protocols include multiple levels of soil and farm activity including no chemical fertilizer or pesticide use, use of compost and crop rotations, as well as animal inclusion for rotational grazing and biodiversity improvement, no till practices, and vegetative cover crops.

APPENDIX IV: FARMWORKER PROTECTIONS

The Agricultural Labor Relations Act was passed in the state of California in 1975. It gave farmworkers the right to bargain collectively with farm employers via a union. It also formed a state agency to oversee elections and investigate if farmworkers wanted union representation. It was the first set of protections for farmworkers established in the United States. The National Labor Regulations Act of 1935 protected the rights of workers to unionize and use collective bargaining but farmworkers were excluded from these protections.

In California, farm labor contractors (FLCs) must register with local, state, and federal governments. Local registration is generally through a county agricultural commissioner. At the state level, FLCs must pass a test demonstrating understanding of safety, labor, and pesticide regulations. They must also submit fingerprints and gain a license from the state. To discourage exploitation of farmworkers, growers are jointly liable for any violations committed by any unregistered FLCs they use for their land. They are not jointly liable for any violations made by registered FLCs.

Safety regulations

The Migrant and Seasonal Agricultural Worker Protection Act (MSPA) began nationally in 1983. It created employment-related protections for farmworkers, that non-exempt FLCs must meet. These include requirements to communicate the terms of employment and required protections; to make sure any housing or transportation meets state and federal requirements; to pay workers as promised. Records must be kept for at least three years.

The Occupational Safety and Health Act, first created in 1970 and amended repeatedly since, includes field sanitation standards employers must meet. Any farm with 11 or more agricultural employees must provide: toilet and handwashing facilities in proximity to the work site and drinking water at drinking temperature. The employer must notify employees of the location of these facilities.

The Environmental Protection Agency oversees the federal Agricultural Worker Protection Standard (AWPS), which regulates pesticides and what is considered safe handling. It was originally established in 1992 and has been revised since. The AWPS also restricts entering fields recently sprayed with pesticides and provides workers with protections if they file violation claims.

California reforms[2]

In California, a series of bills have refined protections for farmworkers within the state. In 2024, Governor Newsom signed two bills to improve housing availability, accessibility, affordability, and safety. California has also created minimum wage and overtime laws for farmworkers that exceed those in other parts of the country. It is only since 2019 that farmworkers have been entitled to overtime pay. Overtime laws have increased in incremental stages. As of 2025, employers must pay farm labor employees overtime after the first eight hours worked per day, as well as for any hours worked above the first 40 per week.

As of 2001, farmworkers must be at least 12 years of age. Hazardous working conditions are only allowed for those aged 16 and older. Those under 18 cannot handle pesticides.

Beginning in 2017, the California legislature passed a series of new bills and amendments designed to create stronger safeguards against harassment in the workplace, as well as protections for those who have already experienced it. All employees are also required to take anti-sexual harassment training, with refresher courses every two years.

Further protections for undocumented workers have also been added in California. Though undocumented workers are technically illegal in the United States, there is also currently no feasible plan to either replace them with documented workers or change the legal status of undocumented workers. To mitigate possible exploitation of undocumented workers in fear of deportation and without legal recourse, the state of California created new protections. In 2013, the state passed a law protecting workers against retaliation from employers for filing claims of unpaid wages regardless of employee legal status. Employers found guilty of not paying wages can lose their business license and/or face fines up to $10,000.[3] As of 2017, employers are prohibited from allowing Immigration and Customs Enforcement (ICE) into work areas without a warrant. ICE is also restricted from entering public buildings with the intention to detain workers while they are pursuing violation claims against employers.[4]

H2A[5]

The Immigration Reform and Control Act (IRCA) established an H2A visa program for farmworkers in 1986. H2B oversees non-agricultural workers. The original H2 program, formed in 1952, did not distinguish between agricultural and non-agricultural workers. The goal of both aspects of the H2 program is to serve the seasonal needs of US-based industries without increasing permanent immigration. It was also formed to oversee international farmworkers entering the country.

The H2A program is regulated by the US Department of Labor, the Department of State, and the Department of Homeland Security. It allows employers to sponsor workers from other countries for agricultural jobs in the United States for a temporary period of less than one year. Sponsorship is only allowed when an employer demonstrates there was inadequate labor available locally. Employers wishing to utilize the H2A program must submit a work order with the State Workforce Agency delineating the conditions of employment. That work order along with an application for certification is then sent to the Department of Labor's Employment and Training Administration (ETA). If accepted, the employer then goes through a training and certification process with the ETA. If approved, the ETA will then post relevant job orders on its registry to recruit possible H2A employees.

Workers admitted under the H2A program are not allowed to apply for visa changes. However, H2A visas can be extended if proven necessary, for no more than three years. If workers wish to return to work in the United States after three years, they must leave the country and reapply for a new H2A visa. For workers, protections under the H2A program include guaranteed employment for hours equivalent to at least 75 percent of what is specified in the contract, wage regulations dependent on location, and housing and transportation that is deemed safe and clean.

APPENDIX V: THE THREE-TIER SYSTEM

The three-tier system is a multi-level distribution regulation system unique to the United States. Its formation began in 1933 following the repeal of the Eighteenth Amendment.

The Eighteenth Amendment instituted federal restrictions on the production and sale of alcohol, making both illegal throughout the entire country. The one exception was that heads-of-households were allowed to make up to 200 gallons (757 liters, or 83 12-bottle cases) of home wine per year. The home wine exception led to a massive increase in planted vineyard area throughout the state of California to provide fruit wanted by home winemakers in the eastern United States.

After 13 years, Prohibition was repealed with the Twenty-first Amendment. The decision to repeal Prohibition came partially from the recognition that making alcohol illegal had not stopped people from drinking. Instead, hidden nightclubs and bars were an open secret. Alcohol was frequently sneaked across the border from Canada and between state lines.

The stronger impetus for repeal was motivated by advocates for states' rights to regulate their own laws against federal oversight. The Twenty-first Amendment was a big win for proponents of states' rights, since part of the agreement of repeal was that regulation of alcohol production and sales should become a state matter, overseen by local laws rather than national ones.

To manage state regulations, a new system for distribution was mandated. The Twenty-first Amendment created a three-tier system of distribution within the country to prevent any one alcohol-related business from becoming a monopoly. The new system prevented the possibility of a producer also serving as wholesaler or retailer. Those three aspects of wine sales were required to stay separate. No producer, for example, can work for or own a business that distributes, or sells wine to consumers.

The three tiers of the system consist of the following:

- Importers or producers. This is essentially the source of the wine, whether the business (or winery) that makes it, or the one that makes it available through import.
- Distributors or wholesalers. These businesses purchase wine from the importer or producer and put it into distribution to become available to retailers.
- Restaurants or retailers. These businesses purchase wine from wholesalers or distributors, and then are able to sell it to the consumer.

The three-tier system oversees all alcoholic beverages, not only wine.

Each state can manage its own laws and regulations to oversee the three-tier system. As a result, each state has its own set of regulations, making it necessary for producers to comply with a unique legal system for each state within which it wishes to sell wine. The federal government oversees the movement of alcohol between states or to other countries.

APPENDIX VI: 154 AVAS OF CALIFORNIA[6]

THE NORTH COAST AVAS

North Coast (1983): includes 6 counties: Mendocino, Lake, Sonoma, Marin, Napa, Solano
Main varieties: incredibly varied

Mendocino County AVAs

Mendocino (1984): reaches both the coastal and inland side of Mendocino
Main varieties: varied

Anderson Valley AVA (1983): the most well-known and planted AVA in Mendocino
Main varieties: Pinot Noir

Cole Ranch (1983): was the smallest AVA in the US until 2021
Main varieties: Riesling, Cabernet Sauvignon

Mendocino Ridge (1997): the only non-contiguous AVA in the United States
Main varieties: Pinot Noir, Zinfandel, Syrah

Yorkville Highlands (1998): connected to and warmer than Anderson Valley
Main varieties: Cabernet Sauvignon, Syrah

Comptche (2024): excluded from the North Coast appellation
Main varieties: Pinot Noir

McDowell Valley (1981): inland Mendocino
Main varieties: Zinfandel, Grenache, Syrah

Potter Valley (1983): unique cooler, elevation section of inland Mendocino
Main varieties: Sauvignon Blanc, Riesling, Chardonnay, Pinot Noir

Redwood Valley (1996): the area where Mendocino's first vineyards were planted
Main varieties: Zinfandel, Petite Sirah

Dos Rios (2005): all terraced slopes for vineyards
Main varieties: Cabernet Sauvignon

Covelo (2006): has four distinct seasons, but no rain in summer
Main varieties: Tempranillo, Grenache, Mencia

Eagle Peak of Mendocino County (2014): much cooler than nearby Redwood Valley
Main varieties: Pinot Noir, Chardonnay

Pine Mountain-Cloverdale Peak (2011): in both Mendocino and Sonoma Counties
Main varieties: Cabernet Sauvignon, Malbec, Sauvignon Blanc

Lake County AVAs

Guenoc Valley (1981): just north of Napa Valley and Mt. St. Helena
Main varieties: Cabernet Sauvignon, Cabernet Franc, Zinfandel, Tempranillo

Clear Lake (1984): completely encircles Clear Lake
Main varieties: Cabernet Sauvignon, Sauvignon Blanc

Benmore Valley (1991): outside the Clear Lake AVA very close to Mendocino
Main varieties: not currently planted; previously Chardonnay

Red Hills (2004): best-known AVA in the region
Main varieties: Cabernet Sauvignon, Cabernet Franc

High Valley (2005): surrounded by mountains and just east of Clear Lake

Main varieties: Cabernet Sauvignon, Syrah, Zinfandel

Big Valley District–Lake County (2013): directly across from High Valley on the west side of Clear Lake
Main varieties: Sauvignon Blanc, Chardonnay, Viognier

Kelsey Bench (2013): directly south of Big Valley with similar climate
Main varieties: Sauvignon Blanc, Chardonnay, Viognier

Upper Lake Valley (2022): on the north side of Clear Lake
Main varieties: Sauvignon Blanc, Cabernet Sauvignon

Long Valley–Lake County (2023): expanded the reach of the North Coast AVA
Main varieties: Cabernet Sauvignon, Cabernet Franc, Petite Sirah

Sonoma County AVAs

Sonoma Valley (1981): the birthplace of fine wine in California
Main varieties: Cabernet Sauvignon, Chardonnay, Zinfandel

Dry Creek Valley (1983): a historic region for old-vine Zinfandel
Main varieties: Zinfandel, Cabernet Sauvignon, Sauvignon Blanc

Knights Valley (1983): hidden in the northeastern corner of Sonoma, west of Napa Valley
Main varieties: Cabernet Sauvignon, Syrah

Chalk Hill (1983): one of the historic homes of Chardonnay
Main varieties: Chardonnay, Cabernet Sauvignon

Russian River Valley (1983): the first area to demonstrate Sonoma's potential for Pinot
Main varieties: Chardonnay, Pinot Noir, Zinfandel

Green Valley of Russian River Valley (1983): the coldest section of the Russian River Valley
Main varieties: Chardonnay, Pinot Noir

Alexander Valley (1984): one of the warmest parts of Sonoma County
Main varieties: Cabernet Sauvignon, Chardonnay, Zinfandel

Sonoma Mountain (1985): the birthplace of organic mountain viticulture, and rise of Cabernet
Main varieties: Cabernet Sauvignon, Chardonnay, Pinot Noir

Sonoma Coast (1987): an enormous AVA that covers the expanse of ocean fog influence
Main varieties: Chardonnay, Pinot Noir

Northern Sonoma (1990): does not include Carneros, Petaluma Gap, or Sonoma Valley
Main varieties: Cabernet Sauvignon

Rockpile (2002): at the northside of Dry Creek Valley and more temperate
Main varieties: Zinfandel, Cabernet Sauvignon

Bennett Valley (2003): hidden at the north side of Sonoma Mountain
Main varieties: Chardonnay, Syrah, Merlot

Fort Ross-Seaview (2012): in the highest part of the West Sonoma Coast
Main varieties: Pinot Noir, Chardonnay

Moon Mountain District of Sonoma County (2013): includes the oldest Pinot Noir and Chardonnay vines in the state
Main varieties: Cabernet Sauvignon, Zinfandel, Pinot Noir, Chardonnay

Fountaingrove District (2015): on the Sonoma side of the Mayacamas mountains
Main varieties: Cabernet Sauvignon, Merlot, Syrah

West Sonoma Coast (2022): encompasses the Coastal Mountain Range of Sonoma
Main varieties: Pinot Noir, Chardonnay, Syrah

Marin County AVAs

Petaluma Gap (2017): defined by wind, straddles Sonoma and Marin Counties
Main varieties: Pinot Noir, Chardonnay, Syrah

Napa County AVAs

Carneros (1983): in both Sonoma and Napa Counties
Main varieties: Chardonnay, Pinot Noir, Merlot, Syrah

Napa Valley (1981): the first AVA in California, and its most famous
Main varieties: Cabernet Sauvignon, Chardonnay, Zinfandel

Howell Mountain (1983): the place the first Cabernet was planted in Napa Valley
Main varieties: Cabernet Sauvignon, Merlot, Zinfandel, Chardonnay, Viognier

Stags Leap District (1989): the origin of the award-winning red in the 1976 Paris Tasting
Main varieties: Cabernet Sauvignon, Merlot, Sauvignon Blanc

Mount Veeder (1990): the foothills of the Mayacamas
Main varieties: Cabernet Sauvignon, Chardonnay, Merlot, Zinfandel

Atlas Peak (1992): rugged slopes on the western side of Napa Valley
Main varieties: Cabernet Sauvignon

Oakville (1993): home to some of the most renowned Cabernet in Napa Valley
Main varieties: Cabernet Sauvignon, Merlot, Sauvignon Blanc, Cabernet Franc

Rutherford (1993): made famous by André Tchelistcheff's love for its soils
Main varieties: Cabernet Sauvignon, Sauvignon Blanc, Merlot, Cabernet Franc

Spring Mountain District (1993): the location of the first white-wine devoted winery in Napa Valley
Main varieties: Cabernet Sauvignon, Merlot, Cabernet Franc, Chardonnay, Zinfandel

St Helena (1995): the northern stretch of the famed Mayacamas bench
Main varieties: Cabernet Sauvignon, Cabernet Franc, Merlot, Syrah, Zinfandel, Sauvignon Blanc

Chiles Valley (1999): tucked into hills east of the Vaca Mountains, near Lake Berryessa
Main varieties: Cabernet Sauvignon, Merlot, Cabernet Franc, Zinfandel

Yountville (1999): the area where the first *vinifera* was planted in Napa Valley
Main varieties: Cabernet Sauvignon, Merlot

Diamond Mountain District (2001): a rugged mountain district above Calistoga

Main varieties: Cabernet Sauvignon, Cabernet Franc

Oak Knoll District (2004): in a cooler section in the south surrounding the town of Napa
Main varieties: Cabernet Sauvignon, Chardonnay, Merlot, Sauvignon Blanc, Riesling

Calistoga (2010): the highest peak daytime temperatures in Napa Valley, with cold nights
Main varieties: Cabernet Sauvignon, Zinfandel, Syrah, Petite Sirah

Coombsville (2011): volcanic soils inundated with cooling winds off the San Pablo Bay
Main varieties: Cabernet Sauvignon, Chardonnay, Merlot, Syrah, Pinot Noir

Crystal Springs (2024): in the foothills of Howell Mountain
Main varieties: Cabernet Sauvignon

Wild Horse Valley (1988): continues into Solano County
Main varieties: Pinot Noir, Chardonnay

Solano County AVAs

Suisun Valley (1982): directly east of Solano County Green Valley AVA
Main varieties: Petite Sirah, a range of small varietal plantings

Solano County Green Valley (1982): abuts the Wild Horse Valley AVA
Main varieties: Syrah, Pinot Noir, Chardonnay, Pinot Gris

SANTA CRUZ MOUNTAINS AVAS

Santa Cruz Mountains (1981): first AVA in the country to be defined by elevation
Main varieties: Cabernet Sauvignon, Pinot Noir, Chardonnay

Ben Lomond (1988): nested within the Santa Cruz Mountains on its western side
Main varieties: Pinot Noir, Zinfandel, Cabernet Sauvignon, Chardonnay

CENTRAL COAST AVAS

Central Coast (1985): includes at least 40 nested AVAs stretching along the coast of California
Main varieties: varied

San Francisco Bay (1999): overlaps at least part of eight counties
Main varieties: varied

Contra Costa County AVAs

Contra Costa (2024): high quality old vine field blends threatened by urban expansion
Main varieties: Zinfandel, Carignane, Mourvèdre

Lamorinda (2016): full of various planting conditions and grape types
Main varieties: varied

Alameda County AVAs

Livermore Valley (1982): one of the founding regions of fine white wine in the state
Main varieties: Cabernet Sauvignon, Petite Sirah, Grenache

Santa Clara County AVAs

Santa Clara Valley (1989): tucked between Santa Cruz Mountains in the west and the Diablo range to the east
Main varieties: Cabernet Sauvignon, Chardonnay, Zinfandel

San Ysidro District (1990): set along the Pajaro River, separates Santa Cruz and Monterey Counties
Main varieties: Chardonnay, Pinot Noir, Merlot

Pacheco Pass (1984): overlapping both the Santa Clara and San Benito Counties
Main varieties: Chardonnay, Merlot, Zinfandel

San Benito County AVAs

San Benito (1987): includes several other nested AVAs
Main varieties: Chardonnay, Cabernet Sauvignon

Chalone (1982): in both the San Benito and Monterey County
Main varieties: Pinot Noir, Chardonnay

Cienega Valley (1982): on the western side of the Gabilan Range
Main varieties: Riesling, Pinot Noir, Chardonnay

Lime Kiln Valley (1982): within the southern part of the Cienega Valley AVA
Main varieties: Mourvèdre, Zinfandel

Mount Harlan (1990): Pinot Noir grown in limestone at high elevation
Main varieties: Pinot Noir, Chardonnay, Aligoté

Pacines (1982): previously for bulk wine, today home to regenerative farming
Main varieties: Chardonnay, Cabernet Sauvignon

Gabilan Mountains (2022): straddles San Benito and Monterey Counties
Main varieties: Pinot Noir, Chardonnay, Syrah, Grenache

Monterey County AVAs

Monterey (1984): runs along the Salinas River, includes several nested AVAs
Main varieties: Chardonnay, Pinot Noir

Carmel Valley (1982): sheltered from the Pacific Ocean, tucked within the Coastal Mountains
Main varieties: Cabernet Sauvignon, Merlot

Santa Lucia Highlands (1992): impressive winds, and relative aridity produce Pinot with power
Main varieties: Pinot Noir, Chardonnay, Syrah

Arroyo Seco (1983): 'dry creek' – named due to the rain shadow produced by mountains
Main varieties: Chardonnay, Riesling, Zinfandel, Merlot

San Antonio Valley (2007): impressively complex soils and warm summer temperatures
Main varieties: hearty reds

San Bernabe (2004): one of the biggest vineyards in the world, a monopole AVA
Main varieties: varied

San Lucas (1987): borders San Bernabe and crosses the Salinas River
Main varieties: varied

Hames Valley (1994): nested within the southern stretch of the Monterey AVA
Main varieties: hearty reds

San Luis Obispo County AVAs

San Luis Obispo Coast, or SLO Coast (2022): includes the coolest, coastal parts of San Luis Obispo
Main varieties: Pinot Noir, Chardonnay, Syrah

Arroyo Grande (1990): a cool section of San Luis Obispo close to the Pacific
Main varieties: Chardonnay, Pinot Noir, Syrah, Grenache

Edna Valley (1982): first vines planted were Chardonnay, close to the Pacific
Main varieties: Chardonnay, Pinot Noir, Syrah

York Mountain (1983): adjacent to, but outside of, the Paso Robles AVA
Main varieties: Grenache, Syrah, Cabernet Sauvignon

Paso Robles (1983): incredibly varied range of microclimates
Main varieties: Cabernet Sauvignon, Syrah, Merlot, Zinfandel

Adelaida District (2014): Paso Robles' westernmost AVA
Main varieties: Bordeaux and Rhône reds

Creston District (2014): east of the Templeton District
Main varieties: Bordeaux reds

El Pomar District (2014): a comparatively cool area for Bordeaux reds
Main varieties: Bordeaux reds

Paso Robles Estrella District (2014): the location of some of the first important Syrah vineyards in the state
Main varieties: Cabernet Sauvignon, Petit Verdot, Merlot, Syrah

Paso Robles Geneseo District (2014): established by German immigrants in the 1880s
Main varieties: Cabernet Sauvignon

Paso Robles Highlands (2014): a higher elevation, and warmer portion in the eastern side of Paso
Main varieties: varied

Paso Robles Willow Creek District (2014): home to some of the most coveted Rhône-style wines in California
Main varieties: Rhône reds, Zinfandel

San Juan Creek (2014): a series of old terraces full of alluvial soils
Main varieties: Cabernet Sauvignon, Zinfandel, Petite Sirah

San Miguel District (2014): named for a Spanish mission in the area
Main varieties: Zinfandel, Sangiovese, Aglianico

Santa Margarita District (2014): the southernmost part of Paso, originally planted by Mondavi
Main varieties: Cabernet Sauvignon, Sauvignon Blanc, Chardonnay

Templeton District (2014): influenced by the winds pushing through the Templeton Gap
Main varieties: Zinfandel, Rhône reds, Mataro

Santa Barbara County AVAs

Santa Maria Valley (1981): in both San Luis Obispo and Santa Barbara Counties, exposed to the Pacific
Main varieties: Chardonnay, Pinot Noir, Grenache, Syrah

Alisos Canyon (2020): cooled by the San Antonio Creek basin
Main varieties: Rhône reds, Tempranillo, Albariño

Santa Ynez Valley (1983): a windy valley on the southern end of Santa Barbara County
Main varieties: varied

Ballard Canyon (2013): limestone soils planted with Rhône varieties
Main varieties: Rhône varieties

Los Olivos District (2016): tucked between Ballard Canyon and Happy Canyon in the Santa Ynez Valley
Main varieties: Sauvignon Blanc, Cabernet Franc, Cabernet Sauvignon

Sta. Rita Hills (2001): one of the state's best-known Pinot Noir regions
Main varieties: Pinot Noir, Chardonnay, Syrah

Happy Canyon of Santa Barbara (2009): the warmer inland side of the county
Main varieties: Cabernet Sauvignon, Sauvignon Blanc

CENTRAL VALLEY AVAS

Yolo County AVAs

Clarksburg (1984): overlaps both Sacramento County and Yolo County
Main varieties: Chenin Blanc, Chardonnay, Petite Sirah

Merritt Island (1983): surrounded by the Elk and Sutter Slough, and Sacramento River
Main varieties: Chardonnay, Chenin Blanc

Capay Valley (2002): especially planted to Iberian varieties
Main varieties: Tempranillo, Cabernet Sauvignon, Petit Verdot

Dunnigan Hills (1983): less prone to frost than other parts of the Sacramento Valley
Main varieties: Cabernet Sauvignon, Merlot, Tempranillo

Winters Highland (2023): overlaps both Solano and Yolo Counties
Main varieties: Petite Sirah, Tempranillo, Zinfandel

Sacramento County AVAs

Lodi (1986): continues into Sacramento and San Joaquin Counties
Main varieties: Cabernet Sauvignon, Chardonnay, Zinfandel

Sloughhouse (2006): the highest elevation and most gravelly area of Lodi
Main varieties: varied

Borden Ranch (2006): volcanic hills and prairie mounds with some alluvial influence
Main varieties: varied

Alta Mesa (2006): a mesa of sandy and gravelly clay loam
Main varieties: varied

Cosumnes River (2006): wetlands and floodplains in Sacramento and San Joaquin Counties
Main varieties: Chardonnay, Sauvignon Blanc, Pinot Gris, Vermentino

Jahant (2006): rolling woodlands with sandy, alluvial soils, Sacramento and San Joaquin Counties
Main varieties: varied

San Joaquin County AVAs

Clements Hills (2006): rolling alluvial hills with rocky clay loam
Main varieties: varied

Mokelumne River (2006): the heart of Lodi, its original population and vineyard center
Main varieties: Cabernet Sauvignon, Chardonnay, Zinfandel

River Junction (2001): a shallow bowl surrounded by rivers
Main varieties: Chardonnay, Cabernet Sauvignon

Tracy Hills (2006): drier and less foggy than surrounding areas
Main varieties: Cabernet Sauvignon, Chardonnay, Montepulciano

Madera County AVAs

Madera (1984): overlaps both Madera and Fresno Counties
Main varieties: French Colombard, Chardonnay, Cabernet Sauvignon

Fresno County AVAs

Squaw Valley-Miramonte (2015): in the Sierra Foothills but south of the Sierra Foothills AVA
Main varieties: varied

Stanislaus County AVAs

Salado Creek (2004): currently no wines are designated with this appellation
Main varieties: Cabernet Sauvignon, Syrah, Sauvignon Blanc

Paulsell Valley (2002): soils are mainly volcanic tuff, sometimes with cobbles
Main varieties: Cabernet Sauvignon, Petite Sirah, Petit Verdot

Diablo Grande (1998): established as a one-winery AVA; the original winery no longer exists
Main varieties: Cabernet Sauvignon, Chardonnay

Kern County AVAs

Tehachapi Mountains (2020): the very southern end of San Joaquin Valley
Main varieties: Zinfandel, Viognier, Tempranillo, Zinfandel

SIERRA FOOTHILLS AVAS

Sierra Foothills (1987): a large AVA crossing eight counties
Main varieties: Zinfandel, Syrah, Cabernet Sauvignon, Barbera

Yuba County AVAs

North Yuba (1985): northmost area in the Sierra Foothills
Main varieties: Cabernet Sauvignon, Syrah, Grenache, Roussanne

El Dorado County AVAs

El Dorado (1983): the northern end of the gold vein found in the mid-1800s
Main varieties: Zinfandel, Cabernet Sauvignon, Syrah

Fair Play (2001): higher-elevation areas fully within El Dorado County
Main varieties: Syrah, Grenache, Mourvèdre, Tempranillo, Barbera

Amador County AVAs

California Shenandoah Valley (1983): continues into both El Dorado and Amador Counties
Main varieties: Zinfandel, Cabernet Sauvignon, Barbera, Petite Sirah

Amador County (1983): contains some of the oldest vineyards in the country
Main varieties: Zinfandel, Barbera, Petite Sirah

Fiddletown (1983): slightly lighter bodied wines than neighboring Shenandoah Valley
Main varieties: Zinfandel, Barbera, Petite Sirah

THE FAR NORTH

Humboldt County AVAs

Willow Creek (1983): overlaps Humboldt and Trinity Counties
Main varieties: Syrah, Cabernet Sauvignon

Trinity County AVAs

Trinity Lakes (2005): remote, mountainous and surrounded by lakes
Main varieties: Syrah, Cabernet Sauvignon

Shasta County AVAs

Manton Valley (2014): overlaps Tehama and Shasta Counties
Main varieties: Syrah, Cabernet Sauvignon

Inwood Valley (2012): in the foothills of Mounts Lassen and Shasta
Main varieties: Syrah, Cabernet Sauvignon

Siskiyou County AVAs

Seiad Valley (1994): a few miles below the Oregon border
Main varieties: Syrah, Cabernet Sauvignon

SOUTHERN CALIFORNIA

South Coast (1985): encompasses five counties in Southern California
Main varieties: A great diversity of varietal plantings

Los Angeles County AVAs

Malibu Coast (2014): overlaps Ventura and Los Angeles Counties
Major varieties: Cabernet Sauvignon, Syrah

Saddle Rock Malibu (2006): along the coast of the Pacific Ocean
Main varieties: Cabernet Sauvignon, Syrah, Sauvignon Blanc

Malibu-Newton Canyon (1987): in the high elevations of the Malibu mountains
Main varieties: Cabernet Sauvignon, Cabernet Franc, Petit Verdot

Palos Verdes Peninsula (2021): hugs the coastline west of Los Angeles
Main varieties: Pinot Noir, Chardonnay, Cabernet Sauvignon, Merlot

Leona Valley (2008): sits within the Sierra Pelona Valley AVA
Main varieties: Chardonnay, Syrah

Sierra Pelona Valley (2010): in the high desert region near Antelope Valley and the Mojave
Main varieties: Zinfandel, Syrah, Tempranillo, Muscat

Kern County AVAs

Antelope Valley of the California High Desert (2010): overlaps Kern and Los Angeles Counties
Main varieties: Tempranillo, Zinfandel, Symphony

San Bernardino County AVAs

Cucamonga Valley (1985): one of the first commercial wine regions of California; overlaps Riverside and San Bernardino Counties
Main varieties: Zinfandel

Temecula Valley (1984): originally named Temecula, renamed 2004
Main varieties: Cabernet Sauvignon, Syrah, Petite Sirah

Yucaipa Valley (2024): high elevation vineyards in the San Bernardino mountains
Main varieties: Cabernet Sauvignon, Merlot, Zinfandel, Syrah

San Diego County AVAs

San Pasqual Valley (1981): showcases well-drained, granitic soils in northern San Diego County
Main varieties: Cabernet Sauvignon, Syrah, Sangiovese

Ramona Valley (2006): a significant diurnal shift helps retain natural acidity
Main varieties: Cabernet Sauvignon, Viognier, Muscat

San Luis Rey (2024): from coastal town Oceanside to the Merriman Mountains
Main varieties: Cabernet Sauvignon, Merlot, Cabernet Franc

GLOSSARY

Alta California. During the occupation of Spain in the western United States, Alta California was part of the larger province of Las Californias. Las Californias included both Alta and Baja California and what today are regarded as the states of California, Nevada, and Utah, as well as portions of Arizona, Wyoming, and Colorado. In the early 1800s, the area became its own province within New Spain. It was named Alta California by Mexico after it gained independence from Spain, and then became part of the United States under the Treaty of Guadalupe Hidalgo. In 1850, the state of California was circumscribed from the larger Alta California.

Angelica. A historically important wine established by the Spanish missions in what is today California. Juice or partially fermented wine from the Mission grape was fortified with brandy or a clear spirit to make a fortified and sweet beverage. A few producers in the state have recently begun to make Angelica again.

Anthocyanin. Antioxidants that protect fruits from ultraviolet radiation are found in the skins of grapes. These are called anthocyanins. In grapes, anthocyanins also provide color to the wine. When the skin of a grape is crushed with its pulp to form juice, the anthocyanins in the grape skin begin to bond with its tannins. They continue to bond during fermentation and aging. This helps retain color in red wines, and changes color as the wine ages.

Appellation. A designated geographical area where fruit is grown. In the United States, for wine, appellations are legally recognized growing regions known as American Viticultural Areas (AVAs). AVAs must be approved by the Alcohol and Tobacco Tax and Trade Bureau (TTB), an aspect of the United States government.

AXR1. Hybrid developed in 1879 and believed to be phylloxera resistant. The vine species *Vitis rupestris*, originating from North America, and the European *Vitis vinifera* were crossed to create AXR1 for use as a rootstock on which to graft other varieties. It was widely planted in California from the 1950s to the 1980s, when it was discovered that it was vulnerable to phylloxera after all. AXR1's vulnerability led to massive replanting in California, especially in Napa Valley in the 1990s.

Back-to-the-land. Significant mid-twentieth century US movement built on the ideals of fostering self-sufficiency while lessening reliance on outside corporate or governmental systems. It generally included growing one's own food, making the living structures, and other goods needed to live off one's own land. Trading between neighbors was also practiced. In the 1960s and 1970s, the back-to-the-land movement helped encourage the rise of organic farming. It also included the most significant period of communes forming through portions of the United States, including California.

Bag-in-box. Colloquial name for a style of packaging where instead of being put into glass bottle for transportation and sale, wine is put into a bag with a sealed tap on one end. The bag is then placed in a cardboard box with the tap accessible through an opening in the box. It is less temperature resilient than glass bottles but as there is no oxygen inside the bag, and no oxygen can enter through the spout, the wine in bag-in-box can stay fresh after opening for months longer than a bottle of wine. As more wine fits into less space and the overall packaging is lighter, bag-in-box has significantly lower carbon emissions compared to glass.

Biodynamics. Organic farming approach that views the entire farm as a living organism, or holistic system. In this way, it is not only the crop that is farmed, but the crop within its overall, surrounding environment. This perspective guides farming decisions. Biodynamics does not allow chemical fertilizers or pesticides and centers health of the soil as a priority.

BIPOC. Acronym that stands for **B**lack, **I**ndigenous, and **p**eople **o**f **c**olor. The term brings focus to the experiences of racial and ethnic groups experiencing prejudice and general resource disadvantage in relation to a mainstream, white society. The term emphasizes solidarity between communities of color, while recognizing there are differences in their experience. Black and Indigenous peoples are specifically named in recognition of their histories of enslavement and genocide.

Bonded winery. A commercially recognized facility that has been registered with and permitted by the TTB. It also pays federal and state taxes on its wine production.

Bordeaux varieties. Varieties originating from the region of Bordeaux in France. When used in winemaking outside of France the wines are described as a Bordeaux-like blend. Such blends can include Cabernet Sauvignon, Merlot, Malbec, Petit Verdot, and Cabernet Franc.

Bracero Program. From 1942 to 1964 the United States and Mexico had an agreement that brought citizens of Mexico into the US for short-term work contracts. The Bracero program was developed to fill labor gaps caused by World War II and was considered Mexico's contribution to the war effort since Mexico did not have soldiers to send to fight in the war. Originally intended to last only during the US efforts in the war, it was extended several times until 1964 after it was realized that white, US citizens returning from the war were able to access better paying jobs outside of agriculture and so another source of laborers was needed.

Bulk wine. Transported in containers larger than 2 liters rather than individual bottles or other types of smaller packages. It is then blended and bottled at its new location. This kind of transportation lowers carbon emissions in comparison to shipping individual bottles the same distance.

Burgundy varieties. Varieties originating from the region of Burgundy in France. When they are grown outside of France, the wines they make are generally said to be Burgundy-like in style, or made with Burgundy varieties. Varieties include Pinot Noir, Chardonnay, and Aligoté.

Carbon footprint. The total amount of carbon dioxide, a greenhouse gas, released into the atmosphere by an individual, business, region, or country. Because an increase of carbon dioxide speeds the process of climate change, the bigger a climate footprint, the larger the environmental impact.

Carbon neutral. An individual, business, region, or country that is carbon neutral sequesters as much carbon as it releases so that the total effect is the equivalent of no carbon emissions.

Carbonic maceration. *See* Whole cluster fermentation.

Certificate of label approval (aka. COLA). Shows that the label on the wine made for commercial sale has been approved by the US government via the TTB. There are also, in far fewer cases, Certificates of Exemption. These are given to wines that are made for special circumstances outside the typical commercial sales route and thereby do not need a COLA.

Chemical farming. Agriculture that relies on synthetic pesticides, fertilizers, and other additions to increase crop yields. It arose significantly following World War II, relying on chemicals developed for the war effort. It is now known that such chemicals cause significant environmental harms such as species die off in plants, animals, insects, and microbes, as well as causing significant human health issues, most especially cancer. Chemical farming continues to be the most common type of farming practiced in the world. It is also known as conventional farming.

Chromatography. Analyzing different chemical components of a wine by separating them from each other. There are multiple types of chromatographers. Paper chromatography helps analyze acid types in wine and is most commonly used to track the progress of converting malic acid to lactic acid in a secondary fermentation known as malolactic fermentation (see below).

Climate action. Any initiative, policy, regulation, or measure that aims to reduce greenhouse gas emissions. Climate action is a process by which greenhouse gas emissions are purposefully reduced through multiple means, most especially moving to renewable forms of energy, mitigating emissions by changing materials, products, or practices, and building structures or doing agriculture in a way that is better adapted to the growing conditions of the place.

Climate change. A long-term shift in weather patterns and temperatures. Climate change can occur naturally due to changes in the earth environment. Large volcanic eruptions, for example, have filled the atmosphere with smoke and ash that obscures the sun leading to lower temperatures and less sun exposure, sometimes over years. Multiple historical famines have occurred in this way. The earth's rotation also changes distance from the sun over long periods of time, changing temperatures and light exposure on the planet. Since the Industrial Revolution of the eighteenth century, climate change has been occurring through human activity. The increase in the greenhouse gases nitrous oxide, methane, and carbon dioxide through human activity has increased temperatures significantly and changed weather patterns worldwide.

Commune. A group of people, sometimes a collection of families, sometimes a collection of individuals or a mix of both, living together, sharing tasks and responsibilities as well as property including land and most other possessions. Communes rose in prominence in California in the 1960s and 1970s. Few existed after that period. Communes continue to be common in other parts of the world such as parts of Europe, Israel, and British Columbia.

Compost. Decayed organic matter such as food waste, autumn leaves, and cuttings from vines collected and mixed with other materials such as livestock feces, herbs, or other plants and allowed to decompose outside. Decomposition of these ingredients is a natural process that creates fertilizer that helps increase soil life and plant health without using chemicals, and without the risk of excess nitrogen emissions caused by chemical fertilizer. Healthy compost includes living earthworms, as well as a range of other insects or bugs, small flora and fauna, and a host of microbes.

Conscientious objector. Person who refuses to serve in the military or use weapons against others based on moral and/or religious beliefs. Conscientious objectors used to face jail time for essentially breaking the nation's laws of military service. The United States now offers what is called a Selective Service Alternative Service Program. The selective service is the organization through which citizens become part of the military. The alternative service program allows individuals to technically be part of the military but to serve their time outside of war or fighting to instead do community service or fulfill other essential needs of the government.

Continental climate. Climate that includes four distinct seasons; dry compared to other climate types. Hot summers and cold winters transition through spring and autumn seasons. Continental climates occur in the center of landmasses usually cordoned off completely from significant bodies of water.

Cover crop. Any plant or mix of plants grown to cover the soil. Cover crops can be planted to serve several different reasons. Cover crops can act as natural soil amendments, helping to increase certain nutrients, repulse various pests, improve water permeability and aid carbon storage. Sometimes cover crops are tilled into the soil at certain parts of the growing season. This is not required, though may be desirable with certain types of cover crops as they fix or draw out different nutrients at different parts of their growing cycle. Sometimes cover crops are tilled into the soil to support vines that may be suffering from reduced water supply. This is controversial as a long-term practice as exposed soil increases in temperature, thus losing microbial life, and so also soil structure over time. Year round or permanent cover crops reduce erosion and soil loss, can improve soil structure as well as moisture permeability, and keep the soil cooler than it would be otherwise. Cover crops are now also being used on rooftops for buildings to help lower the temperature and thus reduce energy needs.

COVID-19 pandemic. At the end of 2019 a unique coronavirus was discovered that quickly led to a global outbreak of an infectious acute respiratory illness. The coronavirus strain that led to the pandemic was named COVID-19. The name comes from **co**rona **vi**rus **d**isease with 19 indicating it was discovered in the year 2019. COVID-19 is still present globally and remains a threat to people with compromised immune systems or other pre-existing conditions. Because of the decline in the number of deaths and hospitalizations caused by COVID-19 since 2022, it is no longer considered to be a pandemic.

DDT. The chemical dichlorodiphenyltrichloroethane, more commonly referred to as DDT, was first synthesized in 1874 as a research experiment. In the late 1930s its insecticidal properties were discovered, and it was developed to fight malaria, dysentery, and other insect-borne illnesses. It was used extensively during World War II and in the Vietnam War to protect soldiers from insects and was used domestically to prevent or lower insect infestations throughout the country. In the 1950s, environmentalists started noticing the decline of various species after the application of DDT to plants. In the late 1950s, two residents of Long Island, Marjorie Spock and Mary Richards, instigated a civil lawsuit against the United States government over the use of DDT in their area. Though they did not succeed in the lawsuit, it is considered the first case of citizens filing suit against the government and set precedent that helped create environmental law. DDT was regulated in California in 1972 thanks to extensive work from farmworkers lobbying the state government. Soon after, it was regulated for the first time by the national government as well.

Direct-to-consumer (DTC). Wineries in many states of the US are allowed to sell directly to consumers wanting to buy their wine. Regulations used to make this practice difficult or impossible as the three-tier system previously demanded wine sales had to happen through wholesalers then retailers or restaurants. Some states still enforce this practice. California wineries benefit significantly from DTC sales and wineries in the state of California successfully sued the government to change their three-tier policy and make it legal to sell DTC across state lines as well. Some states do not allow DTC sales from other states.

Diversity, equity, and inclusion (aka. DEI). Initiatives that strive to welcome people from diverse backgrounds and ensure they are given similar access to resources, including pay, and/or that educate people on how lack of equity can create exclusion of people from differing backgrounds. More recently, DEI programs have added the idea that for people to feel *welcome* they must be treated as if they *belong* in a group. Merely including them depends on inviting them to participate but might not provide the conditions to mean they belong there and have similar access to a platform for decision making and advancement within a company, or organization.

Dry wine. A wine that includes residual sugar (aka. RS) from the juice of grapes that have not fully finished fermentation is considered a sweet wine. A wine that does not have RS is called a dry wine.

Erosion. The geological process of rocks, soils, and other materials from the earth being worn and transported by natural processes such as rain, ice, or wind.

Farm labor contractor (FLC). People or companies that hire farm laborers to then provide labor for various farms and vineyards through contracts made with the FLC rather than each laborer. In the wine industry, FLCs regularly provide additional people to harvest fruit at the end of the season. They sometimes also provide other services like pruning vines in or after winter when the sap flow has significantly slowed, leaf pulling to help change the shape of the canopy on the grapevine, and other needs.

Farm laborer. People who perform planting, cultivating, harvesting, and maintenance on a farm. Essentially any farm or vineyard larger than serving an individual family relies on hiring farm labor. Some rely on labor services through an FLC to reduce the farm's responsibility for the laborers. Others hire people who are employed year-round by the winery or grower directly. In such cases, the year-round employee is better equipped with healthcare, farmworker protections, and other benefits.

Farm-to-table. A system that brings food directly from farms, gardens, and/or growers to nearby restaurants for their menu service to customers.

Field blend. A collection of different grape varieties all grown together in one vineyard. You can also describe vineyards planted with a selection of different vines from the same variety as a varietal selection field blend, but this is less common in regions like California even as it is relatively standard in older vineyards of France. Multi-varietal field blends were the standard technique for planting vineyards before the 1960s. After the rise of so-called fighting varietal wines (a wine made from a single variety) and the increase of consumer protection standards in the 1970s that demanded that what is on the label must reflect accurately what is in the bottle field blends largely surrendered to varietal vineyards. 1970s consumer labeling standards transformed some old-vine field blends into Zinfandel-only plantings, for example. Still, California retains some of the oldest field-blend vineyards in North America, and the world, and was founded on the heritage of these sites.

Fine wine. High-quality wine with craftsmanship and attention to detail. There is no single definition however, so some winemakers ascribe additional interests like single vineyard expressions, or lower intervention in the cellar, or organic farming as requirements of fine wine. Strictly speaking, none of these are standard practice for the category.

Food co-op. Food co-ops emerged in the United States as a way of providing groceries and other household staples that fit the standards and values of a particular community. Food co-ops are owned and managed by members to varying degrees. In some cases co-op members have direct decision-making impact in a democratic fashion. In other cases, members elect a board which makes decisions on products sold and carried based on agreed-upon principles from membership. Food co-ops in the United States have historically been a platform for advancing food-related commitments like organic food as well as support for farmworkers and free trade.

Foot treading (aka. pigeage). Traditional methods for extracting juice from grapes and capturing some portion of tannins and color relied on stepping into a vat of grapes and walking on them. The process of foot treading the grapes helps instill a different sense of structure, aroma and flavor concentration in the resulting wine. Today, basket presses, champagne presses, and bladder presses are all options but some winemakers still prefer to foot tread at least in the initial stages of winemaking when the fruit has just arrived at the winery, or during red wine fermentation instead of doing punchdowns or pump overs. Punch-downs and pump overs are both ways of ensuring that the fruit matter (literally the fruit skins) stays wet from the fermenting juice in order to avoid undue microbial activity (i.e. mold).

Fortified wine. Wines made from a typical juice fermentation that then have brandy or spirits added, i.e. they are fortified with spirits.

Franciscan monks. The St. Francis of Assisi order was founded in 1209 and led to the organization of Franciscan monks, who like their inspirational saint, are committed to principles of obedience, chastity, and poverty. Each was exercised when Spain ordered Franciscan monks to travel to New Spain (which became Mexico) and up into the province that included what is now California to lead their sacred expedition. There was no agriculture in the modern sense in what is now California, though Indigenous people throughout the extended region had complex food systems. Agricultural livestock were not brought until the Spanish pushed their way north. Poverty was certainly an element of their lives in the new colony and they obediently followed the orders of the Spanish government. (How they followed their vows of chastity is unknown.) Franciscan monks helped establish the early colonial impacts on Indigenous communities of California and brought the first *vinifera* vines into the region, thus launching the advent of California wine.

Fruit zone. The area in which a grapevine produces fruit. Above the fruit zone appears the vine's canopy, through which photosynthesis occurs to ripen the fruit. Below the fruit zone the trunk and then roots of the vine grow. At what height and how varied the fruit zone appears depends on how the vine is trained to grow.

Generational demographics. Spans of years are used to denote generational differences between age groups of populations to analyze and describe trends and differences between their experiences based on history in their lifetime, difference in buying habits, influences, and trends. Baby boomers or Boomers were born 1946–1964, Generation X was born 1965–1980, Millennials were born 1981–1996, and Generation Z from 1997–2012.

Grafted vine. *See* Ungrafted vine.

Grape clone. A vine that is genetically identical to a single parent vine from which it was taken as a vine cutting. A grape clone occurs when cuttings of the parent vine are propagated repeatedly, creating a genetically identical collection of vines with the general characteristics of the parent. Some vineyards are instead propagated by cuttings taken from multiple vines so that the new vines are each individually identical to their parent vine, but the vineyard includes cuttings from multiple different parent vines. That is, not all the vines in the site are genetically identical. This style of vineyard relies on selecting vines from which to take cuttings. Each cutting is known as a selection. A vineyard planted with multiple selections is known as a field blend. There are also field blends with multiple grape varieties.

Grape products. During Prohibition some growers and businesses survived by making legally allowed grape products where they could not make wine. These included grape bricks, which were concentrated grape juice, that were then dissolved in water. Consumers were warned not to let it sit too long or it would ferment into wine, thus giving instructions on how to use it to make wine. Other grape products made during Prohibition included wine made as a flavor ingredient for foods like canned soup, and wine used to cure tobacco leaves for cigars.

Greenhouse gases (GHG) – greenhouse gases become trapped in the Earth's atmosphere where they reflect infrared radiation back to the planet's surface rather than releasing it back into space. The result is surface warming, leading to climate change. These gases are formed by both natural and human sources. Some of the gases that form naturally become more impactful due to human activity. GHG include carbon dioxide (CO_2), nitrous oxide (N_2O), methane (CH_4), and water vapor (H_2O). Carbon dioxide is the GHG most increased by human activity, with methane and nitrous oxide next.

Gris grapes. Gris grapes are neither white nor red wine producing grapes but instead mutations that may be grayish-purple, bronze, rose toned, or lightly purple, depending on the parent grape. Gris means gray but implies an in-between color for what in deeper history were called white and black grapes rather than white and red. Pinot Gris (or Pinot Grigio) is one of the most well-known and common gris grapes. Grenache Gris is one of the most exciting.

Growing degree days (GDD). Used in agriculture to estimate when crop development will occur in a growing season. They also help determine the regions best suited to different crops and varieties. GDDs measure heat accumulation on the degree to which a mean daily temperature exceeds the base temperature specific to the species for which GDD is being used.

Head-trained vines. Vines trained to support themselves (once large enough), almost like a small tree, include arms, or vine spurs that are positioned in a circular fashion around the trunk of the vine to form what is called a head. Once the vine becomes thick enough, the arms continue to support themselves and every season grow canopy and a fruit zone. In the beginning, when the vines are younger, they require more maintenance than wire-trained vines but once established head-trained vines are less work to maintain.

High-wire training. A high-wire training system in a vineyard places the fruit zone of the vine much higher than traditional wire training systems. The purpose of high-wire training systems is to make it easier to harvest clusters when ripe, and also to use machinery to harvest as well as perform other essential tasks during the growing period. High-wire training can also be used to allow access for ruminants, especially sheep, beneath the canopy, allowing them to be in the vineyard throughout the season.

Hybrid electric. A hybrid electric motor depends upon both an electric motor and combustion. The goal is to use the best of both methods, reducing carbon emissions and gas needs while ensuring the vehicle can continue operating on gas alone if needed.

Indigenous. People native to a specific region are considered Indigenous to that area. Plants, animals, and ecosystems first formed in a particular area are also indigenous to those areas. The word indigenous is capitalized when referring to groups of people or an individual person.

Interspecific hybrids. *See* text box, page 109.

Judgment of Paris (aka. Paris Tasting). *See* text box, page 164.

Labor rights. Regulations that protect the right of workers to choose their own representatives and collectively bargain for their working conditions and contracts. In the United States, labor rights became an important activist movement beginning in the Great Depression of the 1930s. Farmworkers were excluded from such regulations until the 1970s, when California created the first regulations in the country to enshrine for farmworkers certain basic protections including bathroom breaks, clean water, sun shade, and the right to organize.

LGBTQIA. The initialization LGBTQIA stands for **l**esbian, **g**ay, **b**isexual, **t**ransgender, **q**ueer, questioning, **i**ntersex, and **a**sexual. It often includes a + at the end of the abbreviation to represent evolving understandings of gender and sexual identities. Use of the initials demonstrates coalition building between gender and sexual identities that differ, yet experience marginalization in society for not being heteronormative.

Malolactic fermentation (MLF). A secondary fermentation that converts naturally occurring malic acid (which is sharply acidic, much like the character of green apples) into lactic acid (which is creamier like yogurt). Especially essential in red wines but can be advantageous in some white wines, depending on stylistic goals. For example, it is more likely to find MLF in Chardonnay than Sauvignon Blanc.

Manong. Filipino word meant to convey respect to a Filipino man. It often conveys the notion of a beloved uncle, or person who acts as an elder, or parental presence beyond the literal birth parents of the person. In a US context, the first wave of Filipino immigrants (who were almost entirely men) and who usually worked as migrant laborers in harvest industries such as salmon, fruit, produce, and wine grapes, have been affectionately named as the Manong to denote their beloved elder status in the Filipino American community.

Mass spectrometry. In wine, mass spectrometry is used to separate and analyze the main components of wine at very close levels. Aroma and flavor compounds are especially measured.

Mediterranean climate. Warm to hot dry summers, and cooler, damp winters characterize Mediterranean climates. They generally occur through the Mediterranean region of Europe and Northern Africa, as well as on the western side of many continents, including North America, specifically California. Broadly speaking, a Mediterranean climate includes two seasons: summer and winter.

Mediterranean diet. An emphasis on fruits, vegetables, nuts, seeds, beans, and legumes as well as plenty of olive oil, occasional fish and very little red meat is considered a Mediterranean diet. Such a diet gained prominence when it was touted as a greater proactive treatment of heart disease than American diets. In such a case, it also importantly includes red wine.

Mega Purple. A concentrated as well as sweet grape juice made from the hybrid variety Rubired, bred by Harold P. Omo of UC Davis. Rubired is a teinturier grape (q.v.) that when concentrated can be used to add deeper color while also increasing a *grapey* flavor to wine. It tends to be used by winemakers trying to add extra oomph or color, or to fill the midpalate of Cabernet Sauvignon, especially in cooler vintages.

Metamorphic soils. Rocks exposed to high temperatures and pressure beneath the earth's surface and thereby chemically transformed are known as metamorphic rocks. Such rocks, when eroded and weathered, create metamorphic origin soils.

Migrant worker. A person who migrates for work between countries or regions for the sake of temporary employment in various industries. Rather than settling in one area and working from there, their income depends on moving between employment opportunities in different regions.

Mission era. The period between the arrival of Spain's sacred expedition to what is now California via San Diego, and the secularization of missions by Mexico. The mission era was enacted by Indigenous neophyte servants and Spanish military but led by Franciscan monks sent by Spain. It included the first known arrival of *Vitis vinifera* in what is now California.

Mission grape. A variety of *Vitis vinifera* brought from Spain to North and South America as the first wine variety introduced to the so-called new world.

Neophytes. In the mission era, a neophyte referred to an Indigenous person who was a new convert to Catholicism. To become a neophyte depended only on taking Communion, whether the Indigenous person knew the meaning of Communion or conversion. Once converted, neophytes were no longer allowed to leave the mission. This was enforced by Spanish military and neophytes worked in service of the mission and sometimes also the pueblo.

Neo-prohibitionism. The goal of reducing alcohol consumption not only through personal choice but also via legislation, changes and restrictions in sale, and tarnishing the reputation of alcohol. This is a direct reference to Prohibition in the United States when alcohols really were made fully illegal in the country.

Old vine. Vineyard sites that are older in relation to what is normal for the region. The OIV defined old vines in 2024 for the first time as a plant (or vineyard) of 35 years or older. Wineries like Turley look for vineyards at least 50 years old. The Historic Vineyard Society of California has played an important role in building recognition for older vineyards throughout California.

Organic. Organic practices rely on naturally derived substances and biological or mechanical rather than chemical methods to grow crops and livestock.

Organic farming. Relies on farming with naturally derived substances and biological or mechanical practices rather than chemical ones to produce crops including livestock and grapes. Importantly, organic farming can include naturally *derived* products rather than the plant or farm product itself. That is, organic farming need not depend on direct application of milk itself, or insecticide plants, as two examples, to count as organic, but might utilize treatment applications derived from those sources.

Organophosphates. A collection of human-made chemicals originally developed as nerve gas, herbicides, and insecticides. They were used as an alternative to DDT initially because their lifespan in soils is shorter, but it turns out their initial impact on animal species is far more severe. Organophosphates operate as poison to the human species and also continue to be toxic in foods farmed with them. Some organophosphates damage the nervous systems in vertebrates (including humans and other mammals). Some organophosphates that act more strongly as nerve agents were used in chemical warfare, especially in Vietnam. Their use in agriculture became more common in wine as DDT became less common. Today, most organophosphates are regulated in California.

Overhead sprinklers. There are multiple forms of irrigation systems, one of which is overhead sprinklers. This relies on high-pressure sprinklers just above the canopy level that spray water through the canopy and fruit zone. The effect is to reduce overall temperatures within the vineyard during heat events, and to protect vines against extreme cold in frost events.

Own rooted vine. *See* Ungrafted vine.

Ozone disinfectant. Ozone is a gas used in wineries (as well as home pools) to disinfect water, surfaces, and equipment. When it is sprayed on equipment or surfaces it kills bacteria, fungi, and viruses. It is now used for cleaning wine barrels, which reduces water usage and prevents organic buildup within the vessel. Because ozone destroys organic compounds, its treatment of water makes water already used in winery settings reusable for other applications such as rinsing and winery cleaning.

Parent material. In geology the weathered rock or geological deposit of rock or bedrock forms the parent material of what becomes soil in the region. Chemically it is considered the primary contribution to resulting soil. How a parent material weathers and erodes depends on the structure of the parent material itself and leads to a combination of rocks, gravels, soils, silts, or windblown loess in a region depending on the interaction between a parent material's structure and the region's weather and climate.

Pesticides. Chemical compounds (i.e. human-made through the application of chemistry) used to ward off fungi, bacteria, viruses, pests, or to feed plants chemically. These occur significantly in chemical farming for produce, livestock, or viticulture, as well as in various household goods like cleaners and solvents.

Phenols. Aromatic chemicals that are easily made airborne and form naturally within smoke, herbs, and other plants. They are intentionally used to create plastics, medicine, and household disinfectants. In wine, they can be found as volatile phenol compounds that contribute to smoke impact or smoke taint. *See also* Smoke impact.

Phylloxera. A pest impactful of vineyards (as well as in some other plant species, though usually due to a different species of the pest) that in the right growing conditions causes damage to the vine root, allowing viruses to enter the plant that over time change chemistry in the fruit, and eventually destroy the vine. Interestingly, phylloxera is common in a multitude of soil types but in sandy soils the virus that enters vines via phylloxera's impact cannot proliferate. Thus, older own-rooted vines in sandy soils survive without the impact of phylloxera.

Pierce's disease. Sap-feeding insects penetrate vines and introduce the bacteria that causes Pierce's disease. *Xylella*, the bacteria that causes Pierce's, is hard to combat as it lives in the water conductive system of the vine. Leafhoppers and sharpshooters move from vine to vine spreading the bacteria. Pierce's disease arrived in California as early as the 1880s and continues to impact vineyards today.

Plate tectonics. The movement of large plates of rock in response to planet activity, largely based in the dynamics of molten rock (to put it simply, underground lava). These geological plates provide the earth's surface on which all species live. As geological plates move along the earth's surface they impact each other through various types of movement which leads to the formation of mountains, volcanoes, inland valleys, and other topographical features.

Presidio. A fortified military settlement under the history of Spain and Spanish colonized America.

Private label wine. Private label wine is made by a company different from the one that sells it under its own name. Private label wines are often made for restaurants or retailers to have their own cuvée. Sometimes these are grown and made for the company that will sell it. In other instances, private label wines are made by blending bulk wines purchased from a bulk wine broker rather than a specific winery producer.

Pueblo. A town or village; a modest-sized settlement in a region.

Rain cycle (aka. the water cycle). The circular movement of water from the earth's surface to the air as gas through evaporation and back to the ground through precipitation either in the liquid form of rain or the solid form of snow or hail. That water then evaporates into gas, or water vapor, again and perpetuates the cycle.

Rainwater catchment. Capturing the run-off of rain from a building or other structure that does not absorb water but channels it, and then storing that water for use later.

Raisin grape. *See* Wine grape.

Ready-to-drink (RTD). An RTD is a beverage already mixed and packaged so that it can be opened and enjoyed instantaneously.

Refractometer. A device to measure the sugar content of grapes as a way to help determine the potential alcohol level of the resulting wine. Refractometers can take multiple forms and for the most part are only estimates of resulting alcohol but even so can help decide the best time for picking.

Regenerative. Describes techniques that help grow or restore aspects of a habitat, biology, or species previously damaged.

Regenerative organic viticulture. A system of farming that seeks to strengthen the ecosystem as a whole, especially via soil health. Biodiversity, carbon sequestration, elimination of chemicals, and encouragement of carbon sequestration are all core to the principles of regenerative organic viticulture.

Remote sensing technologies. Aircraft, satellites, and other equipment that provide the means to digitally scan or photograph a vineyard and analyze its health via various standards. Examples include: multispectral imagery that analyzes disease formation within the vineyard; sensing equipment that examines water stress; mapping that can assert the differences in electroconductivity (and thereby moisture) in the soils, or health of the vines.

Repeal. Legislative action that eliminated Prohibition in 1933 and changed the laws, allowing production and sales of alcohol again on a state-by-state basis.

Reverse osmosis (RO). Filtering process that separates the water, alcohol, flavor, aromatics, and structural aspects of wine from each other in order to remove parts of the wine no longer desired. RO can be used to lower resulting alcohol levels, temporarily remove smoke impact, intensify color, and eliminate undesired compounds.

Rhône varieties. Cultivars that originated from the Rhône Valley of France, regardless of where they are grown now. These include Syrah, Grenache, Mourvèdre, Counoise, Cinsault, Viognier, Roussanne, Marsanne, Clairette, Picpoul (or Piquepoul), Muscadin, Vaccarese, Terret Noir, Bourbelenc, and Picardin. Some include noir (red), blanc (white), and gris expressions.

Rootstock. The roots of a plant, which are grafted to another species or variety to increase the top variety's yields, size, or disease resistance. In wine, rootstock was initially developed to create resistance to phylloxera. It has since also been used to increase drought resistance or heighten yields.

Sacramental wine. Grape wine used for consecration declaring something sacred, including bread and wine in the Eucharist. Sacramental wine is used for Communion, and in New Spain was used to convert Indigenous peoples into neophytes overseen by and working for the church.

San Bernabe. The largest vineyard in North America, planted in Monterey. Also considered one of the three largest vineyards in the world.

Seasonal worker. In California, seasonal workers are essential to production and harvesting for wine as well as innumerable other crops. Farms tend to need more people to complete essential tasks during certain parts of the season such as pruning and harvest. Farms remain the most common source of seasonal employment. It is most common that seasonal laborers are non-citizens or immigrant labor as it is easier for most citizens to find higher paying work outside of agriculture.

Sedimentary soils. Soils formed by processes occurring due to rivers, streams, or the sedimentation of lakes, seas, or oceans are considered sedimentary soils. Parent material for sedimentary soils are rock and bedrock formations such as sandstone, siltstone, mudstone, and calcareous rock.

Shade cloth. A knitted fabric wrapped around the fruit zone of vineyards (and more recently suspended above vineyards in some parts of the world), providing sun protection from ultraviolet radiation while still allowing photosynthesis to occur. At its best it lowers temperatures and creates more manageable light distribution for vines.

Sharecroppers. Farmers who did not own land but were able to farm it in lease from landowners by trading a portion of the crop or proceeds of its sale. Sharecroppers were especially common in the post-Civil War south, where Black farmers were common but not allowed equal land ownership to pre-Civil War landowners.

Sikhism, or Sikhi. Sikhi is a religion and philosophy from the country of India that originates from around the fifteenth century. The central principles of Sikhi are love and oneness. A disciple or learner of Sikhism is known as a Sikh.

Skin contact maceration. Maceration is the process of making wine by soaking the skin (and sometimes also the stems) in the fermenting juice and then wine for some period of time. Maceration has become more common for red wine in recent decades when plushness, velvety texture, and impact have been desired. The phrase skin contact maceration generally refers to white wines that include maceration of the skins in juice or fermented wine. This is less common as most white wines are made by removing the juice quickly from the skins. Colloquially, white wines that have gone through skin contact maceration (aka skin contact) are sometimes called orange wines.

Smoke impact (aka smoke taint). Volatile phenol compounds in wildfire smoke are active in the atmosphere within the first couple of days after a wildfire is active. Because phenols are volatile, they are less impactful the older the smoke becomes. Volatile phenol compounds in wildfire smoke cause problems when they encounter grapes still on the vine and also have permeable skins. In such cases, phenol compounds bond with the sugar in grape skins and lead to smoke impact.

Soil orders. In soil taxonomy, a grouping of soils with shared or similar dominant characteristics. A soil order is a broad category and is the most general classification of soils.

Soil series. Soil grouping classified by similar or shared physical, chemical, and biological properties. There are 30,000 soil series in the world, and around 17,000 in the United States.

Soil vs. dirt. Soil is considered to be a living organism that includes microbial life as well as smaller flora, fauna, and fungi that together contribute to the structure, permeability, and water absorbing qualities of the soil, as well as its ability to support plant life such as vines. Dirt, on the other hand, is soil that has lost its structure and been relocated by tractor, shovel, hand, flooding, erosion, or wind. Essentially, dirt is inert, without structure or productive life. Soil is located in a particular place, is structural in a way that depends on the life within it, and offers great moisture retention and permeability.

Solera. A solera system depends on aging that includes multiple vintages held over time and over time blended together. For barrel-aged soleras, barrels are arranged in rows with the bottom-most barrel including the longest aged wine. Barrels of wine are mixed over time and the oldest vintages remain preserved within the solera at least in part. The benefits of solera aging include increased depth of character as well as complexity, greater resistance to oxidation, and development of savory character. Champagne, whiskey, sherry, and balsamic vinegar all utilize a solera method.

Spring frost. Spring frost occurs in vineyards where temperatures fall below freezing after winter has ended and the vine has begun its sap flow again. Usually spring frost refers to periods after bud break has occurred and vine shoots have begun appearing. The freezing temperatures damage the young growth of the vine and can lead to reduced yields, a seasonal disruption, or the death of a vine, depending on its severity.

St. George rootstock. An older rootstock option for grating *Vitis vinifera* to a North American species, specifically *Vitis rupestris* that is resistant to phylloxera as well as various diseases. St George is an advantage, even in modern viticulture, because it obtains deep root depth suited for a range of viticultural challenges including drought-prone areas, shallow soils, and even salinity content. In the AXR1 crisis, vineyards planted with St. George remained resistant to phylloxera.

Sweet wine. *See* Dry wine.

Table grape. *See* Wine grape.

Table wine. Broadly speaking, table wine is wine suited for drinking with a meal. It tends to be below 14 percent alcohol and is differentiated from sparkling, sweet, or fortified wines. Table wine is generally, though not always, made with wine grapes.

Tannin. An astringent and bitter compound that provides texture, structure, and mouthfeel in wines, tannin occurs naturally in nature. Outside of wine, it is most readily experienced in black tea, where the experience of tannins feels slightly drying or grippy in the mouth. In wine, the effect can be similar, though it is possible to have very smooth tannins that nevertheless feel structural in the wine. Tannins occur in fruits, seeds, woods, walnuts, cinnamon, tea, coffee, grape skins and seeds, and more. Tannins also bind to anthocyanins in grape skins, helping to fix the color of the wine, smooth the tannins, and make the wine more ageable.

Tariffs. A tax or duty applied to goods imported from other countries. Tariffs are generally paid by consumers in the country receiving the goods, even though they are often interpreted as a restriction on the exporting country. The exporting country might experience a reduction in export sales, but the tax is paid by consumers in the country of import.

Teinturier grapes. Most grapes have color in their skins, thanks to anthocyanins, but do not have color in the pulp of the grape. When pressed, the grape juice is uncolored like the pulp. Teinturier grapes have anthocyanins in both the skin and the pulp, creating color through both and when pressed their juice also has natural color.

Tilling. Soil is overturned, mixed with other ingredients, and manipulated to create a texture and consistency desired by the farmer. Traditionally, tilling has been referred to as preparing or cultivating the soil, as if it is an essential part of farming. There are instances, like hard clays for example, when breaking up the soil before planting can make that area of land more prepared for growing crops. Studies have shown that ongoing tilling decreases microbial life in the soil, increases erosion, and releases carbon into the atmosphere. In some cases, some areas between vine rows benefit from occasional tilling to improve short-term water availability to a vine. In some regions, ground rodents become a greater problem without tilling. In such cases, tilling under the vine but not between vine rows can help. Unturned and undisturbed soils that allow year-round cover crops are untilled soils.

Trellis-trained. *See* Wire-trained vines.

Ultraviolet (UV) shortwave. Shortwave radiation refers to a portion of the ultraviolet spectrum that has greater energy compared to other forms of radiation but also shorter wavelength. That means that it is on the higher end of the energy levels of the light spectrum. UV rays are known to contribute to various forms of cancer. The ozone layer and the obstruction layer of greenhouse gases filters out some UV radiation.

Undocumented worker. In the United States, undocumented workers remain a highly controversial issue. Undocumented workers are non-citizen, non-visa carrying, foreign-born workers who, according to US law, do not have the legal right to remain in the United States long-term. Undocumented workers are not entitled to worker protection, employee benefits, or even consistent wages. Because undocumented workers are considered illegal within the United States, they have few means to complain about mistreatment from employers. The associated risk is deportation or imprisonment.

Ungrafted vine (aka. own rooted). Grapevines planted directly into the soil so that their own roots descend into the earth, feeding the canopy and fruit of the vine directly. In other words, the top of the plant and the roots are not only the same species, but one unified plant. Vines are also commonly *grafted*. The top of a desired grape variety is spliced into the roots of another grape variety to create a single growing vine that is made from two conjoined vines. In viticulture, it is common for plants to be grafted to avoid pests like phylloxera, or to help the vine increase its yields, speed its ripening cycle, or make the vine more resistant to drought.

Untilled soils (aka. Zero tillage). *See* Tilling.

US generations. *See* Generational demographics.

Vapor pressure deficit (VPD). The amount of moisture the air can hold before reaching saturation. The vapor pressure deficit increases as temperatures increase. An increase in the vapor pressure deficit effectively makes the air drier, or more absorbent of water, thereby causing greater aridity in soils as well as plants.

Viña Madre. *See* text box, page 290.

Vine canopy. The vine's collection of leaves above the roots and fruit zone that photosynthesize and thereby create the energy needed for ripening and eventually harvest.

Vine training. *See* Wire-trained vines and Head-trained vines.

Viticulture. The study of cultivating grapes and producing a fruit crop from them, generally though not only for making wine, is known as viticulture.

Vitis californica. A wild species of grapevine native to the western parts of the United States overlapping what is now southern Oregon and the northern half of California. Although native to the United States, the species is not immune to phylloxera as the louse developed east of the Rockies, beyond the range of *Vitis californica.*

Vitis girdiana. A species of grapevine native to the southern parts of California and the Baja Peninsula of Mexico, *Vitis girdiana* grows wild along riverbeds and streams. It has also naturally crossed with *Vitis vinifera* in the form of the Mission grape to create the hybrid now known as Viña Madre.

Vitis vinifera. The species *Vitis vinifera* is native to Europe and the most prominent, but not the only, grapevine species for making wine, especially fine wine.

Weather station. An electronic device placed within vineyards to measure multiple aspects of weather and climate data in the vineyard. Weather stations record different levels of complexity and information depending on how advanced the system is. The most basic weather stations record temperatures over the course of a day (and night) as well as wind speeds. As they gain in complexity they may also record humidity, wind direction, precipitation levels and timing, solar radiation levels and timing of sun exposure (sometimes as well as direction), soil temperature, and soil moisture. At their optimum, weather stations guide viticultural decisions and provide long-term insight into climate change.

Weathering of parent material. Rocks and minerals break down on the earth's surface thanks to rain, sun exposure, freezing, air movements, and also the activity of living organisms. This process is called weathering and creates a transformation over time of rocks and bedrock into soils.

Wente clone. A dominant type of Chardonnay, expecially in California but actually the foundation of many types of Chardonnay worldwide. The clone originated in the multi-generational Wente vineyards of Livermore. It is highly regarded in wine growing throughout the world.

Whole cluster fermentation. Fermentation that occurs with fruit still on the stems, to use the entire cluster for fermentation, with the fruit on the cluster generally left intact. This can lead to fermentation ocurring inside the berry without exposure to oxygen, unlike when a berry is broken and the juice ferments with oxygen. When fermentation happens within an intact berry that process is called *carbonic maceration.* Inclusion of stems for fermentation creates a more apparent texture in the resulting wine, increases tannins while decreasing the perception of acidity, and usually also changes the aromatics or flavor of the resulting wine. Fermentation inside an intact berry through carbonic maceration generally reduces tannins and often color in the resulting wine and increases its overall sense of fruitiness.

Wind turbines. A turbine captures the movement generated by wind moving over its rotors and converts it into electricity. Wind turbines look like giant circular fans, broadly speaking. Multiple turbines built through an extended area can all together be called a wind farm.

Wine auctions. Auctions for trade, where wine shops or retailers can bid on wine to resell in their business. Trade auctions can include buying wine that is still in barrel before it has been bottled. Doing so can mean that barrel of wine is exclusive to your business. Auctions also exist for consumers or collectors who want to purchase harder to find wines, rare or old vintages, or wine no longer available within the standard wine market. Auctions can occur both online or live.

Wine brokers. Brokers serve as an intermediary between a buyer and seller of wine and can broker wine at any level of the wine industry. They might help collectors sell their wine to new buyers. They also might perform sales between wineries and wine shops or restaurants, or wineries and collectors.

Wine collectors. Wine lovers who not only appreciate wine but also buy it as a long-term investment strategy. Collectors purchase wine with the intention of reselling it later, after it has increased in resale value.

Wine cooperative. A group of wine growers that has created a wine-making system in which growers' fruit is purchased in total and given a standardized price by volume according to the pricing standards of the cooperative. The cooperative uses the grapes to make various types of wine and sells them. This approach provides a reliable buyer for their vineyards. Individual growers generally do not taste the wine made from their plot as it is generally part of a larger blend.

Wine grape. The intended use of a grape helps determine its type. Generally, the characteristics of certain grapes have helped define how they are used. Grapes made into raisins generally have more fiber, as well as more iron, in comparison to other grape types, and are higher in sugar, tasting sweeter than other grape types. These are dried to make raisins, which become more shelf stable and have been used to make rations during food shortages or wartime efforts. *Table grapes* tend to be larger, with firmer pulp and thinner skins. They are less acidic and have lower sugar than other grape types, most especially wine grapes. They can still be used to make raisins or wine, which will differ from what is typically used for either. Wine grapes are smaller than table grapes, with a higher juice content than either raisin or table grapes, and have comparatively thicker skins than either. Wine grapes are grown to have more flavor and concentration, as well as more aroma than either raisin or table grapes.

The thickness of the skins is an important part of the wine's resulting aroma.

Winemaking consultant. Experienced winemakers work with clients to offer guidance and/or advice on each stage of the winemaking process. They can assist in starting a new winemaking project, improving quality in an established one, identifying vineyards to fulfill the winery owners' stylistic preferences, and more. Consultants generally have multiple clients. Since the 1990s, there has been a global phenomenon known as the flying winemaker, that is, a consultant that has clients in multiple parts of the world.

Winter freeze. Liquids inside a grapevine can freeze, bursting cells inside the vine when temperatures drop below freezing. Such freeze can significantly damage or kill the vine. It can also damage the vine without killing it but eliminate its ability to produce fruit.

Wire-trained vines. In modern viticulture, vines are most often trained in relation to a system of wires and posts that guide the structure of the vine as it grows. This helps determine the structure of the canopy, the fruit zone, and the overall growing area of the vine. Vines trained to wire are also sometimes called trellised vines. There are many shapes for training the trellis of the vine. The trellis is essentially the support system for the vine's growth through a wire method. Wires for growing vines are generally thin but strong, used to catch the vine canopy as it grows and guide its shape. The alternative to wire trained vines, broadly speaking, is head-trained vines. *See also* Head-trained vines.

COMPANY AND ORGANIZATION ABBREVIATIONS

AAAV. Association of African American Vintners
AWOC. Agricultural Workers Organizing Committee
CCOF. California Certified Organic Farmers
CCSW. California Code of Sustainable Winegrowing
CWA. California Wine Association
EPA. Environmental Protection Agency
FDA. Federal Department of Agriculture
HVS. Historic Vineyard Society
ICE. Immigration & Customs Enforcement

IPOB. In Pursuit of Balance
IWCA. International Wineries for Climate Action
LAVA. Los Angeles Vintners Association
NFWA. National Farm Workers Association
NOAA. National Oceanic and Atmospheric Association
OIV. International Organization of Vine and Wine
OSHA. Occupational Safety and Health Administration
ROA. Regenerative Organic Alliance
ROC. Regenerative Organic Certification
TTB. Alcohol and Tobacco Tax and Trade Bureau
UFW. United Farm Workers Union
USDA. United States Department of Agriculture
WAB. Wine Advisory Board
WHO. World Health Organization

NOTES

Chapter 1

1. Atkins and Bauer, 15–16.
2. The first documented explorers were from Spain both from the south via the Gulf of California, and overland from what is now Arizona. By 1579, England responded to Spain's exploits in North America by sending their own ship to the coast of California. Other explorers also landed along the shores of California until Spain then laid claim to the region. Numerous accounts exist of these European and Indigenous contacts. Many though not all were cordial, with Indigenous people sometimes offering foods and trading with explorers.
3. The influx of outsiders into California because of the gold rush and then statehood thrust non-Natives into Native territories, usually with the result of violence. The first Governor of California stated in 1851 that extermination was the only option to fix the problem.
4. California became a US territory in 1848. Statehood followed in 1850. Immediately following statehood, the California legislature passed the Act for the Government and Protection of Indians. It continued the policy established by Mexico of instituting forced labor of Native people who the state saw as breaking the law, while also creating laws specific to Natives-only that restricted their behavior far more than non-Natives. Natives were denied citizenship in the United States until 1924. It was finally granted in recognition of the high rate of enlistment of Indigenous peoples to serve the United States in World War I.
5. Russian explorers first landed in Alaska in 1741, and then colonized the extended area. From there, they traveled south and east through what is now British Columbia, Washington, Oregon, and northern California. Spain worried they would continue south. In 1811, Russians built settlements in what are

now Bodega, California and the coast of Sonoma. The purpose was to grow food for their settlements in Alaska. Previously, Spain had only ventured as far north as modern-day San Francisco, but in 1823 built a mission in what is now the town of Sonoma. Historians disagree as to whether this new mission was in response to the Russian advance or not.

6. In 1579, Sir Francis Drake was sent by England to explore the coast of California. Spanish explorers had already toured the area and England was seeking potential resources. Drake hammered a stake into the coast claiming the coast north of San Francisco for England, but England did not return to the area until the 1800s and so did not maintain the claim.
7. Bolton, 49.
8. Pinney, *City of Vines*, 1.
9. Pinney, *City of Vines*, 10. At the end of the mission era, mission lands universally ended up in the hands of wealthy non-Native families.
10. Atkins and Bauer, 69.
11. Ornelas-Higdon, 17.
12. Atkins and Bauer, 3.
13. Atkins and Bauer, 65.
14. Atkins and Bauer, 67.
15. Ornelas-Higdon, 14.
16. Ornelas-Higdon, 27.
17. Though the exact number is not known, it is believed that North America has more species of wild grapevine than any other part of the world, followed by eastern Asia. In California, *Vitis girdiana* is endemic to the southern part of the state; *Vitis californica* grows through the northern part of California and into southern Oregon.
18. Ornelas-Higdon, 18.
19. California's Mission grape is known in Chile as Pais, or Criollo Chica in Argentina. DNA-typing has revealed it to be the Spanish cultivar, Listán Prieto. It can be found in the Canary Islands but essentially nowhere else in Europe.
20. Brady, 12.
21. Pinney, *City of Vines*, 22.
22. Ornelas-Higdon, 26.
23. Pinney, "The Early Days in Southern California," 2.
24. Pinney, *City of Vines*, 18.
25. Pinney, *A History of Wine in America, Vol. 1*, 240.
26. Pinney, *City of Vines*, 20.
27. Ornelas-Higdon, 15.
28. ibid., 30.
29. Pinney, *A History of Wine in America, Vol. 1*, 240. Ornelas-Higdon, 36.

30. Ornelas-Higdon, 43–4.
31. Pinney, *A History of Wine in America, Vol. I*, 242.
32. Pinney, *City of Vines*, 25. Ornelas-Higdon, 32.
33. Pinney, *City of Vines*, 13.
34. Ornelas-Higdon, 30.
35. Pinney, *City of Vines*, 33.
36. ibid., 36.
37. Ornelas-Higdon, 53.
38. ibid., 52.
39. Pinney, *City of Vines*, 46.
40. Ornelas-Higdon, 44.
41. ibid., 67.
42. ibid., 59.
43. ibid., 69.
44. Pinney, *City of Vines*, 61.
45. ibid.
46. Pinney, *A History of Wine in America, Vol. 1*, 255.
47. ibid., 257.
48. Pinney, *City of Vines*, 27.
49. Madley, 11, 351. Johnston-Dodds, 1.
50. Hannickel, 154.
51. Johnston-Dodds, 2.
52. Cook, 44. This figure of 150,000 is disputed by historians as being too low.
53. Madley, 3.
54. Phillips, 292.
55. United States Department of State, Office of the Historian. https://history.state.gov/milestones/1866-1898/chinese-immigration. Access date: 21 March 2024.
56. Pinney, *A History of Wine in America, Vol. 1*, 292.

Chapter 2

1. Teiser and Harroun, "The Volstead Act," 51.
2. Pinney, *A History of Wine in America, Vol. 1*, 218.
3. Lapsley, 99.
4. Donovan.
5. Phylloxera in California is also further discussed in chapter 7.
6. Walker, 209.
7. Unfortunately, the newly imported vines also brought with them phylloxera, and it is likely they were also the cause of Pierce's disease. Phylloxera did not exist in California prior to the 1850s. Gale, 213.

8. Pinney, *A History of Wine in America, Vol. 1*, 351–4.
9. Pinney, *The Makers of American Wine*, 92.
10. Pinney, *A History of Wine in America, Vol. 2*, 8.
11. Pinney, *A History of Wine in America, Vol. 1*, 359.
12. Also known as the Panama-Pacific International Exposition.
13. Hannickel, 182–4.
14. Pinney, *A History of Wine in America, Vol. 1*, 369.
15. Teiser and Harroun, "The Volstead Act," 56.
16. Pinney captures these figures from the the US Department of Agriculture, Bureau of Agricultural Economics, Crop Reporting Board, *Fruits and Nuts Bearing Acreage*, 1919–1946 (Washington, D.C.: GPO, 1949), p. 29. *A History of Wine in America, Vol. 2*, 19.
17. Pinney, *A History of Wine in America, Vol. 2*, 19.
18. Sullivan, *Napa Wine*, 262.
19. Ornelas-Higdon, 196.
20. California sparkling wine was still called champagne at the time.
21. Teiser and Harroun, "The Volstead Act," 54.
22. Pinney, *A History of Wine in America, Vol. 2*, 9.
23. Burnham.
24. Lerner.
25. ibid.
26. Teiser and Harroun, "The Volstead Act," 54.
27. ibid., 69.
28. ibid. 73.
29. Pinney, *A History of Wine in America, Vol. 2*, 66.
30. Teiser and Harroun, "The Volstead Act," 70–1.
31. ibid., 71.
32. ibid.
33. Sullivan, *Napa Wine*, 247.
34. ibid.
35. Teiser and Harroun, "The Volstead Act," 73.
36. Teiser and Harroun, *Winemaking in California*, 211.
37. ibid.
38. Pinney, *City of Vines*, 249.
39. Pinney, *A History of Wine in America, Vol. 2*, 138.
40. Pinney, *City of Vines*, 250.
41. ibid.
42. Pinney, *A History of Wine in America, Vol. 2*, 134.
43. Liquor: California Invasion.

44. ibid.
45. Pinney, *A History of Wine in America, Vol. 2*, 127–8.
46. Pinney, *A City of Vines*, 251.
47. Elaine Chukan Brown, "How The Gallos Did It," 190.
48. United States Department of Defense.
49. Pruitt, "The Post World War II Boom".
50. Lawrence.
51. Rothstein, 77.
52. ibid., ix–x.

Chapter 3

1. Pinney, *A History of Wine in America, Vol. 2*, 140.
2. Vankin.
3. Olmstead and Rhode, 18.
4. Sassen, 39.
5. Olmstead and Rhode, 3.
6. Villarejo, 3.
7. "The San Joaquin Valley: Past, Present, Future."
8. Olmstead and Rhode, 13.
9. ibid., 6.
10. Mabalon and Romasanta, 15.
11. Both are discussed in chapter 1.
12. Krebs.
13. Olmstead and Rhode, 17.
14. ibid.
15. ibid., 18.
16. Bulosan, 121.
17. Pulido, 63.
18. Mabalon and Romasanta, 27.
19. ibid., 16.
20. Sowers, 24.
21. Mabalon and Romasanta, 30.
22. ibid., 33.
23. Araiza, 160–1.
24. Shaw, 37.
25. ibid., 132.
26. Rosenfeld, 427.
27. Shaw, 44.
28. ibid.

29. Araiza, 170–1.
30. Shaw, 43.
31. ibid.
32. Garcia, 75–7.
33. Olmstead and Rhode, 17.
34. ibid.
35. ibid.
36. Garcia, 22.
37. Governor Reagan, the California Board of Agriculture, and President Nixon all supported the growers and did publicity events to support them. Shaw, 43–4.
38. Olmstead and Rhode, 17.
39. Flores, 46.
40. Postel, 2–4.
41. ibid., 3.
42. Gordon, 57.
43. Shaw, 122.
44. Grier, 1232.
45. Kobilinsky.
46. Kauffman, 113–14.
47. Postel, 6.
48. ibid., 15.
49. Lear, 359–60.
50. Pulido, 86 and Shaw, 136.
51. Kauffman, 253.
52. Shaw 124.
53. ibid.
54. Kauffman, 252.
55. Shaw, 123.
56. ibid., 128–33.
57. ibid.
58. ibid., 123.
59. Moses, 166, and Pruitt "The Nazis Developed Sarin Gas."
60. Shaw, 133.
61. ibid.
62. Pinney, *A History of Wine in America, Vol. 2*, 351–2.
63. Williams, et al.
64. This will be addressed further in chapter 4.

Chapter 4

1. Amerine and Singleton, 287.
2. Pinney, *A History of Wine in America, Vol. 2*, 225.
3. The connections and impact of the farmworker labor struggle, and the rise of environmentalism are explored in chapter 3.
4. MLK, Jr. was known to carry Thurman's book, *Jesus and the Disinherited*, with him during civil rights activities as a source of inspiration. Thurman's principles played a significant role in the civil rights movement, which then gave guiding force to many of the other social initiatives of the time.
5. Chapter 3 discusses the efforts to ban the use of DDT and related chemicals from agriculture in the 1960s and early 1970s.
6. Burns, 143.
7. Bank of America.
8. Martin.
9. Berkeley Library.
10. Gilder Lehrman Institute of American History.
11. Kitchell.
12. This is further discussed in chapter 2.
13. Waters, 295.
14. ibid., 299.
15. The personal as political was an idea starting to take hold among young people in the late 1960s, including in regards to food. Waters did not invent the idea but captured that spirit and helped give it a tangible focus through her work at Chez Panisse and broader attention to local farming. The book by Frances Moore Lappé, *Diet for a Small Planet*, published the same year Chez Panisse opened, helped spur the idea on in relation to food.
16. Tower.
17. Goldstein, 43.
18. Raskin.
19. Waters, 241.
20. Goldstein, 265–6.
21. ibid., 296.
22. ibid., 265–6.
23. ibid., 267.
24. ibid., 265–6.
25. Kauffman, 115–16.
26. ibid., 113.
27. Dona Brown, 2017.

28. Gardner, 7.
29. Kauffman, 183.
30. ibid., as well as Reed.
31. Gardner, 3.
32. Kauffman, 186.
33. Based on personal interviews with Sanford. He later partnered with Michael Benedict and so today the site of Sanford's planting is known as the Sanford-Benedict vineyard, one of the jewels of the Sta. Rita Hills.
34. Kauffman, 186.
35. ibid., 190.
36. Townsend.
37. All quoted in Kauffman, 200.
38. Kauffman, 180.
39. ibid., 190.
40. Alan Chadwick Archive, "Covelo Village Garden Project."
41. Dan White.
42. Cole-Johnson.
43. By refusing the draft, Muhammad Ali lost his heavyweight boxing title, was given significant fines, and also handed prison time. He was one of the best-known individuals to not just avoid the draft by leaving the country, but to do so publicly and face punishment by the federal government.
44. York, 19–21.
45. Nigro.
46. Beck.
47. ibid.
48. From a personal interview with Coturri.
49. Beck.
50. Bank of America, 1970.
51. Pinney, *A History of Wine in America, Vol. 2*, 240–1.
52. *Time*.
53. Franson.
54. Bank of America, 1973.
55. Pinney, *A History of Wine in America, Vol. 2*, 240–1.
56. Winiarski, 53. The Bank of America predictions especially are mentioned in numerous interviews with other winemakers of the time as well.
57. Franson.
58. ibid.
59. Alston, et al., 4.
60. Delfino.

61. Sonderling.
62. Lander and Mobley.
63. ibid.
64. Winiarski, 10, 18, 24, 38.
65. ibid., 10.
66. In 2000, the Freedom of Information Act was used to solicit previously classified documents from Nixon's presidency. As a result, it was disclosed that Nixon and his advisor Henry Kissinger had made plans to drop an atomic bomb on North Vietnam, with multiple military exercises done in advance as preparation for the possibility. It is unclear if Nixon truly would have dropped a nuclear bomb, but many now believe that even if the peace protests did not immediately end the war, they did dissuade Nixon from escalating it or using the nuclear arsenal. Burr and Kimball.
67. Nixon.
68. Blumenthal.
69. Green.
70. McCraw.
71. Parry.
72. McCoy, 66.
73. ibid., 67.
74. Agnew, 216.
75. CBS News.

Chapter 5

1. The influence of Parker is also discussed in chapter 4.
2. This is discussed further in chapter 4.
3. Hoemmen, et al. 68.
4. Wine Institute, "California Wine Sales."
5. "Y2K".
6. Insert history footnote
7. Giliberti.
8. Ferguson.
9. "Y2K".
10. Salvucci.
11. Davis.
12. Fish.
13. Corie Brown.
14. California Agriculture.
15. Popp.
16. CBS News.

17. Kelli White.
18. Gaffney.
19. Cueller.
20. McHugh.
21. FDIC.
22. Asimov, "Where Anxiety is All That's Flowing."
23. Fabrikant.
24. Wine Cap.
25. *Wine Spectator,* "2011 Vintage Report."
26. *Wine Spectator,* "2010 Vintage Report."
27. UC Davis, "Economy and weather."
28. ibid.
29. Weed.
30. Halverson.
31. Examples and general knowledge of the trend from personal interviews with winemakers.
32. There are numerous other examples. This information is based on personal interviews with winemakers.
33. Based on personal interviews with Clendenen.
34. Bonné.
35. Gibb.
36. The original piece from Parker is no longer available online. Quoted from Yarrow, "Debating Robert Parker."
37. The original interview is no longer available online. Quoted from Roberts.
38. Robinson, "Bottle fight."
39. Goode.
40. Gibb.
41. Yarrow, "Debating Robert Parker".
42. Adams.
43. California State Assembly.
44. Romano.
45. Wildfires and climate change are discussed further in chapter 15.
46. Erica Duecy has done a market survey that shows these statistics. It is available for purchase via her website. She also presented it at the 2024 Wine Writer Symposium in Napa Valley.
47. Leporati.

Chapter 6

1. California Department of Food & Agriculture.
2. Shah.

3. Yarrow, "The Silicon Valley effect on wine."
4. California Department of Food & Agriculture.
5. Caparoso.
6. Prothero, 134–48.
7. Public Policy Institute of California.
8. Amerine and Winkler; A. J. Winkler.
9. Shabam.

Chapter 9

1. Winetitles Media.

Chapter 10

1. USGS.

Chapter 13

1. Winiarski, 1–2, and 38.
2. Personal interviews with Richard Sanford.
3. Club Oenologique.
4. Bakshi.
5. Wise.
6. https://www.vitaevino.org/
7. https://www.comeoveroctober.com/
8. Willcox.
9. https://www.280project.com/
10. Cosme, et al., 1.
11. All monetary references in US dollars unless otherwise noted.
12. Dunham.
13. Dunham.
14. Winkler and Nicholas.

Chapter 14

1. Liv-ex
2. Meningers International.
3. Rumgay, et al.
4. Jones, et al.
5. Twenge.
6. PennState Extension
7. World Health Organization.
8. Canadian Centre on Substance Use and Addiction.
9. Holland.

10. Alaska House.
11. Stockwell, et al.
12. United States Department of Health and Human Services.
13. Carter, "ICCPUD Report Says Any Drinking Increases Mortality Risk."
14. Chapter 2 further examines this.
15. https://madd.org/
16. Code of Federal Regulations.
17. CBS News.
18. Rumgay, et al.
19. Mukamal and Rimm.
20. Ortolá, et al.
21. World Health Organization.
22. CBS News.
23. Tian, et al.
24. Azizov and Zaiss.
25. Rumgay, et al.
26. Northwest Medicine.
27. Selley.
28. Carter, "ICCPUD Report Says Any Drinking Increases Mortality Risk."
29. NIH.
30. Carter, "How Neo-Prohibitionists Came to Shape Alcohol Policy."
31. Esser, et al.
32. Discussed in chapter 2.
33. Carter, "The Fight Over Moderate Drinking."
34. Austen.
35. Mathieu.
36. Carter, "How Neo-Prohibitionists Came to Shape Alcohol Policy."
37. Carter, "Wine Industry Fires Back."

Chapter 15

1. Burke, et al.
2. Thordarson and Self.
3. D'Arrigo, et al.
4. Hall and Manabe.
5. NASA.
6. The introduction and section on climate change are partially taken from my chapter "Drought, Fire and the Future" in *On California*.
7. Kerlin.
8. This is discussed further on page 426, under Climate Action and Farming.
9. Abatzoglou, and Williams.

10. Red wine can also be added back to pressed wine to provide additional color and flavor.
11. This does not necessarily mean the smoke impact will be perceptible already in the wine, just that the bonds have been extracted from the juice.
12. Herve, et al.
13. Australian Wine Research Institute, "Treating smoke-affected juice.".
14. Australian Wine Research Institute, "Smoke taint.".
15. Huo, et al.
16. In 2020, a ranking of the largest wildfires in state history showed that six out of ten occurred in 2020. A fire in 2024 has since surpassed one of the 2020 fires. Swanson.
17. Lansing.
18. Wine Institute, "California Wine 2017 Harvest Report."
19. Until 2017, most wildfires that affected wine quality had occurred within forested or agricultural areas. In 2017, fires burned through urban centers surrounding vineyards as well. It was not only woods and other plant life that burned but buildings, vehicles, and other man-made materials. As a result, there was also a different range of volatile phenols within the smoke. The 2017 fires opened the possibility to study these as well.
20. The volatile phenols guaiacol and 4-methylguaiacol have the greatest impact on wine character; 4-ethylguaiacol, 4-ethylphenol, eugeneol, o-cresol, p-cresol, furfural, and syringol are secondary markers that can be tested in an expanded panel and are now believed to have an important though less prominent impact on wine character. More research is needed to determine what compounds cause what sensory characteristics.
21. Becker and Knoche.
22. Ristic, et al.
23. Baumgartner.
24. Comménil, et al.
25. Martine.
26. https://californiasustainablewine.com/
27. Systembolaget.
28. ibid., 37–42.
29. https://www.portoprotocol.com/about-us/
30. https://sustainablewine.co.uk/about/
31. Doyle.
32. UC Davis, "Economy and weather".
33. Stormwater Report.
34. Doyle.
35. Sonoma County regulations include standards that determine safe drinking water. Doyle.

36. Rieger, "UC Davis Winery."
37. Portillo, et al.
38. Clarke.
39. EPA.
40. This is considered further in the section on farming, p. 348. IPCC.
41. Smart.
42. Rieger, "Gallo Glass."
43. Verallia.
44. International Aluminium Organization.
45. Plastics Engineering.
46. Smith.
47. Sette.
48. Snowden-Seysses.
49. Bailey.
50. Snowden-Seysses.
51. ibid.
52. Wasilewski, et al.
53. Paso Robles Daily.
54. Boone.
55. https://living-future.org/lbc/
56. D'Angelo.
57. Thach.
58. Booker and Weber.

Epilogue

1. Menegat, et al.
2. Rodale Institute "Regenerative Organic Viticulture".

Appendices

1. https://www.ttb.gov/wine/labeling
2. https://calmatters.org/newsletter/california-farmworkers-housing-laws/
3. https://leginfo.legislature.ca.gov/faces/billNavClient.xhtml?bill_id=201320140AB263
4. https://leginfo.legislature.ca.gov/faces/billNavClient.xhtml?bill_id=201720180AB450
5. https://www.dol.gov/agencies/whd/agriculture/h2a#:~:text=Workers%20employed%20under%20the%20H,period%20specified%20in%20the%20contract.
6. As of 2024, TTB.

REFERENCES

Abatzoglou, John T. and A. Park Williams. "Impact of anthropogenic climate change on wildfire across western US forests." *PNAS.* Vol. 113, No. 42, 18 October 2016, 11770–5.

Adams, Andrew. "2020 Fires Caused $3.7 billion in losses for Wine Industry." *Wine Business Monthly.* Published 20 January 2021. Accessed: 5 November 2022. https://www.winebusiness.com/news/article/240575

Agnew, Eleanor. *Back from the Land.* Chicago: Ivan R. Dee, 2004.

Alan Chadwick Archive Living Library and Archive. "UC Garden Project: University of California, Santa Cruz Garden Project." Accessed: 10 September 2023. https://chadwickarchive.org/garden-projects/ucsc/

_____. "Covelo Village Garden Project." Accessed: 10 September 2023. https://chadwickarchive.org/garden-projects/covelo/

Alaska House Coalitions. "Law Requiring Warning that Alcohol Causes Breast and Colon Cancers Passes the Alaska State Legislature." Published: 17 May 2024. Accessed: 8 September 2024. https://akhouse.org/2024/05/17/law-requiring-warning-that-alcohol-causes-breast-and-colon-cancers-passes-the-alaska-state-legislature/

Alston, Julian M., James T. Lapsley, and Olena Sambucci. "Grape and Wine Production in California." In Philip L. Martin, Rachael E. Goodhue, and Brian D. Wright, eds, *California Agriculture: Dimensions and Issues.* Giannini Foundation of Agricultural Economics, 2019.

American Cancer Society. "Health Risks of Smoking Tobacco." Accessed: 8 September 2024. https://www.cancer.org/cancer/risk-prevention/tobacco/health-risks-of-tobacco/health-risks-of-smoking-tobacco.

html#:~:text=eyes%2C%20and%20bones.-,How%20smoking%20 tobacco%20affects%20your%20cancer%20risk,people%20in%20 the%20United%20States

Amerine, Maynard A. and Vernon Singleton. *Wine: An Introduction for Americans.* Berkeley: University of California Press, 1965.

Amerine, M. A. and A. J. Winkler. "Composition and quality of musts and wines of California grapes." *Hilgardia.* Vol. 15, Issue 6, February 1944, 493–675.

AP. "Liquor Consumption in U.S. at a 3-Decade Low." *New York Times.* Published: 25 November 1989. Accessed: 8 September 2024. https://www.nytimes.com/1989/11/25/us/liquor-consumption-in-us-reported-at-a-3-decade-low.html

Araiza, Lauren. *To March for Others: The Black Freedom Struggle and the United Farm Workers.* Philadelphia: University of Pennsylvania Press, 2014.

Arsenault, Raymond. *Freedom Riders: 1961 and the Struggle for Racial Justice.* Oxford: Oxford University Press, 2006.

Asimov, Eric. "When Wine Becomes Crucial to Cultural Identity." *New York Times.* Published: 5 June 2023. Accessed: 4 September 2024. https://www.nytimes.com/2023/06/05/dining/drinks/wine-cultural-identity.html

_____. "Where Anxiety Is All That's Flowing." *New York Times.* Published: 28 July 2009. Accessed: 3 October 2024. https://www.nytimes.com/2009/07/29/dining/29pour.html

Atkins, Damon B. and William J. Bauer, Jr. *We Are the Land: A History of Native California.* Oakland: University of California Press, 2021.

Austen, Ian. "Yukon Government Gives in to Liquor Industry on Warning Label Experiment." *New York Times.* Published: 6 January 2018. Accessed: 8 September 2024. https://www.nytimes.com/2018/01/06/world/canada/yukon-liquor-alcohol-warnings.html

Australian Wine Research Institute. "Smoke taint: practical management options for grapegrowers and winemakers." December 2018.

_____. "Treating smoke-affected juice or wine with activated carbon." June 2020.

Azizov, Vugar and Mario M. Zaiss. "Alcohol Consumption in Rheumatoid Arthritis: A Path through the Immune System." *Nutrients.* Vol. 13, Issue 4, 1324. Published: 16 April 2021. Accessed: 10 September 2024. https://pmc.ncbi.nlm.nih.gov/articles/PMC8072698/

Bailey, Pat. "UC Davis launches world's first LEED Platinum winery." UC Davis. Published: 5 October 2010. Accessed: 10 September 2024. https://www.ucdavis.edu/news/uc-davis-launches-world%E2%80%99s-first-leed-platinum-winery#:~:text=Additionally%2C%20the%20winery%20has%20been%20designed%20to,building's%20energy%20requirements%20for%20air%20quality%20and

Bakshi, Henna. "Making Fine Wine in Palestine, Despite it All." *Wine Enthusiast*. Published: 17 June 2024. Accessed: 4 September 2024. https://www.wineenthusiast.com/culture/travel/palestinian-wine/?srsltid=AfmBOopH2IMgOVZzeEflev4F_AyQtWxZwGmjT0fVlSgXLAz5law3pGfN

Bank of America. *California Wine Outlook: An Economic Study Prepared by Bank of America*. San Francisco: Bank of America, 1973.

Bank of America National Trust and Savings Association. *Outlook for the California Wine Industry*. San Francisco: Bank of America, 1970.

Baumgartner, Kendra. "Measuring Grape Smoke Exposure Impact." Crops Pathology and Genetics Research; Davis, CA, U.S. Department of Agriculture, Agriculture Research Service. Research duration: 1 August 2020–31 July 2025. Accessed: 23 October 2024. https://www.ars.usda.gov/research/project/?accnNo=438756

Beck, John. "Meet 'The Godfather' of Sonoma Wine Country." *Sonoma Magazine*. Published: October 2023. Accessed: 8 November 2023. https://www.sonomamag.com/meet-the-godfather-of-sonoma-wine-country/

Becker, T. and M. Knoche. "Water induces microcracks in the grape berry cuticle." *Vitis*. Vol. 51, No. 3, January 2012, 141–2. (PDF) Accessed: 23 October 2024. https://www.researchgate.net/publication/285947540_Water_induces_microcracks_in_the_grape_berry_cuticle

Berkeley Library. *Free Speech Movement*. Accessed: April 2024. https://www.lib.berkeley.edu/visit/bancroft/oral-history-center/projects/free-speech-movement

Blumenthal, Seth E. "Children of the silent majority: Nixon, new politics and the youth vote, 1968–1972." Boston University Doctoral Student Dissertation, 2013.

Bolton, Herbert Eugene. "The Mission in the Spanish-American Colonies." *American Historical Review* 23 (1917–18): 42–61.

Bonné, Jon. "Cathy Corison – Chronicle's Winemaker of the Year 2011." *San Francisco Chronicle*. Published: 1 January 2012. Accessed:

1 October 2024. https://www.sfgate.com/food/article/Cathy-Corison-Chronicle-s-Winemaker-of-the-Year-2433541.php

Booker, Christopher and Sam Weber. "Cow burps are a significant contributor to climate change." *PBS News*. Published: 6 May 2022. Accessed: 1 September 2024. https://www.pbs.org/newshour/show/cow-burps-are-a-major-contributor-to-climate-change-can-scientists-change-that#:~:text=Yes%20Not%20now-,Cow%20burps%20are%20a%20major%20contributor,change%20%E2%80%94%20can%20scientists%20change%20that%3F&text=Livestock%20production%E2%80%94primarily%20cows%E2%80%94produce,how%20some%20livestock%20process%20food

Boone, Virginie. "The Wild History of Lytton Springs." *Sonoma County Winegrowers*. Published: 14 August 2024. Accessed: 10 September 2024. https://sonomawinegrape.org/the-wild-history-of-lytton-springs/

Brady, Roy. "Alta California's First Vintage." In Muscatine, Amerine, and Thompson, eds., *Book of California Wine*, 10–15.

Brain, Matt. "The Winkler Index is an elegant relic." Brain on Winemaking: JamesSuckling.com. Published: 8 May 2024. Accessed: 8 August 2024. https://www.jamessuckling.com/wine-tasting-reports/brain-on-winemaking-the-winkler-index-is-an-elegant-relic/

Brown, Corie. "California Wine Industry Turns Sour." *Los Angeles Times*. Published: 5 January 2003. Accessed: 10 August 2024. https://www.latimes.com/archives/la-xpm-2003-jan-05-me-wine5-story.html

Brown, Dona. *Back to the Land: The Enduring Dream of Self-Sufficiency in Modern America.* Madison: University of Wisconsin Press, 2011.

Brown, Elaine Chukan. "California wine without boundaries." JancisRobinson.com. Published: 23 August 2022. https://www.jancisrobinson.com/articles/california-wine-without-boundaries

______. "Harlan Makes a Change." JancisRobinson.com. Published: 18 May 2022. https://www.jancisrobinson.com/articles/harlan-makes-change

______. "Dewilding California vine history." JancisRobinson.com. Published: 26 March 2022. https://www.jancisrobinson.com/articles/dewilding-california-vine-history

______. "Winiarski's new climate index." JancisRobinson.com. Published: 21 March 2022. https://www.jancisrobinson.com/articles/winiarskis-new-climate-index

______. "J. Lohr Wildflower Valdiguié." JancisRobinson.com. Published: 11 March 2022. https://www.jancisrobinson.com/articles/j-lohr-wildflower-valdiguie

_____. "Coturri appoints a successor." JancisRobinson.com. Published: 1 March 2022. https://www.jancisrobinson.com/articles/coturri-appoints-successor

_____. "Jasmine Hirsch takes the lead." JancisRobinson.com. Published 15 February 2022. https://www.jancisrobinson.com/articles/jasmine-hirsch-takes-lead

_____. "Randall Grahm's Gallo wine." JancisRobinson.com. Published: 9 February 2022. https://www.jancisrobinson.com/articles/randall-grahms-gallo-wines

_____. "Farewell Kita." Published: 27 January 2022. JancisRobinson.com. https://www.jancisrobinson.com/articles/farewell-kita

_____. "Raj Parr's chilly, rustic, new chapter." JancisRobinson.com. Published: 22 January 2022. https://www.jancisrobinson.com/articles/raj-parrs-chilly-rustic-new-chapter

_____. "Next steps in diversifying the wine industry." JancisRobinson.com. Published: 13 January 2022. https://www.jancisrobinson.com/articles/next-steps-diversifying-wine-industry

_____. "Drought, Fire and the Future." In Susan Keevil, ed., *On California: From Napa to Nebbiolo... Wine Tales from the Golden State*. Academie du Vin Library, 2021, 117–23.

_____. "How the Gallos Did It." In Susan Keevil, ed., *On California: From Napa to Nebbiolo... Wine Tales from the Golden State*. Academie du Vin Library, 2021, 189–93.

_____. "Surveying the Santa Cruz Mountains: Vineyard Manager Prudy Foxx." Hawk Wakawaka Wine Reviews. Published: 20 October 2014. Accessed: 1 December 2024. https://wakawakawinereviews.com/2014/10/20/surveying-the-santa-cruz-mountains-vineyard-manager-prudy-foxx/

Bulosan, Carlos. *America is in the Heart*. Seattle: University of Washington Press, 1993.

Burke, Andrea, Helen M. Innes, Laura Crick, and Rob Wilson. "High sensitivity of summer temperatures to stratospheric sulfur loading from volcanoes in the Northern Hemisphere." *PNAS*. Vol. 120, No. 47, 21 November 2023.

Burnham, Kelsey. "Prohibition in Wine Country." In *Napa Valley Register*. Archived from the original on 20 April 2010. Retrieved 26 March 2024. https://napavalleyregister.com/lifestyles/real-napa/prohibition-in-wine-country/article_ed8bdf22-4a81-11df-bb7d-001cc4c002e0.html.

Burns, John F. "Back in the World: Vietnam Veterans, California, and the Nation." In Marcia A. Eymann and Charles Wollenberg, *What's Going On?: California and the Vietnam Era*. Berkeley: University of California Press, 2004, 129–51.

Burr, William and Jeffrey Kimball. "Nixon White House Considered Nuclear Options Against North Vietnam Documents Reveal." *The National Security Archive*. Published: 31 July 2006. Accessed: 3 January 2024.

Byrne, Brendan, Junjie Liu, Kevin W. Bowman, et al. "Carbon emissions from the 2023 Canadian wildfires." *Nature*. Vol. 633, 2024, 835–9. Published: 26 September 2024. Accessed: 5 October 2024. https://www.nature.com/articles/s41586-024-07878-z

California Agriculture. "California's Wine Industry Enters New Era." Published: 1 January 2003. Accessed: 10 May 2023. https://calag.ucanr.edu/Archive/?article=ca.v057n03p71

California Department of Food & Agriculture (CDFA). "California grape acreage report, 2023 crop." Published: 30 April 2024. Accessed: 9 September 2024. https://www.nass.usda.gov/Statistics_by_State/California/Publications/Specialty_and_Other_Releases/Grapes/Acreage/2024/grpacDETAILED2023Crop.pdf

California State Assembly. "The Impact of Wildfires on California Agriculture Report." Assembly held 18 November 2020. Accessed: 8 October 2024. https://agri.assembly.ca.gov/sites/agri.assembly.ca.gov/files/The%20Impact%20of%20Wildfires%20on%20California%20Agriculture%20Informational%20Hearing%20Report.pdf

Canadian Centre on Substance Use and Addiction. "Canada's Guidance on Alcohol and Health." https://www.ccsa.ca/canadas-guidance-alcohol-and-health

Caparoso, Randy. "Latest update on the 100+ grapes grown in Lodi." Lodi Wine: Letters from Lodi. Published: 20 March 2024. Accessed: 21 March 2024. https://www.lodiwine.com/blog/Latest-update-on-the-over-100-wine-grapes-grown-in-Lodi#:~:text=See%20our%20September%206%2C%202022,European%20family%20of%20wine%20grapes.

Carter, Felicity. "ICCPUD Report Says Any Drinking Increases Mortality Risk." *Wine Business Monthly*. Published: 15 January 2025. Accessed: 15 January 2025. https://www.winebusiness.com/news/article/297184

Carter, Felicity. "The Fight Over Moderate Drinking: Why Studies on Effects are Unlikely to Happen." *Wine Business Monthly*. Published: 1

April 2024. Accessed: 8 September 2024. https://www.winebusiness.com/wbm/article/284945

_____. "How Neo-Prohibitionists Came to Shape Alcohol Policy." *Wine Business Monthly.* Published: 25 March 2024. Accessed: 8 September 2024. https://www.winebusiness.com/wbm/article/284944

_____. "Wine Industry Fires Back Against the WHO's 'Zero Safety.'" *Wine Business Monthly.* Published: 21 October 2023. Accessed: 10 September 2024. https://www.winebusiness.com/news/article/278363

CBS News. "How Morley Safer Convinced Americans to Drink More Wine." Published: 28 August 2016. Accessed: 22 September 2023. https://www.cbsnews.com/news/how-morley-safer-convinced-americans-to-drink-more-wine/

Christensen, L. P., N. K. Dokoozlian, M. A. Walker, and J. A. Wolpert. *Wine Grape Varieties in California.* Publication 3419. Oakland: University of California, Agriculture and Natural Resources, 2003.

Clarke, Jim. "The Science Behind UV Light in Vineyards." *SevenFiftyDaily.* Published: 14 November 2022. Accessed: 10 September 2024. https://daily.sevenfifty.com/the-science-behind-uv-light-in-vineyards/#:~:text=Vines%2C%20mildews%2C%20and%20human%20beings,and%2C%20potentially%2C%20disease%20free

Club Oenologique. "Chateau Musar in the war years – a vintage report with a difference." Published: 23 May 2020. Accessed: 4 September 2024. https://cluboenologique.com/story/chateau-musar-in-the-war-years-a-vintage-report-with-a-difference/

Code of Federal Regulations. "Part 16—Alcoholic Beverage Health Warning Statement." Published: 14 February 1990. Accessed: 8 September 2024. https://www.ecfr.gov/current/title-27/chapter-I/subchapter-A/part-16

Cole-Johnson, Samantha. "Biodynamic viticulture in the Americas." JancisRobinson.com. Published: 28 February 2023. Accessed: 12 September 2023. https://www.jancisrobinson.com/articles/biodynamic-viticulture-americas

Comménil, Pascal, Loïc Brunet, and Jean-Claude Audran. "The development of the grape berry cuticle in relation to susceptibility to bunch rot disease." *Journal of Experimental Botany.* Vol. 48, No. 313, August 1997, 1599–1607.

Cook, Sherburne Friend. *The Population of the California Indians, 1769–1970.* Berkeley: University of California Press, 1976.

Cory, Kenneth and John A. Nejedly. *A Report of the Joint Committee on Public Domain on Crude Oil Pipelines in California.* Sacramento: California State Lands Commission, 1974.

Cosme, Fernanda, Luís Filipe-Ribeiro, and Fermando M. Nunes. "Impact of Climate Changes on Grapes and Grape Products." *Global Warming and the Wine Industry.* London: IntechOpen, 2024, 1–12.

Cueller, Steven S. "The 'Sideways' Effect." *Wines & Vines.* Published: January 2009. Accessed: 3 October 2024. https://winebusinessanalytics.com/sections/printout_article.cfm?content=61265&article=feature

D'Angelo, Madeleine. "Silver Oak Alexander Valley Winery Becomes the World's Largest Living Building." *Architect Magazine.* Published: 22 April 2020. Accessed: 10 September 2024. https://www.architectmagazine.com/design/silver-oak-alexander-valley-winery-becomes-the-worlds-largest-living-building_o

D'Arrigo, Rosanne, Richard Seager, Jason E. Smerdon, et al. "The anomalous winter of 1783–1784: Was the Laki eruption or an analog of the 2009–2010 winter to blame?" *Geophysical Research Letters.* Vol. 38, Issue 5, 16 March 2011.

Davidson, William. "The Great Plains: America's Carbon Vault." *Energy and the Environment: Economics and Policy.* University of Nebraska-Lincoln, Department of Agricultural Economics, Fall 2016.

Davis, Marc. "How September 11 Affected the U.S. Stock Market." *Investopedia.* Published: 11 September 2023. Accessed: 4 March 2024. https://www.investopedia.com/financial-edge/0911/how-september-11-affected-the-u.s.-stock-market.aspx#:~:text=On%20the%20first%20day%20of,during%20the%20global%20coronavirus%20pandemic

Delfino, Aldo. *1970 Napa County Agricultural Crop Report.* Napa County Department of Agriculture, 1971.

Donovan, Tristan. "The Four Horsemen May Charge Over the Earth—but Coca-Cola Will Remain." *The Atlantic*, November 8, 2013. Accessed: March 31, 2024. https://www.theatlantic.com/health/archive/2013/11/the-four-horsemen-may-charge-over-the-earth-but-coca-cola-will-remain/281178/

Doyle, Sarah. "Why isn't rainwater capture more popular in Sonoma County's wine industry?" *The Press Democrat.* Published: 10 January 2023. Accessed: 10 September 2024. https://www.pressdemocrat.com/article/lifestyle/why-isnt-rainwater-capture-more-popular-for-our-trademark-industry-2/

Draper, Paul. "Zinfandel." In Muscatine, Amerine, and Thompson, eds., *Book of California Wine*, 223–34.

Dunham, John and Associates. "United States Economic Impact Study 2022." *Wine America*. Published: 21 September 2022. Accessed: 4 September2024.https://wineamerica.org/economic-impact-study/2022-american-wine-industry-methodology/#:~:text=Winery%20%7C%20Supplier%20%7C%20Association-,2022%20Economic%20Impact%20Study%20of%20the%20American%20Wine%20Industry%20Methodology,publications%2C%20and%20Data%20Axle.%C2%B9

EPA. "Wastewater Technology Fact Sheet Ozone Disinfection." (PDF) Published: September 1999. Accessed: 10 September 2024. https://www3.epa.gov/npdes/pubs/ozon.pdf

Esser, Marissa B., Adam Sherk, Yong Liu, and Timothy S. Naimi. "Deaths from Excessive Alcohol Use – United States, 2016–2021. *MMWR Morbidity Mortality Weekly Report*. (PDF) Col. 73, 154–61. Published: 9 February 2024. https://www.cdc.gov/mmwr/volumes/73/wr/pdfs/mm7308a1-H.pdf

Eymann, Marcia A. and Charles Wollenberg. *What's Going On?: California and the Vietnam Era*. Berkeley: University of California Press, 2004.

Fabrikant, Geraldine. "They're Pinching Hundred-Dollar Bills." *New York Times*. Published: 3 October 2008. Accessed: 3 October 2024. https://www.nytimes.com/2008/10/04/business/04luxury.html

FDIC. "Chapter 1." Crisis and Response: An FDIC History, 2008–13. FDIC, 2019, 3–32.

Ferguson, Scott. "Sparkling Wine." *Wine Business Monthly*. Published: 10 September 2002. Accessed: 8 August 2024. https://www.winebusiness.com/wbm/article/18995

Fish, Tim. "California Winery Woes: From Boom to Bankruptcy." *Wine Spectator*. Published: 18 September 2003. Accessed: 10 August 2024. https://video.winespectator.com/articles/california-winery-woes-from-boom-to-bankruptcy-21753

Flores, Lori A. *Grounds for Dreaming: Mexican Americans, Mexican Immigrants, and the California Farmworker Movement*. New Haven: Yale University Press, 2016.

Franson, Paul. "Class of 1972 wine." *Decanter Magazine*. Published: 1 January 2002. Accessed: 8 September 2022. https://www.decanter.com/features/class-of-1972-wine-249048/

Gaffney, Jacob. "California's 2000 Vintage Sets Record for Size of Grape Crush." *Wine Spectator*. Published: 9 March 2001. Accessed: 3 March 2024. https://www.winespectator.com/articles/californias-2000-vintage-sets-record-for-size-of-grape-crop-20953

Gale, George D. *Dying on the Vine*. Berkeley: University of California Press, 2011.

Garcia, Matt. *From the Jaws of Victory: the triumph and tragedy of Cesar Chavez and the Farm Worker Movement*. Berkeley: University of California Press, 2012.

Gardner, Hugh. *The Children of Prosperity*. New York: St. Martin's Press, 1978.

Gibb, Rebecca. "Should Robert Parker Have Listened to Disraeli?" *Wine-Searcher*. Published: 29 January 2014. Accessed: 2 October 2024.https://www.wine-searcher.com/m/2014/01/should-robert-parker-have-listened-to-disraeli

Gilder Lehrman Institute of American History. "History Resources: Ronald Reagan on the unrest on college campuses, 1967." Accessed: May 2024. https://www.gilderlehrman.org/history-resources/spotlight-primary-source/ronald-reagan-unrest-college-campuses-1967

Giliberti, Ben. "Bubbling Over." *The Washington Post*. Published: 28 December 1999. Accessed: 8 August 2024. https://www.washingtonpost.com/archive/lifestyle/food/1999/12/29/bubbling-over/913776d1-0f5a-45f4-ba6e-42c9500d4890/

Goldstein, Joyce. *Inside the California Food Revolution*. Berkeley: University of California Press, 2013.

Goode, Jamie. "Novelty at the Expense of Quality?" *Wine Anorak*. Published: 15 February 2014. Accessed: 8 October 2024. https://www.wineanorak.com/wineblog/uncategorized/novelty-at-the-expense-of-quality

Gordon, Robert. "Poisons in the Fields: The United Farmworkers, Pesticides, and Environmental Politics." *Pacific Historical Review*. Vol. 68, No. 1, February 1999, 51–77.

Green, Mark. "How Ralph Nader Changed America." *The Nation*. Published: 1 November 2015. Accessed: 3 March 2024. https://www.thenation.com/article/archive/how-ralph-nader-changed-america/

Grier, James W. "Ban of DDT and Subsequent Recovery of Reproduction in Bald Eagles." *Science*. New Series, Vol. 218, 17 December 1982, 1232–5.

Hall, Alex and Syukuro Manabe. "The Role of Water Vapor

Feedback in Unperturbed Climate Variability and Global Warming." *American Meterological Society.* (PDF) Published: 1 August 1999. Accessed: 4 September 2024. https://journals.ametsoc.org/configurable/content/journals$002fclim$002f12$002f8$002f1520-0442_1999_012_2327_trowvf_2.0.co_2.xml?t:ac=journals%24002fclim%24002f12%24002f8%24002f1520-0442_1999_012_2327_trowvf_2.0.co_2.xml

Halverson, Nathan. "Weather, economy bring flood of bad news; best to be said about 2010 is it's almost over." *The Press Democrat.* Published: 17 October 2010. Accessed: 8 October 2024. https://www.pressdemocrat.com/article/news/weather-economy-bring-flood-of-bad-news-best-to-be-said-about-2010-is-it/

Hannickel, Erica. *Empire of Vines, Wine Culture in America.* Philadelphia: University of Pennsylvania Press, 2013.

Harding, Julia and Jancis Robinson, with Tara Q. Thomas. *The Oxford Companion to Wine, Fifth edition.* Oxford: Oxford University Press, 2023.

Hawkin, Paul. *Regeneration: Ending the Climate Crisis in One Generation.* New York: Penguin Books, 2021.

Hayes, Sue Eileen. "Those Who Worked the Land." In Muscatine, Amerine, and Thompson, eds., *Book of California Wine*, 25–9.

Herve, Eric, Steve Price, and Gordon Burns. "Smoke Markers in Red Wines and Maceration Times." ETS Laboratories, 2019.

Hoemmen, Garrett, C. Matthew Rendleman, Brad Taylor, et al. "Analysis of structural changes on grape grower's return per ton: A case study of developing American Viticultural Areas." *Wine Economics and Policy.* Vol. 2, Issue 2, December 2013, 67–75.

Holland, Fiona. "Japan releases first official alcohol consumption guidelines." *Just Drinks.* Published: 20 February 2024. Accessed: 12 September 2024. https://www.just-drinks.com/news/japan-official-alcohol-consumption-guidelines-published/

Houriet, Richard. *Getting Back Together.* London: Abacus, 1973.

Huo, Yiming, Renata Ristic, Carolyn Puglisi, et al. "Amelioration of Smoke Taint in Wine via Addition of Molecularly Imprinted Polymers during or after Fermentation." *Journal of Agriculture and Food Chemistry.* Vol. 72, Issue 32, 2024, 18121–31.

Hutchison, John N. "Northern California from Haraszthy to the Beginnings of Prohibition." In Muscatine, Amerine, and Thompson, eds., *Book of California Wine*, 30–48.

Hyslop, Stephen G. *Contest for California: From Spanish Colonization to the American Conquest.* Norman: University of Oklahoma Press, 2012.

International Aluminium Organization. "A Circularity Case for Aluminum Compared with Glass and Plastic." (PDF) Published: January 2023. Accessed: 24 September 2024. https://international-aluminium.org/wp-content/uploads/2022/02/Aluminium-vs-glass-and-plastic-FINAL-Information-Sheet-1.pdf

IPCC. Emissions Trends and Drivers. In IPCC, 2022: Climate Change 2022: Mitigation of Climate Change. Contribution of Working Group III to the Sixth Assessment Report of the Intergovernmental Panel on Climate Change. Cambridge University Press, Cambridge, UK and New York, NY, USA, 2022.

Johnson, Hugh and Jancis Robinson. *The World Atlas of Wine, 8th edition.* London: Octopus Publishing, 2019.

Johnston-Dodds, Kimberly. "Early California Laws and Policies Related to California Indians." (PDF) Sacramento: California State Library, California Research Bureau, September 2002.

Jones, Nicholas, Rachel Marks, Roberto Ramirez, and Merarys Ríos-Vargas. "2020 Census Illuminates Racial and Ethnic Composition of the Country." *Census.gov.* Published: 12 August 2021. Accessed: 10 January 2024. https://www.census.gov/library/stories/2021/08/improved-race-ethnicity-measures-reveal-united-states-population-much-more-multiracial.html#:~:text=to%20self%2Didentify.-,Multiracial%20Population,people%2C%20a%20230%25%20change

Kauffman, Jonathan. *Hippie Food: How back-to-the-landers, longhairs, and revolutionaries changed the way we eat.* New York: HarperCollins Publishers, 2018.

Keevil, Susan, ed. *On California: From Napa to Nebbiolo… Wine Tales from the Golden State.* London: Academie du Vin Library Ltd, 2021.

Kerlin, Kat. "California's 2020 Wildfire Season." *UC Davis News.* Published: 4 May 2022. Accessed: 8 December 2024. https://www.ucdavis.edu/climate/news/californias-2020-wildfire-season-numbers#:~:text=Quick%20Summary&text=Just%20over%209%2C900%20wildfires%20burned,of%20acres%20burned%20in%20California

Kitchell, Mark. *Berkeley in the Sixties: A film by Mark Kitchell.* California Newsreel, 1990.

Kobilinsky, Dana. "After 50 Years, DDT Still Impacts Lake Ecosystem." *The Wildlife Society.* Published: 25 June 2019. Accessed: 1 October

2024. https://wildlife.org/after-50-years-ddt-still-impacts-lake-ecosystem/#:~:text=They%20also%20determined%20a%20shift,as%20well%2C%E2%80%9D%20he%20said

Krebs, A. V. "Bitter Harvest." *Washington Post.* Published: 2 February 1992. Accessed: 16 December 2024. https://www.washingtonpost.com/archive/opinions/1992/02/02/bitter-harvest/c8389b23-884d-43bd-ad34-bf7b11077135/

Lancaster University. "Global study reveals time running out for many soils, but conservation measures can help." *ScienceDaily*. Published: 14 September 2020. Accessed: 4 May 2024. https://www.sciencedaily.com/releases/2020/09/200914115905.htm

Lander, Jess and Esther Mobley. "Meet Napa's 'class of 1972,' the cool kids who changed American wine forever." *San Francisco Chronicle.* Published: 9 June 2022. Accessed: 7 September 2023. https://www.sfchronicle.com/food/wine/article/napa-valley-1972-caymus-silver-oak-cakebread-17228780.php

Lansing, Roger. "Special Report: Understanding the Effect of Smoke Taint in Grapes and Wine." *Wine Business Monthly.* Published: February 2010. Republished online: 1 December 2017. Accessed: 2 November 2023. https://www.winebusiness.com/wbm/article/192827

Lapsley, James T. *Bottled Poetry: Napa Winemaking From Prohibition to the Modern Era.* Berkeley: University of California Press, 1996.

Lavin, Kate. "Almonds reach 1-million bearing acres." *Wine Business Analytics.* Published: 28 February 2018. Accessed: 1 October 2024. https://winebusinessanalytics.com/news/article/196283/Almonds-Reach-1-Million-Bearing-Acres#:~:text=%E2%80%9CThe%20only%20reason%20to%20plant,%2C%20if%20not%2050%25.%E2%80%9D

Lawrence, Quil. "Black vets were excluded from GI bill benefits – a bill in congress aims to fix that." *All Things Considered.* NPR. Published: 18 October 2022. Accessed: 14 April 2024. https://www.npr.org/2022/10/18/1129735948/black-vets-were-excluded-from-gi-bill-benefits-a-bill-in-congress-aims-to-fix-th

Lear, Linda. *Rachel Carson: Witness for Nature.* New York: Henry Holt and Company, 1997.

Leporati, Gregory. "The NBA Is Obsessed with Wine. Here's How That Happened." *Wine Enthusiast.* Published: April 2024. Accessed: 8 October 2024. https://www.wineenthusiast.com/culture/nba-wine/#:~:text=The%20NBA%20Officially%20Partners%20with %20Kendall%2DJackson%20Wines&text=Kendall%2D Jackson%2C%20though%2C%20is,have%20their%20own%20 wine%20labels

Lerner, Michael. "Unintended Consequences of Prohibition." *Prohibition: A film by Ken Burns and Lynn Novick.* PBS. Accessed: 31 March 2024. https://www.pbs.org/kenburns/prohibition/unintended-consequences/

"Liquor: California Invasion." *Time Magazine.* Published: 8 February 1943. Accessed: 1 April 2024. https://content.time.com/time/subscriber/article/0,33009,774205,00.html

Live-ex. "Liv-ex 100 down another 2.5%, California bounces back." Published: 6 December 2024. Accessed: 18 January 2025. https://www.liv-ex.com/2024/12/liv-ex-indice/

Mabalon, Dawn B. and Gayle Romasanta. *Journey for Justice: The Life of Larry Itliong.* Stockton: Bridge and Delta Publishing, 2018.

Madley, Benjamin. *An American Genocide, The United States and the California Catastrophe, 1846–1873.* New Haven: Yale University Press, 2016.

Martin, Michael T. "'Buses Are a Comin'. Oh Yeah!': Stanley Nelson on *Freedom Riders.*" *Black Camera.* Vol. 3, No. 1, 2011, 96–122. *Project MUSE.* https://dx.doi.org/10.2979/blackcamera.3.1.96

Martine, Katherine. "Developing Wildfire Resilience: Oregon State University Researchers Discuss 'Exciting Avenue of Research' into Smoke Protective Coating for Grapes." *Wine Business Monthly.* Published: 23 October 2024. Accessed: 23 October 2024. https://www.winebusiness.com/news/article/294211

Martyris, Nina. "GIs Help Bring Freedom to Europe, and a Taste for Oregano to America." *The Salt: What's On Your Plate. NPR.* Published: 9 May 2015. Accessed: 2 April 2024. https://www.npr.org/sections/thesalt/2015/05/09/405302961/gis-helped-bring-freedom-to-europe-and-a-taste-for-oregano-to-america

Mathieu, Henry. "More Americans view alcohol consumption as unhealthy – survey." *Just Drinks.* Published: 16 August 2024. Accessed: 8 September 2024. https://www.just-drinks.com/news/more-americans-view-alcohol-consumption-as-unhealthy-survey/#:~:text=Americans%20are%20increasingly%20viewing%20 alcohol,is%20bad%20for%20one's%20health

McCoy, Elin. *The Rise of Robert M. Parker, Jr., and the Reign of American Taste.* New York: HarperCollins, 2005.

McCraw, David E. "The 'Freedom from Information' Act: A look back at Nader, FOIA, and what went wrong." Published: 21 November 2016. Accessed: 3 January 2024. https://www.yalelawjournal.org/forum/the-freedom-from-information-act-a-look-back

McHugh, Adam. "20 Years Ago, This Movie Changed a Tiny California Wine Region Forever." *Food & Wine*. Published 18 September 2024. Accessed: 3 October 2024. https://www.foodandwine.com/sideways-movie-changed-santa-ynez-valley-8714407#:~:text=The%20impact%20Alexander%20Payne's%20Oscar,movie%20theater%20in%20October%202004

Mendelson, Richard. *Appellation Napa Valley: Building and Protecting an American Treasure*. Napa: Val de Grace Books, Inc., 2016.

Menegat, Stefano, Alicia Ledo, and Reyes Tirado. "Greenhouse gas emissions from global production and use of nitrogen synthetic fertilisers in agriculture." *Scientific Reports*. Issue 12, 14490, 2022. Published: 25 August 2022. Reviewed: 10 December 2024. https://www.nature.com/articles/s41598-022-18773-w

Moses, Marion. "Farmworkers and Pesticides." *Confronting Environmental Racism*. Ed. Robert Bullard. Boston: South End Press, 1992, 161–81.

Mukamal, Kenneth and Eric B. Rimm. "Is alcohol good or bad for you? Yes." *Harvard Public Health*. Published: 22 August 2024. Accessed: 10 September 2024. https://harvardpublichealth.org/policy-practice/is-alcohol-bad-for-you-or-is-alcohol-good-for-you-yes/

Muscatine, Doris, Maynard A. Amerine, and Bob Thompson, eds. *The University of California / Sotheby Book of California Wine*. Berkeley: University of California Press, 1984.

Nabuurs, G-J., R. Mrabet, A. Abu Hatab, et al. "Chapter 7: Agriculture, Forestry and Other Land Uses (AFOLU)." (PDF) *Climate Change 2022: Mitigation of Climate Change*. https://www.ipcc.ch/report/ar6/wg3/downloads/report/IPCC_AR6_WGIII_FullReport.pdf

NASA Science Editorial Team. "Steamy Relationships: How Atmospheric Water Vapor Amplifies Earth's Greenhouse Gas Effect." Published: 8 February 2022. Accessed: 4 September 2024. https://science.nasa.gov/earth/climate-change/steamy-relationships-how-atmospheric-water-vapor-amplifies-earths-greenhouse-effect/

Nigro, Dana. "Paul Dolan, Champion of Green Grapegrowing in California, Dies at 72." *Wine Spectator.* Published: 29 June 2023. Accessed: 10 September 2023. https://www.winespectator.com/articles/paul-dolan-champion-of-green-grapegrowing-in-california-dies-at-72

NIH. "Incorporating harm reduction into recovery." Published: 3 October 2023. Accessed: 1 September 2024. https://www.niaaa.nih.gov/news-events/research-update/incorporating-harm-reduction-alcohol-use-disorder-treatment-and-recovery

Nixon, Richard. "The Silent Majority Speech." Delivered: 3 November 1969. Accessed: 10 October 2023. https://watergate.info/1969/11/03/nixons-silent-majority-speech.html/

Northwest Medicine. "Cancer Care: Prevention & Early Detection: Reduce Your Risk." Accessed: 8 September 2024. https://www.cancer.northwestern.edu/cancer-care/prevention-early-detection/reduce-your-risk.html#:~:text=%22Food%2C%20Nutrition%20and%20the%20Prevention,directly%20linked%20to%20dietary%20choices

Olmstead, Alan L. and Paul W. Rhode. *A History of California Agriculture*. Oakland: the Regents of the University of California, 2017.

Ornelas-Higdon. Julia. *The Grapes of Conquest: Race, Labor and the Industrialization of California Wine 1769–1920*. University of Nebraska Press, 2023.

Ortolá, Rosario, Mercedes Sotos-Prieto, and Esther Garcia-Esquinas. "Alcohol Consumption Patterns and Mortality Among Older Adults With Health-Related or Socioeconomic Risk Factors." *JAMA*. Published: 12 August 2024. Accessed: 10 September 2024. https://jamanetwork.com/journals/jamanetworkopen/fullarticle/2822215

Parry, Manon. "Ralph Nader: Public Health Advocate and Political Agitator." *American Journal of Public Health*. Published: February 2011. Accessed: 3 January 2024. https://www.ncbi.nlm.nih.gov/pmc/articles/PMC3020208/

Paso Robles Daily News. "Winery celebrates six years of wind-powered energy." Published: 22 March 2024. Accessed: 10 September 2024. https://pasoroblesdailynews.com/winery-celebrates-six-years-of-wind-powered-energy/189858/

PennState Extension. "Alcoholic Beverage Consumption Statistics and Trends 2023." Published: 14 February 2023. Accessed: 10 September 2024. https://extension.psu.edu/alcoholic-beverage-consumption-statistics-and-trends-2023

Phillips, George H. *Vineyards and Vaqueros: Indian Labor and the Economic Expansion of Southern California, 1771–1877*. Norman: University of Oklahoma Press, 2010.

Pinney, Thomas. *The City of Vines: A History of Wine in Los Angeles*. Berkeley: Heyday; San Francisco: California Historical Society, 2017.

________. *The Makers of American Wine: A Record of Two Hundred Years.* Berkeley: University of California Press, 2012.

________. *A History of Wine in America: From Prohibition to the Present, Vol. 2.* Berkeley and Los Angeles: University of California Press, 2005.

________. *A History of Wine in America: From the Beginnings to Prohibition, Vol. 1.* Berkeley and Los Angeles: University of California Press, 2007.

________. "The Early Days in Southern California." In Muscatine, Amerine, and Thompson, eds., *Book of California Wine,* 2–9.

Planet Tracker. "Valuing the Global Food System: Food Systems, Circularity, Equity." Published: 21 February 2023. Accessed: 4 May 2024. https://planet-tracker.org/valuing-the-global-food-system/#:~:text=Moreover%2C%20further%20down%20the%20supply,16%20and%2020%25%20of%20GDP

Plastics Engineering. "Flat, Recycled PET Bottles Look to Disrupt Wine Packaging." Published: April 2024. Accessed: 24 September 2024. https://www.plasticsengineering.org/2024/04/flat-recycled-pet-bottles-look-to-disrupt-wine-packaging-004484/#!

Popp, Jamie. "2003 Wine Report." *Bev Industry.* Published: 1 April 2004. Accessed: 8 August 2024. https://www.bevindustry.com/articles/83546-2003-wine-report#:~:text=In%202003%2C%20wine%20shipments%20in%20the%20United,up%205%20percent%20to%20627%20million%20gallons

Portillo, Maria C., Erick Alberto Tena-Garcia, Eugenia Vila, et al. "Sanitisation of wine-ageing barrels using near ultraviolet radiation." *IVES: Technical Reviews, Vine & Wine.* Published: 29 March 2023. Accessed: 10 September 2024. https://ives-technicalreviews.eu/article/view/7467

Postel, Sandra. "Marjorie Spock: An Unsung Hero in the Fight Against DDT and in the Rise of the Modern Environmental Movement." *Nassau County Historical Society Journal.* Issue 75, 2020, 1–19.

Prothero, Donald R. *California's Amazing Geology, Second Edition.* Boca Raton: CRC Press, 2024.

Pruitt, Sarah. "The Post World War II Boom: How America Got into Gear." *History.* Published: 14 May 2020. Updated: 10 August 2023. Accessed: 2 April 2024. https://www.history.com/news/post-world-war-ii-boom-economy

_____. "The Nazis Developed Sarin Gas During WWII, but Hitler was Afraid to Use It." *History.* Published: 1 April 2019. Accessed: 2 April

2024. https://www.history.com/news/the-nazis-developed-sarin-gas-but-hitler-was-afraid-to-use-it

Public Policy Institute of California (PPIC). "Droughts in California." Published: April 2021. Accessed: 1 March 2024. https://www.ppic.org/publication/droughts-in-california/#:~:text=California%20has%20the%20nation's%20most,09%2C%20and%202012%E2%80%9216

Pulido, Laura. *Environmentalism and Economic Justice: Two Chicano Struggles in the Southwest.* Tucson: University of Arizona Press, 1996.

Raskin, Jonah. "Saved the Farm." *Pacific Sun.* Published: 28 April 2021. Accessed: 4 May 2024. https://pacificsun.com/saved-the-farm/

Reed, Roy. "Back-to-Land Movement Seeks Self-Sufficiency." *New York Times.* Published: 9 June 1975. Accessed: 8 May 2024. https://www.nytimes.com/1975/06/09/archives/backtoland-movement-seeks-selfsufficiency-the-growing-backtotheland.html

Ridder, M. "Sales volume of wines in Systembolaget stores in Sweden from 1st quarter of 2020 to first quarter of 2024." *Statista.* Published: 26 June 2024. Accessed: 18 August 2024. https://www.statista.com/statistics/756877/quarterly-sales-volume-of-wines-in-systembolaget-stores-in-sweden/#:~:text=Published%20by%20M.,Systembolaget%20in%20Sweden%202009%2D2023

Rieger, Ted. "Gallo Glass Plans to Install New Hybrid Electric Furnace to Decarbonize Glass Production." *Wine Business Monthly.* Published: 19 September 2024. Accessed: 20 September 2024. https://www.winebusiness.com/news/article/292719

_____. "UC Davis Winery Starts-Up New Clean-in-Place System." *Wine Business Monthly.* Published: 9 September 2024. Accessed: 10 September 2024. https://www.winebusiness.com/news/article/292197

Ristic, R., Anthea Fudge, Kerry A. Pinchbeck, and Robert De Bei. "Impact of grapevine exposure to smoke on vine physiology and the composition and sensory properties of wine." *Theoretical and Experimental Plant Physiology.* Vol. 28, No. 1, February 2016.

Roberts, Joe. "Uber-critic Robert Parker Drops the Gloves in Sommelier Journal." *1 Wine Dude.* Published: 2 May 2012. Accessed: 8 October 2024. https://www.1winedude.com/uber-critic-robert-parker-drops-the-gloves-in-sommelier-journal-interview/

Robinson, Jancis. "Bottle fight: Novelty v classic wines." *Financial Times.* Published: 14 February 2014. Accessed: 8 October

2024. https://www.ft.com/content/31c989da-8829-11e3-8afa-00144feab7de#axzz2tVQmm7pC

Robinson, Jancis, Julia Harding, and José Vouillamoz. *Wine Grapes*. New York: HarperCollins Publishers, 2012.

Rodale Institute. "Regenerative Organic Viticulture and the Soil Carbon Solution." Published: 2020. Accessed: 1 September 2024. https://rodaleinstitute.org/wp-content/uploads/Rodale-Soil-Carbon-White-Paper_v11-compressed.pdf

____. "Regenerative Organic Agriculture and Climate Change." Published: 2014. Accessed: 1 September 2024. https://rodaleinstitute.org/wp-content/uploads/rodale-white-paper.pdf#:~:text=With%20the%20use%20of%20cover%20crops%2C%20compost%2C,needle%20past%20100%%20to%20reverse%20climate%20change.&text=Simply%20put%2C%20recent%20data%20from%20farming%20systems,to%20widely%20available%20and%20inexpensive%20management%20practices

Romano, Aaron. "How Did 2020's Wildfires Impact California Wine?" *Wine Spectator*. Published: 23 March 2021. Accessed: 8 October 2024. https://www.winespectator.com/articles/how-did-2020-s-wildfires-impact-california-wine

Rosenfeld, Seth. *Subversives: the FBI's War on Student Radicals, and Reagan's Rise to Power*. New York: Farrar, Straus and Giroux, 2012.

Rothstein, Richard. *The Color of Law: A Forgotten History of How Our Government Segregated America*. New York: Liveright Publishing Corporation, 2017.

Rumgay, Harriet, Neil Murphy, Pietro Ferrari, and Isabelle Soerjomataram. "Alcohol and Cancer: Epidemiology and Biological Mechanisms." *Nutrients*. Vol. 13, Issue 9, 3173. Published: 11 September 2021. Accessed: 8 September 2024. https://pmc.ncbi.nlm.nih.gov/articles/PMC8470184/#:~:text=More%20than%2030%20years%20ago,available%20epidemiological%20and%20experimental%20evidence

Salvucci, Jeremy. "What Was The Dot-Com Bubble, And Why Did It Burst?" *The Street*. Published: 12 January 2023. Accessed: 8 August 2024. https://www.thestreet.com/dictionary/dot-com-bubble-and-burst#:~:text=In%20the%20early%2090s%2C%20the,that%20was%20sweeping%20Wall%20Street

"The San Joaquin Valley: Past, Present, Future and from the air." *On the Road, the California Swing*. Engineering with Nature.

Published: 20 January 2022. Accessed: 9 September 2024. https://ewn.erdc.dren.mil/the-san-joaquin-valley-past-present-future-and-from-the-air/#:~:text=The%20first%20irrigation%20canal%20in,aquatic%20ecology%20as%20a%20child)

Sassen, Saskia. *The Mobility of Labor and Capital: A Study in International Investment and Labor Flow.* Cambridge: Cambridge University Press, 1988.

Schwartz, Ariel. "Winery's Solar Tracking System Makes for a Greener Tipple." *Fast Company.* Published: 20 February 2009. Accessed: 10 September 2024. https://www.fastcompany.com/1173458/winerys-solar-tracking-system-makes-greener-tipple

Selley, Chris. "A Scorching New Critique of Canada's 'Pseudoscientific' Alcohol Guidelines." *National Post.* Published: 27 February 2023. Accessed: 8 September 2024. https://nationalpost.com/opinion/critique-of-canadas-alcohol-guidelines

Sette, Samantha. "11 Boxed Wines Industry Insiders Love." *Wine Enthusiast.* Published: 28 July 2022. Accessed: 24 September 2024. https://www.wineenthusiast.com/ratings/best-boxed-wine/?srsltid=AfmBOooo07RMNiSJZ6XBwrzPqCEQXuExHPrXM7ezOySDAku-k73a86N2O

Shabam, Patrick L. "The limitations of the Winkler Index." *Wines & Vines.* January 2019.

Shah, Vikas. "The reality of Silicon Valley." *Thought Economics.* Published: 19 October 2023. Accessed: 1 October 2024. https://thoughteconomics.com/the-reality-of-silicon-valley/#:~:text=The%20San%20Francisco%20Bay%20Area%20(more%20commonly,a%20little%20smaller%20than%20Turkey%20and%20Indonesia.&text=It's%20now%20multi%2Dbillion%20dollar%20businesses

Shaw, Randy. *Beyond the Fields: Cesar Chavez, the UFW, and the Struggle for Justice in the 21st Century.* Berkeley: University of California Press, 2008.

Smart, Richard. "Carbon footprints, wine and the consumer." *JancisRobinson.com.* Published: 4 December 2019. Accessed: 10 September 2024. https://www.jancisrobinson.com/articles/carbon-footprints-wine-and-consumer#:~:text=Studies%20in%20Australia%20and%20elsewhere,than%20some%20other%20wine%20producers)

Smith, Barbara. "Bonny Doon Vineyard – First US Winery to Launch

Low Carbon Footprint Paper Bottle." *Wine Business Monthly.* Published: 18 April 2024. Accessed: 24 September 2024. https://www.winebusiness.com/news/article/286294

Snowden-Seysses, Diana. "Carbon Capture During Fermentation Could Make Wine a Negative-Emission Industry." *Seven Fifty Daily.* Published: 12 July 2021. Accessed: 10 September 2024. https://daily.sevenfifty.com/carbon-capture-during-fermentation-could-make-wine-a-negative-emission-industry/

Sonderling, Don. "Then … and Now." *Wine Country This Week.* Published: 20 June 2021. Accessed: 3 March 2024. https://napagreen.org/40-years-of-wine-country-this-week/

Sowers, Stacey K. *¡Si, Ella Puede!: The Rhetorical Legacy of Dolores Huerta and the United Farm Workers.* Austin: University of Texas Press, 2019.

Stockwell, Tim, Robert Solomon, Paula O'Brien, et al. "Cancer Warning Labels on Alcohol Containers: A Consumer's Right to Know, a Government's Responsibility to Inform, and an Industry's Power to Thwart." *J Studies Alcohol Drugs.* Vol. 81, Issue 2, 284–92. Published: March 2020. Accessed: 8 September 2024. https://pubmed.ncbi.nlm.nih.gov/32359059/

Stormwater Report. "UC Davis, GE, and Winesecrets Turn Stormwater into Wine." Published: 4 April 2017. Accessed: 10 September 2024. https://stormwater.wef.org/2017/04/uc-davis-ge-winesecrets-turn-stormwater-wine/

Sullivan, Charles. *A Companion to California Wine.* Berkeley: University of California Press, 1998.

______. *Napa Wine: A History.* San Francisco: The Wine Appreciation Guild Ltd, 1994.

Swanson, Charles. "10 largest wildfires in California history by acreage." *The Press Democrat.* Published: 30 July 2024. Accessed: 10 August 2024. https://www.pressdemocrat.com/article/news/10-largest-wildfires-in-california-history-by-acreage/#:~:text=The%20Creek%20Fire%2C%20which%20started,View%20Image%20Gallery

Systembolaget. "Different for a reason: Responsibility Report 2023." (PDF) Published: 21 March 2023. Accessed: 8 May 2024. https://www.omsystembolaget.se/globalassets/pdf/ansvarsredovisning/systembolaget-responsibility-report-2023.pdf

Teiser, Ruth and Catherine Harroun. "The Volstead Act, Rebirth, and Boom." In Muscatine, Amerine, and Thompson, eds., *Book of California Wine*, 50–81.

________. *Winemaking in California*. New York: McGraw-Hill Book Company, 1983.

Thach, Liz. "Calculating the ROI of Sustainability at Jackson Family Wines." *Forbes*. Published: 28 November 2023. Accessed: 24 September 2024. https://www.forbes.com/sites/lizthach/2023/11/28/calculating-the-roi-of-sustainability-at-jackson-family-wines/

Thordarson, Thorvaldur and Stephen Self. "Atmospheric and environmental effects of the 1783–1784 Laki eruption: A review and reassessment." *Journal of Geophysical Research: Atmospheres*. Vol. 108, Issue D1, 16 January 2003.

Tian, Yalan, Jiahuui Liu, Yue Zhao, et al. "Alcohol consumption and all-cause and cause-specific mortality among U.S. adults: prospective cohort study. *BMC Medicine*. Vol. 21, Article 208, 2023. Published: 7 June 2023. Accessed: 14 September 2024. https://bmcmedicine.biomedcentral.com/articles/10.1186/s12916-023-02907-6#citeas

Time. "Industry: American Wine Comes of Age." Published: 27 November 1972. Accessed: 5 November 2021. https://time.com/archive/6877663/industry-american-wine-comes-of-age/

Tower, Jeremiah. "James Beard: America's First Foodie." *American Masters Digital Archive* (WNET). Released: March 7, 2014. Accessed: 1 April 2024. https://www.pbs.org/wnet/americanmasters/archive/interview/jeremiah-tower/

Townsend, Peggy. "The birth of UC Santa Cruz: audacious and academic." *UC Santa Cruz Newscenter*. Published: 1 January 2015. Accessed: 14 October 2023. https://news.ucsc.edu/2014/12/birth-of-ucsc.html

TTB. "Established American Viticultural Areas: California." Updated: 29 October 2024. Accessed: 20 November 2024. https://www.ttb.gov/regulated-commodities/beverage-alcohol/wine/established-avas#California

Twenge, Jean M. "How Gen-Z Changed its Views on Gender." *Time*. Published: 1 May 2023. Accessed: 1 September 2024. https://time.com/6275663/generation-z-gender-identity/

UC Davis. "Economy and weather put the squeeze on wine grape supply, survey finds." Published: 23 September 2011. Accessed: 3 October 2024. https://phys.org/news/2011-09-economy-weather-wine-grape-survey.html

______. "The LEED Platinum Teaching and Research Winery." Accessed: 24 September 2024. https://wineserver.ucdavis.edu/about/facilities/leed-platinum-teaching-and-research-winery#:~:text=Early%20

in%202011%2C%20UC%20Davis,totaling%20more%20than%20%2420%20million

United States Department of Defense. "75 Years of the GI Bill: How Transformative It's Been." Published: 9 January 2019. Accessed: 14 October 2023. https://www.defense.gov/News/Feature-Stories/Story/article/1727086/75-years-of-the-gi-bill-how-transformative-its-been/

United States Department of Health and Human Services. *Alcohol and Cancer Risk: The U.S. Secretary General's Advisory, 2025*. Published: 3 January 2025. Accessed: 5 January 2025. https://www.hhs.gov/surgeongeneral/priorities/alcohol-cancer/index.html

United States Department of State, Office of the Historian, Foreign Service Institute. "Chinese Immigration and the Chinese Exclusion Acts." *Milestones: 1866–1898.* Accessed: 21 November 2023. https://history.state.gov/milestones/1866-1898/chinese-immigration

United States Department of Treasury, Alcohol and Tobacco Tax and Trade Bureau. *American Viticultural Area (AVA) Manual for Petitioners*. (PDF) P 5120.4, September 2012.

USGS. *California Water Science Center.* "California's Central Valley: Regional Overview." Accessed: 1 October 2024. https://ca.water.usgs.gov/projects/central-valley/about-central-valley.html

Vankin, Jonathan. "How California Agriculture Has Shaped the State and the Country." *California Local.* Published: 8 May 2023. Accessed: 15 May2023.https://californialocal.com/localnews/statewide/ca/article/show/36707-california-agriculture-dairy-wheat-fruit-vegetables/

Verallia. "Verallia inaugurates its first 100% electric furnace." Published: 10 September 2024. Accessed: 9 December 2024. https://www.verallia.com/en/communique-de-presse/cognac-electric-furnace-inauguration/

Villarejo, Don. *New Lands for Agriculture: The California State Water Project.* Davis: The California Institute for Rural Studies, 1981.

Walker, M. Andrew. "UC Davis' Role in Improving California's Grape Planting Materials." In *Proceedings of the ASEV 50th Anniversary Meeting, Seattle, Washington, June 19–23, 2000,* 209–15.

Wasilewski, Tomasz, Zofia Hordyjewicz-Baran, Magdalena Zarębska, et al. "Sustainable green processing of grape pomace using micellar extraction for the production of value-added hygiene cosmetics." *Science & Wine*. Published: 19 June 2022. Accessed: 24 September 2024.https://www.ciencia-e-vinho.com/2022/06/19/sustainable-green-processing-of-grape-pomace-using-micellar-extraction-for-the-production-of-value-added-hygiene-cosmetics/#:~:text=The%20study%20showed%20that%20grape%20pomace%20produced,cosmetics.%20Professor%20Tomasz%20Wasilewski%2C%20PhD%20DSc.%20Eng

Waters, Alice. *Coming to My Senses: the Making of a Counterculture Cook.* New York: Clarkson Potter/Publishers, 2017.

Weed, Augustus. "Is California Preparing an Internet Wine Crackdown?" *Wine Spectator.* Published: 21 July 2009. Accessed: 10 October 2023. https://www.winespectator.com/articles/is-california-preparing-an-internet-wine-crackdown-40283

White, Dan. "Slow Food Nation: Conversation with Alice Waters." UC Santa Cruz Newscenter. Published: 5 August 2015. Accessed: 3 September 2023. https://news.ucsc.edu/2015/08/alice-waters-q-and-a.html

White, Kelli. "Merlot: Rise, Demise, and Rebirth." *GuildSomm.* Published: 24 October 2019. Accessed: 3 October 2024. https://www.guildsomm.com/public_content/features/articles/b/kelli-white/posts/california-merlot

Willcox, Kathleen. "The Astounding Impact of Come Over October: Demanding Attention and Opening Minds." *Wine Industry Advisor.* Published: 28 October 2024. Accessed: 30 October 2024. https://wineindustryadvisor.com/2024/10/28/the-astounding-impact-of-come-over-october-demanding-attention-and-opening-minds/#:~:text=An%20Island%20of%20Positivity%20in,advertising%20and%20social%20media%20campaigns

Williams, David R., Jourdyn Lawrence, and Brigette Davis. "Racism and Health: Evidence and Needed Research." *Annual Review of Public Health*, Vol. 40, 2 February 2019, 105–25.

Wine Cap. "Banking failures and the fine wine market: Performance during economic uncertainty." Accessed: 3 October 2024. https://winecap.com/learn/banking-failures-and-the-fine-wine-market-performance-during-economic-uncertainty#

Wine Institute. "California Wine Sales Reach $43.6 Billion in U.S. Market in 2019." Published: 2020. Accessed: 8 August 2024. https://discovercaliforniawines.com/news/california-wine-sales-reach-43-6-billion-in-u-s-market-in-2019/

_____. "California Wine 2017 Harvest Report." (PDF) Published: 2018. Accessed: 8 August 2023. https://wineinstitute.org/wp-content/uploads/2020/07/CaliforniaWineHarvestReport2017_CalifWines.pdf

Wine Spectator. "2011 Vintage Report: California." Published: 17 November 2011. Accessed: 3 October 2024. https://www.winespectator.com/articles/is-california-preparing-an-internet-wine-crackdown-40283

_____. "2010 Vintage Report: United States." Published: 17 November 2010. Accessed: 3 October 2024. https://www.winespectator.com/articles/2010-vintage-report-united-states-44066

Winetitles Media. "Grapevine research solves Chardonnay clonal mystery." Published: 4 August 2020. Accessed: 1 November 2024. https://winetitles.com.au/grapevine-research-solves-chardonnay-clonal-mystery/

Winiarski, Warren. "Creating Classic Wines in the Napa Valley," an oral history conducted in 1991, 1993 by Ruth Teiser, Regional Oral History Office, The Bancroft Library, University of California, Berkeley, 1994.

Winkler, A. J. *General Viticulture.* Berkeley: University of California Press, 1965.

Winkler, Klara J. and Kimberly A. Nicholas. "More than wine: Cultural ecosystem services in vineyard landscapes in England and California." *Ecological Economics.* (PDF) Vol. 124, April 2016, 86–98. https://www.sciencedirect.com/science/article/pii/S0921800916301227

Wise, Jason, director. *SOMM: Cup of Salvation.* Documentary. 2023.

World Health Organization News Release. "No Level of Alcohol Consumption is Safe for Our Health." *World Health Organization,* 2023. https://www.who.int/europe/news/item/04-01-2023-no-level-of-alcohol-consumption-is-safe-for-our-health

"Y2K." *Smithsonian Spotlight.* Published: 11 September 2019. Accessed: 8 August 2024. https://www.si.edu/spotlight/y2k

Yarrow, Alder. "Shake Ridge Ranch in all its glory." JancisRobinson.com. Published: 19 June 2024. Accessed: 8 August 2024. https://www.jancisrobinson.com/articles/shake-ridge-ranch-all-its-glory

_____. "Debating Robert Parker." *Vinography.* Published: 27 January 2024. Accessed: 10 October 2024. https://www.vinography.com/2014/01/debating_robert_parker

_____. "The Silicon Valley effect on wine." JancisRobinson.com. Published: 21 March 2017. Accessed: 1 March 2023. https://www.jancisrobinson.com/articles/the-silicon-valley-effect-on-wine

York, Alan. "About the Author." *Unpublished manuscript.* Courtesy of Rodrigo Soto, 2013.

INDEX

Page numbers in *italic* refer to pictures

Abba, Louie Jr and Joe *262*
Accendo winery 160, 161
Adams, Mark 234
Adelaida District 231, 234, 235, 240, 383
Adelaida winery 236–7
agrichemicals 48–52, 348–9
Agricultura General 17
Agricultural Labor Relations Act 52–3, 370
Alameda County (Central Coast) 52, 212–14, 381
Alaska 45, 101, 113, 115, 177, 285, 316
Alban, John 232, 233
Alcohol and Tobacco Tax and Trade Bureau 211, 250, 361–4, 391, 392, 394
alcohol content 361–4
Alexander Valley 136–7, *137,* 143, 376
  wineries 156, 157, 200, 338, 343
Alisos Canyon 244
Allegro 307
Almaden winery 52, 60, 137, 192, 221, 261
Alta California *see* Mexican Alta California; Spanish Las Californias
Alta Mesa 260, 264, 385
Amador County (Sierra Foothills) 207, 274–6, 278–81, 387
  wineries 284
Amista winery 344
Ancient Peaks winery 237
Anderson Valley 112, 117, 118, 119, 120–23, 129, 375
  wineries 124, 151, 199,
Antelope Valley of the California High Desert 291–2, 294, 388
anthocyanins 327, 390, 409
Araujo winery 125, 160, 165, 167, 174
Armington, Bob 45
Arnot Roberts winery 108, 128, 144, 278
Arroyo Seco 102, 213, 216, 217, 218, 237, 382
  wineries 213, 216, 237
Asseo, Stephen 238
Association of African American Vintners 6, 87
Atlas Peak 168, 170, 379
Au Bon Climat winery 81–2, 241, 245–6, 249
Australia 7, 102, 126, 154, 166, 208, 213, 268, 270, 328, 329
AVAs (American Viticultural Areas) listed
  Central Coast 381–4
  Central Valley 385–6
  Far North 387–8
  North Coast 168, 375–80
  Santa Cruz Mountains 380
  Sierra Foothills 387
  Southern California 388–9
  see also individual AVAs
AVAs, designation of origin system 361–2
AXR1 rootstock 165–6, 351, 391, 408

B Corp 184, 365, 367
Babcock, Brian 228
Baby boomer generation (Boomers) 86, 310–13, 316, 318, 321, 357–8, 399
Bacigalupi family 141
Back-to-the-Land 52, 61–3, 65–6, 126, 283, 391
Bag-in-a-box 205, 240, 314, 316, 339, 391
Ballard Canyon 101, 244, 248, 384
Baxter winery 121, 123–4
Beaulieu Vineyards 29, 33, 68, 127, 159, 162–3, 166

Beckmeyer, Hank 265, 283
Beckstoffer, Andy 127, 166
Bedrock winery 107, 135, 138, 144–5, 210, 280
Beinstock, Gideon 276, 282, 283
Ben Lomond 198, 380
Benmore Valley 128, 129, 376
Bennett Valley 143, 157, 378
Benzinger Family 65, 125, 139, 141–2
Beringer 127, 162
Bien Nacido winery 241, 246
Big Valley District 129, 377
biodynamics 48, 63–4, 65, 117, 119, 120, 132, 138, 139–42, 148, 154, 155, 169, 176, 180, 181, 184, 191, 197, 224, 227, 238, 355, 365, 368, 369, 391
BIPOC 7, 87, 88, 208, 300, 391–2
  see also Black Americans; Indigenous people
Birichino winery 198, 222
Bisceglia Brothers 42
Black Americans 6–7, 38–9, 44, 45–6, 314–5, 391–2, 407
  Association of African American Vintners 6, 87
  and civil rights 38, 46, 53, 56, 57–8
  internships 88
  winery owners 115, 124, 171, 247–8
  Two Eighty Project 208, *209,* 300
Black Panther party 46, 59
Bohan family 132
Boisset, Jean-Charles 138
Bokisch Family winery 265
bonded wineries 31, 34, 68, 77, 196, 392
Bonny Doon 154, 190, 195–6, 197, 198, 204–5, 232
Bonterra (earlier Fetzer) 65, 119–20, 139, 261
Borden Ranch 260, 264, 265, 385
Boulton, Dr. Roger 336–7
Bracero program 44–5, 48, 50, 392
brand performance 311–12
Brecher, John 96
Brick & Mortar winery 307
Broc Cellars 60, 207
Brown Estate winery 171
Brown, Jerry 337
Bruce, David 194
Buena Vista winery 132, 137–8, 162
building design 336–7, 342–3, 366
bulk table wines 41–2
Bulosan, Carlos, *America is in the Heart* 45
Burgess Cellars 67
Busch, Pamela 87
Byron Blatty winery 291, 295

Cabernet Sauvignon
  history and significance 2, 78, 96, 176–7, 212, 231–2
  plantings by AVA 380–89
  plantings by hectare and county 111
  vineyards *95, 116*
Cain (Napa Valley) 160, 330
Cakebread, Rosemary 160
Calaveras County 3, 175, 275–6, 280–81, 282
Calera winery 81, 101, 222, 223, 224
California Agricultural Relations Act (1975) 52, 370
California as US territory 1, 20–22
California Certified Organic Farmers 125, 368
California Certified Sustainable 368
California climate
  climate map 100
  droughts *versus* floods 103, 105–6
  Mediterranean climate 93–4, 101–2
  mesoclimates 106
  variability 324–5
  winds 102–3
  see also climate change; growing conditions
California Delta 3, 32, 103, 207, 251–4, *252,* 257–9, *258,* 265, 276
California Department of Agriculture 33, 357
California farming
  1920s–1975 timeline 40
  big volume businesses 41–2
  infrastructure development 43–4
  innovation and diversity 358–60
  regenerative 302–3, 343–9
  see also California wine, biodynamics, historical events; climate change; immigrant farmworkers; sustainability
California grape varieties 106–111
California Land Stewardship Institute 367
California soils 94
California topography and geology *92,* 93, 97–101
California wine, historical events
  Eighteenth Amendment and Volstead Act 23–4, 28, 373
  global forces and
    financial crises (21st century) 78–9
    historic 5–7
  Granholm *vs.* Heald 261
  grape varieties 2–4, 106–111
    hectares and counties 94, 111
    wild grapevines 16–17
  history overviews 5–7, 115, 289–90, 351–3
  history timelines
    (1769–1899) 12
    (1900–1967) 25
    (1920–1975) 40
    (1960–1991) 54
    (1991–2025) 74
  movements
    Back-to-the-Land 61–3, 391

Demeter Biodynamic Certification 368–9
land back 356–7
natural wine 73, 75, 86–8
In Pursuit of Balance 81–2
Regenerative Organic Certification 369
organizations
California Land Stewardship Institute 367
California Wine Association 27, 29, 33, 351
California Wine Institute 85
Historic Vineyard Society 135
Wine Advisory Board 33
Wine Institute 32–3 32–4, 35, 85, 334, 351, 357, 368
regeneration and innovation 350–51
demographics/buying habits 357–8
land back movement 356–7
lessons from farming 353–5
lessons from history 351–3
vineyards *see* individual AVAs; individual wineries
see also California farming; cultural significance of wine
California Wine Association 27, 29, 33, 351
California Wine Institute 32–4, 35, 85, 334, 351, 357, 368
Calistoga 162, 168, 169, 170, 174–5, 179, 183, 379, 380
Cambie, Philippe 157
Camins2Dreams 205, 222, 230, 246–7
Canada 15, 89, 116, 316, 318, 319, 328, 335, 373
Cannard, Bob 59–60, 66, 141
Cantisano, Amigo Bob 125
Caparoso, Randy 263–4
Capay Valley 256–7, 385
Cappiello, Patrick 263
Caraccioli winery 205, 218
Carmel Valley 216, 217, 219, 220, 382
Carneros 67, 125, 130, 140–41, 143, 159, 168–9, 173, 178, 183, 213, 378
Carson, Rachel, *Silent Spring* 48–49, 50, 64
Cathiard winery 161, 171–2
Caymus Cellars 67–8, 156, 306
Central Coast
AVAs 362, 381–4
map and known details 201, 204
Rhone Rangers 232–3
topography, climate, wines 202–4, 256–7
wineries 205–6
Central Valley 24, 28, 52
AVAs 259, 264, 271, 385–6
geology 99, 101, 251–5
irrigation, farming, climate 40, 42, 43–4, 103
map 272
sustainability 251–2
Chadwick, Alan 63–4, 120, 139
Chalk Hill 68, 141, 143, 157, 357
Chalone 216, 217, 220, 222, 381
Chappellet winery 108, 159, 168, 172
Chardonnay
history 2, 126, 140–41, 161, 193, 202–3, 212–13
plantings, hectares and counties 111
Vintner's Reserve 112, 126
Charles Krug winery 34, 159, 162, 163, 172, 179, 380
Chateau Montelena 67, 140, 162, 163, 164, 176, 183
Chateau Musar winery 299
Chavez, Cesar 45–6, 270, 352
chemicals 48–52, 64, 348–9
Cherry, Cris 238
Chez Panisse 58–61, 66, 81–2, 178, 228, 352
Chiles Valley 168, 170, 379
Chinese Exclusion Act 22, 26, 44
Chinese immigrants 22, 26, 256
Christian Brothers 52, 232
Cienega Valley 220, 222, 223–4, 382
Clarksburg 254, 256–7, 258–9, 385
Clear Lake 112, 116, 126, 128, 129, 376, 377
Clements Hills 260, 264, 265, 386
Clendenen, Jim 82, 241, 245–6, 249
climate action 150, 159, 335, 337, 344–9
certifications 365–9
see also climate change
climate, Californian *see* California climate
climate change 7, 301
bottling and packaging 338–40
farming practices and 302–3, 324–5, 338
chemicals 48–52, 348–9
land management 346–8, 355
farming, regenerative 343–9, 356–60
greenhouse gases 323–5
livestock 348
natural disasters 322–3
renewables 342–3
state and federal governance 334–5
water 336–8
wildfires 85–6, 325–31
adjusting approach to winemaking 331–2
winemaking, effects on 340–41
see also California climate; climate action; sustainability
Clos Saron 275, 276, 282
Cole Ranch 108, 121–2, 129, 375
Come Over October 300, 320–21
Comptche 122, 123, 376
Concannon family 3, 212, 213
Continuum winery 159, 163, 168, 172–3, 214
Contra Costa (Central Coast) 145, 203, 204, *207,* 209, 210–11, 362, 381

Coombsville 161, 168, 169, 171, 173, 175, 182, 213, 380
cooperative wineries 32, 34, 134, 261, 350, 412
Corison winery 159, 169, 173
Corison, Cathy 82, *115*, 159, 221
Corra winery 160
Cosumnes River 260, 264, 385
Coturri, Phil 65–6, 141–2, 156, 157
Covelo 64–5, 120, 122, 123, 139, 376,
Covid pandemic 75, 85–6, 302, 304–8, 318–19
Crabb, H.W. 162, 360
Creston District 236, 383
Cruse, Michael 157
Crystal Springs 168, 171, 380
Cucamonga Valley (Southern California) 291, 292, 293, 294, 296, 389
cultural significance of wine
  changing spending habits 318–19, 357–8
    generational changes 311–15
    harnessing trends 315–16
    two-tier spending 305–9
    Covid and 318–19
  economic impact 300–301
  egalitarianism 75
  employment figures 301
  health issues 316–20
  market future 309–11
  market investors 308–9
  market volatility 302, 304–6
  movie *Somm* 75, 83
  online and social media marketing 75–6, 82–5
  pro- and anti-wine lobbies 298–300
  production, environmental impact 302–3
  workers *see* immigrant farmworkers
  see also California wine, historical events; Prohibition

Dalla Valle winery 2, 160, 167, 183
Daniel, John 34, 163, 177
Darling winery 145
DDT and organophosphates 48–52, 64, 360, 395–6, 403
DEI (diversity, equity, inclusion) 88, 314–315, 396
Delta breezes 103, 210, 254, 258, 260
Demeter Biodynamic Certification 368–9
Di Giorgio Wine Company 42
Diablo Grande 271, 386
Diamond Mountain District 168, 170, 379
DiCostanzo winery 161, 173
direct-to-consumer 60, 71, 76, 80, 159, 261, 305, 396
distillers' buy-out 36, 41
distribution 22, 373–4
Dolan, Paul 65, 120, 139
Domain Dujac 340
Domaine Carneros 159, 174
Dos Limones 66
Dos Rios 122, 123, 376
Draper, Paul 60, 194, *195*, 199
Drew winery 81, 121, 124
Dry Creek River 344
Dry Creek Valley 65, 135–6, 138, 141, 142
  wineries 145, 200, 342, 344
Dry Creek winery 67, 68, 145–6, 344
DuMol winery 80, 146
Dunnigan Hills 256–7, 385
DuPratt *121*
Dust Bowl 345–6, 348, 353–4

Eagle Peak of Mendocino County 122, 123, 376
Eberle, Gary and Marcy 231–2
economic depressions 31–4, 47, 78–80
Eden Rift winery 98, 221, 223–4
Edmeades, Donald 68, 120–21
Edmonds St. John 60, 232, 277, 278
Edna Valley 102, 203, 225, 226, 383
Eighteenth Amendment and Volstead Act 23–4, 28, 373
Eisele winery 161, 174–5, 179
El Dorado County (Sierra Foothills) 275, 277–8
El Pomar District 236, 383
Empting, Cory 161, 169, 176, 181
Enfield Wine Co. winery 82, 132, 142, 146, 161, 210, 262, 280
enslavement 13–22, 392
environmental issues *see* sustainability
Environmental Protection Agency 371
Enz, Bob 221–2
Epoch Estate winery 204, 228–9, 230
Erickson, Andy 160, 175
Ernest and Julio Gallo 36–7, 145, 269, 360
Estrella Vineyard 231, 232
Estrella District 236, 283
Extradimensional Wine Co. Yeah! 121, 142, 146–7, *147*, 210, 280

Fair Play 278, 387
Far Mountain winery 142
Far North 108, 285–289, 387–8
farm-to-table movement 57, 58–61, 84, 397
farming *see* biodynamics, California farming; climate change; immigrant farmworkers; sustainability
Favia, Annie 160, 175, 208
Favia winery 160, 175, 280
Federal Department of Agriculture 49, 51, 64

fermentation, research into 27, 326–7, 328, 340
Fernandez, Elias 184
Fetzer (now Bonterra) 65, 119, 120, 125, 139, 262
Fetzer Food and Wine Centre 120
Fiddletown 279, 280, 387
Filipino workers 44–5, 53, 401
Filoli vineyard 110, 208–9
financial (housing) crisis 78–9
Fiorentini, Jordan 229, *229*
First Nation *see* Indigenous people
Fish Friendly Farming 117, 119, 159, 178, 367
FitVine 307
Flowers winery 132, 142, 147–8, 198,
Forlorn Hope winery 82, 275, 280, 282
Forrestel, Beth 208, 300
Fort Ross-Seaview 63, 98, 113, 132, 141, 143, 378
fortified wine 17, 30, 39, 41, 46, 55, 108, 279, 291, 293, 390, 398, 408
Fountaingrove District 143, 330, 378
Foxx, Prudy 190–91
Franciscan Sacred Expedition 15–16
Franscioni, Rosella and Gary 219
freedom movements 44–7, 53, 57–8
French Paradox 71, 78, 83, 166–7, 305, 317, 350
French Camp Vineyards 234
Frenchtown Farms 276, 282–3
Fresno County (Central Valley) 42, 386
Frey Vineyards 64–5, 119, 120, 139
Frog's Leap winery 125, 159, 161, 165, 169, 175–6
From, Vailia *229*

Gabilan Mountains 218, 220, 221, 222, 223, 382
Gallagher, Patricia 68, 164
Gallica winery 160
Gallo 36–7, 42, 52, 88, 125, 138, 196, 199, 261, 269, 339
Galloni, Antonio 84
Generation X (Gen-X) 86, 310–13, 316, 318, 321, 357–8, 399
Generation Z (Gen-Z) 86, 90, 302, 311–13, 316, 321, 357–8, 399
genocide 4, 13–22, 392
geology and topography of California *92,* 93, 97–101
GI Bill 37, 38, 41
Gibb, Rebecca 85
Gimelli, Ken 222
Gines, Ben 45
Glaab, Ryan and Megan *147,* 155–6
glass 270, 314, 333, 338–9, 341, 391
global financial crisis (21st century) 78–80, 82
global vineyard acreage (2022) 301
gold rush 1, 21, 115, 118, 206, 207, 259, 267, 273–4, 276–7, 290
Goldstein, Joyce 60, 61, 84
Gomez, Tara *205,* 246–8
Goode, Jamie 85
Graciano grape 228, 230, 250, 279, 295
Graf, Richard 222
Graham, Kate 142, 146
Grahm, Randall 98, 153, 190, 195–7, *196,* 198–9, 204, 205, 221, 224, 232
Granholm *vs.* Heald 261
grapes and smoke taint 86, 326–30
grapes, most planted 2–4, 106–11
  as of 2023, by county 111
  Cabernet Sauvignon *see* Cabernet Sauvignon
  Carignan 136–7
  Chardonnay *see* Chardonnay
  French Colombard 107
  Graciano 230
  Merlot 78
  Mission Grape 5, 17
  Napa Gamay (Valdiguié) grape 3
  Petite Sirah 3–4, 213
  Pinot Gris 224, 254, 258
  Pinot Noir 131, 222
  Sauvignon Blanc 161
  Syrah 231, 232–3
  Touriga Nacional 108
  Viña Madre grape 110, 290
  Zinfandel *see* Zinfandel
  see also planting; vineyards
Great Depression 31–4, 47
Green & Red 60
Green Valley of Russian River Valley 141, 143, 146, 377
Greenwood Ridge winery 121
Grgich Hills winery 165, 169, 176, 348
Grgich, Miljenko "Mike" 140, 163, 165, 176, 348
growing conditions
  California 93–4
  Lodi 259–60
  Napa Valley 168–9
  North Coast 116–17
  Paso Robles 233–5
  San Joaquin Valley 268, 270
  Santa Barbara 242–4
  see also California climate
Guenoc Valley 126, 128, 129, 376

H2A visa program 372
Hames Valley 217, 218, 383
Handley winery 121
Hanzell winery 81, 140, 148, *195,* 165, *195,* 198, 213

Happy Canyon of Santa Barbara 244, 245, 250, 384
Haraszthy, Agoston 137, 352
Harken wines 307
Harlan, Bill 176–7, 181
Harlan, Will 161, 169
Harlan winery 2, 167, 176–7, 181, 309
Harvey, Kevin 197
Hawk & Horse winery 130
@hawk_wakawaka 83, *140, 200*
health issues 316–20
Herrmann York 291, 295–6
High Valley 129, 376
Hill, Julia Butterfly 118, 285
hippie movement 65–6
Hirsch, David 63, 132, 148, 151
Hirsch Vineyard 81, 98, 132, 147
Hirsch, Jasmine 81, 148
Historic Vineyard Society 131, 135, 136, 145, 210, 402
Hochar, Serge 299, 300
Hochschule Geisenheim University 310
Hoffman, Dr. Stanley 237
Holland, Ashley 154–5
Howell Mountain 34, 37, 163, 168, 169, 171, 379, 380
Huerta, Dolores 45–6, 52
Humboldt County (Far North) 98, 286–287, 387
Husch winery 121

Illumination winery 161
immigrant farmworkers
  Acts 371
    Agricultural Labor Relations (1975) 370
    Immigration Reform and Control (1986) 372
    Migrant and Seasonal Agricultural Worker Protection (1983) 370–71
    National Labor Relations (1935) 47–8
    Occupational Safety and Health (1970) 51, 371
  cheap labour negotiations 47–8
  Chinese 22, 26
  collaboration 44–47, 53
  DDT and 48–52, 50
  Filipino 44–47, 401
  H2 visa program 372
  labor movement 41–2, 44–7
  Manong 44–5, 53, 401
  Manongs and (Braceros) Mexicans 44–5, 48
  organizations
    Agricultural Farm Workers Organizing Committee 45–6
    National Farmworkers Association 45–6
    United Farm Workers 40, 45–6, 50–53, 270
  shortages 22, 26, 42, 44, 47, 90, 305, 310
  strikes and boycotts 43, 45–7, 50–53, 269–70
  undocumented 89, 301, 371, 410
  see also California farming
immigration 4, 22, 26, 41–53, 89–90, 302, 345, 370–2
Immigration Reform and Control Act 372
Imutan, Andy 46
In Pursuit of Balance (IPOB) 81–2, 84
Indian Indenture Act (1850) 21, 26, 44
Indigenous people 5, 7, 13–22, 26, 87, 118, 208, 293, 391–2, 400
  apprenticeships 87, 208, 300
  bounty killings of 21–22
  and European influx 5, 13–16, 20–21, 255, 267–8, 293, 398
  enslavement, indentured servitude 5, 13–22, 289
  Indian Indenture Act (1850) 21, 26, 44
  internships 208, 300
  land, and land back 345, 356–7
  reservations 256, 345
  wineries 246–8, 356–7
industrial agriculture 61, 64, 323
industrialization 323–4
Industrial Revolution 322–4, 393
Inglenook winery 34, 68, 159, 162, 163, 177
Instagram 83, 85, 131, 209
International Organization of Vine & Wine 135, 402
International Wineries for Climate Action 150, 151, 184, 335–6, 341, 365–6
internet 1, 75–6, 77, 80, 83, 84, 312, 313
interspecific hybrids 27, 73, 87, 105, 109–10, 152
investment
  and adaptation 77–8
  celebrity investors/growers 129
  distillers' buy-out 36, 41
  financial (housing) crisis 78–9
  and tax 66–8
  wine investors 308–9
Inwood Valley 286, 288, 388
Iruai winery 108, 288
Italian Americans 117, 134
Italian-Swiss Colony 134
Itliong, Larry 45–6, 270

J Winery 149
J. Lohr winery 216, 232, 237–8, 342
Jackson family 88, 150, 151, 157, 335, 343, 345
Jackson, Jess 126–7, 360
Jahant 260, 264, 265, 385
James Berry Vineyard 232, 242

Jardine, Jason 147, 197–8
Japanese people
 anti-immigration laws 44
 internment 44
 laborers 45, 53
Jensen, Josh 101, 222–3
Jimenez, Ivo 176, 348
Johnson, Eric 225
Johnson, Hugh 96
Johnson, Julie 160, 175
Jolie-Laide 82, 108, 121, 149, 278
Jonata and The Hilt winery 248
Jordan Winery 67, 149, 156
Jordan, Judy 149
Jordan, Tom 67
Joseph Swan 60, 131, 149–50

Kandler, Nathan 262
Keller winery 138, 150, 185
Kelsey Bench 129, 377
Kendall-Jackson 88, 112, 126, 127, 151, 338, 360
Kern County (Central Valley) 42, 388
King, Martin Luther 56, 65
Kistler, Steve 147, 148, 152, *153*
Klein, Jill 160, 179
Knights Valley 143, 157, 377
Koshitzky, Maayan 161
Kraemer, Ann 279–80
Krankl, Manfred and Elaine 249–50
Krug, Charles 34, 159, 162, 163, 172, 179, 360

L'Aventure winery 238
La Clarine Farm winery 275, 283
La Cresta Blanca 202–3
La Belle winery 161
La Questa 192, 194
Labbé, Justine 171
labor shortages 22, 26, 42, 44, 47, 90, 305, 310
labelling 361, 363–4
Lake County (North Coast) 116, 126–9
 AVAs 129, 326–7
 known details of 126
 wineries 126, 127–8, 129, 130
Lamorinda 210, 381
land back movement 356–7
LangeTwins winery 261, 265, 343
Langtry, Lillie 129
Lazy Creek winery 121
Ledge winery 234
Lee, Theodora 81, *115,* 124
LEED 336, 343, 365, 366
Lemon, Ted 151
Leona Valley (Southern California) 292, 294, 388
Leviathan winery 160
Lifestyle, Diet, Wine & Health congress 320
Lime Kiln Valley 180, 221–2, 382
Lindquist winery 227–8
Lindquist, Bob 204, 228, 232–3, 241, 245
Littorai winery 81, 121, 148, 151
Livermore Valley 3, 36, 202, 208–9, 212, 213, 214, 215, 381, 411
Living Building 156, 342–3, 366
Lockwood, John 82, 132, 146,161, 262, 280
Lodi *253,* 259–64
 AVAs 264, 385–6
 growing conditions 259–60
 known details of 259
 planted area 260–61
 varieties 108, 264
 wineries 32, 166, 261–3
  listed 265–7
Lodi Native 263–4
Lodi Rules 334, 368
Lodi Winegrape Commission 334
logging 118, 285
Lohr, Jerry 237
Lone Star winery 344
Long Valley-Lake County 129, 377
Long, Zelma 159, 163
Los Angeles 13, 19–22, 61, 206, 290–91
 producers 110, 291
Los Angeles County (Southern California) 294, 295
Los Angeles Vintners Association 291
Los Olivos District 244, 384
Luna, Miguel 161
Lynch, Kermit 61

MacDonald, Graeme and Alex 161, 179
MacDonald wines 159, 161, 179
MacNeil, Karen 96, 299–300
Madera County (Central Valley) 111, 268, 271, 386
MAHA winery 238
Mahle, Pax *147,* 152–3, 191, 263
Malibu (Southern California) 240, 289, 291, 294, 295, 388
Mangaoang, Ernesto 45
Manong 44–5, 53, 401
Manton Valley 286, 288, 388
Manuel, Pete 45
Marin and Solano Counties 112, 186–7, 378, 380
market volatility 55–6, 302
Martin, Neil 85
Martin Ray winery 193, 195, 199
Martini, Louis M. 34, 138, 140, 163, 213
Martini, Louis M. Winery 34, 127, 138, 140, 159, 160, 163, 171–2, 178–9

Martini, Louis P. 34, 127, 138
Massican winery 160
Masson, Paul 29, 52, 193, 194, 211
Matthiasson winery 81, 160, 179, 191
Matthiasson, Steve 87, 165, 208, 262, 300
Mayacamas winery 66, 141, 160, 163, 180
McCoy, Carlton 88
McDowell Valley 122, 123, 376
Mendocino County (North Coast) 117–25
  AVAs 122–3, 375–6
  known details of 117
  wineries 119–24
Mendocino Ridge 121, *121*, 123, 375
Meredith, Carole 96, *115*, 233
Meridian winery 232
Merlot 78, 83, 107, 157, 169, 233, 235, 244, 257
Merritt Island 257, 258, 259, 385
Mexican Alta California 5, 13–14, 18–21, 161–2, 255, 390
  land management 255–6
  wine 19–20
Mexican-American workers 44–5
Meyer Family winery 344
Michael David Winery 267
Migrant and Seasonal Agricultural Worker Protection Act 370–71
migrant workers *see* immigrant farmworkers
Milliken, Beth Novak 184
Millennial generation 86, 90, 302, 311–13, 316, 321, 357–8, 399
Mission Grape 5, 13, 17, 94, 107, 110, 115, 259, 273, 278, 279, 289, 290, 390, 402, 411
Mokelumne River *207*, 259, 260–61, 264, 265, 386
Monarch Challenge 214–15
Monarch Tractor 1, 6, 203, *214*, 214–15
Mondale, Walter 50
Mondavi family 163, 172–3
Mondavi, Carlo 214–15
Mondavi, Peter 34, 163, 172
Mondavi, Robert 34, 66, 140, 148, 163, 166, 176, 179, 180, 235, 236, 237, 261, 265, 352, 384
Monte Rosso 138
Monterey 4, 5, 18, 26, 93, 102, 106, 111, 127, 139, 203–4, 215–20
  AVAs 382–3
  vineyards (1970s) 216, 217–8, 268
  wineries 218–20
Moon Mountain District of Sonoma County 138, 141–2, 143, 179, 198, 378
Moore, Alecia 250
Mount Eden vineyard 193, 199
Mount Harlan 220, 223–4, 382
Mount Veeder 60, 68, 162, 163, 168, 170, 180, 379
Mulville, Kelly 222

Nader and General Motors 70
Napa Gamay grape 3, 277
Napa Green 368
Napa Valley *95*, 158, *160*
  AVAs 169–71, 378–80
  brand 308
  economic significance 158–9
  growing conditions 168–9
  initiatives and research groups 6, 33–4, 159, 352
  Monarch Challenge 215
  Napa River 105, 159
  producers 159–61
  soils and climate 105–6, 168–9
  vineyard history
    first vines 161–3, 165
    (1970s–1980s) 68, 274
  phylloxera 162–3, 165–8
  wildfires 330, 332
  wineries 65, 66, 67–8, 81, 88, 127, 159–60, 306
    listed 171–86
    new wave 159–61, 167–8, 169, 177, 208, 335, 340, 348, 360
Napa Valley Enological Research Laboratory 33–4
National Labor Relations Act (1935) 47–8
natural wine movement 73, 75, 86–8
Navarro winery 60, 121
NBA players 88
neo-prohibitionism 316–20
New Clarivaux Vineyard 288–9
Newfound winery 180
Newsom, Gavin 356
Newton, Jeff 243
Neyers, Bruce 60–61
Nielsen, Uriel 241
Niess, Matt 87, 142
Nixon presidency 57, 69–70
North Coast 112–17
  2017 wildfires 328
  AVAs 375–80
  growing conditions 116–17
  known details of 126
  map *114*
  regions and counties 112
  wineries, listed 123–4
Northern California AVAs Map 287
Northern Sonoma 117, 122, 127, 130, 134, 143, 378
Novak, Mary 68, 184

O'Neill vinters 205, 307
Oak Knoll District 161, 168, 170, 380
Oakville 66, 162, 168, 170, 215, 233, 379
  wineries 173, 176, 179, 180, 181, 183
Obsidian Ridge winery 130
Occidental winery 152
Occupational Safety and Health Act (1970) 51, 371
1% for the Planet 184, 367
Opus One winery 180
Ordinaire, Oakland 87
organic farming, 63–6, 119, 125, 128, 139, 151, 169, 368
Osborne, Angela 245
Ovid winery 181

Pacheco Pass 212, 220, 381
packaging 339–40
Paderewski, Ignacy 231
Padilla, Gilbert 46
Paicines vineyard 222–3
Palos Verdes Peninsula (Southern California) 292, 294, 388
Parducci 66, 119, 127
Paris Exposition (1889) 162, 203
Paris Tasting (1976) 68, 164–5
Parker, Robert 2, 57, 70–2, 82–5, 183, 249–50
Parr, Raj 81, 225–6, 249
Parra, Pedro 139, 142, 147
Paso Robles 98, 101, 105, 108, 203–4, 225–6, 228–9, 230, 231–40, 241, 277, 325, 342, 354
  AVAs 235–6, 384
  growing conditions 233–5
  wineries listed 236–40, 284
Passalacqua, Tegan 82, *133*, 135, 211, 262, 266–7, 275, 279
  see also Turley; Sandlands
Paul Masson Mountain Winery 52, 193
Paulsell Valley 271, 386
Pax winery 152–3
Peace Movement 56–7
Peschon, Françoise 160, 165, 174
Petaluma Gap 102, 103, 138, 144, 150, 157, 185, 187, 378
Peterson, Joel 135, *136*, 144–5
Petroski, Dan 160
Petite Sirah grape 3–4, 111
Phelan Farms winery 108, 226–7
Phelps 60, 232
Philips, Kevin *266*
Phillips, Jean 183
Phillips, Michael and David 266
Pinney, Thomas 8
phylloxera 26–7, 162, 165–8
Pierce's Disease 22, 110
Pine Mountain-Cloverdale Peak 122, 123, 141, 144, 376
Pisoni winery 218–19
planting
  field blend, cultivars, hybrids 107–11
  and growing conditions 93–4
  hectares and counties 94
  replanting and marketing (late 19th century) 26–7
  surge (1970s) 68
  and Winkler Index 104–5
  see also grapes, most planted; vineyards; individual AVAs; individual wineries
Popelouchum 98, *196*, 197, 199, 224
Potter Valley 108, 122, 123, 376
pre-industrial winemaking 60, 189, 193, 194, 243
Prohibition 23–4, 28–30, 360, 373
  market post-Prohibition 30–31
  neo-prohibitionism 316–20
  see also cultural significance of wine
Promontory winery 161, 181
Puckett, Madeline 83

Quintessa winery 181–2
Quivira 65, 139, 344
Qupé 60, 228, 232, 241, 245

Radio-Coteau winery 121, 154
Ramey winery 154
Ramona Valley (Southern California) 295, 389
Ravenswood winery 135, *136*, 138
Raw Wine fair 87
Read Holland winery 154–5
Reagan, Ronald 49, 56, 58–9, 71, 231
Red Hills 63, 127, 129, 130, 376
Redwood Valley 120, 122, 123, 376
regenerative farming 302–3, 343–9, 353–7
  innovation and diversity 358–60
Regenerative Organic Certification 359–60, 369
Renaissance winery 283
Renfro, Christopher 87, 111, 208–9, *209*, 300
research 6, 33–4, 94–7, 104–5, 327–30, 352, 360, 393
  at UC Davis 2, 96–7, 110, 148, 212–13, 233, 268, 352
Rhys winery 197, 199
Richmond, Pete 161
Rideau, Iris 247, *258*
Ridge Geyserville 60, 135–7, 200
Ridge Lytton Springs 135, 342
Ridge Monte Bello winery 60, 95, 101, 194, *195*, 199–200, 306
Riesling 108, 118, 121, 132, 155, 167, 173, 198, 221, 237, 244, 274, 286

River Junction 271, 386
ROAR winery 219
Robert Biale winery 161, 182
Robert Mondavi winery 60, 66, 68, 161, 163, 193, 265
Roberts, Arnot 108, 128, 144
Robinson, Jancis 85, 96
Rochioli winery 155
Rockpile 135, 141, 143, 378
Rodale Institute 346–7, 354–5
Roederer 119, 124, 204
Rombauer winery 182–3
Rootdown 121–2
Roots Fund 88
Rorick, Matthew 280, 282
rosé 108, 128, 327
Rothschild, Baron Phillippe de 180
Royal Nonesuch Farm, The 230
RTDs (ready-to-drink beverages) 314
Rued winery 344
Russian colonists 113
Russian River Valley 131, 138, 141, 143, 377
  wineries 149, 155, 158, 195
Rutherford 161, 166, 168, 170, 175–6, 177, 379
Ryder Estate Winery 342
Ryme winery 155–6

Sacramento Valley (Central Valley) 252–3, 255–7, 270, 385
Saddle Rock Malibu (Southern California) 388
Salado Creek 271, 386
Salmon Safe 367
San Andreas Fault 98–101, 116
San Antonio Valley 217, 218, 382
San Benito County 220–25, 381–2
San Bernabe 216, 217, 218, 382, 406
San Bernardino County (Southern California) 292, 294, 389
San Diego County (Southern California) 5, 87, 293, 295, 389
San Francisco 7, 60, 87, 110, 202
San Francisco Bay 206–10, 381
San Francisco world fair 28
San Gabriel 17, 19, 110
San Joaquin Valley
  agriculture/worker relations 42–3, 46, 51–2
  AVAs 271, 386
  growing conditions 268, 270
  industries 270–71
  known details of 267
  varieties 267–8, 270
  Yokut people 267–8
San Juan Creek 234, 236, 384
San Lucas 217, 218, 382
San Luis Obispo County 225–8, 383
San Luis Rey (Southern California) 295, 389
San Miguel District 234, 236, 384
San Pasqual Valley (Southern California) 295, 389
San Ysidro District 212, 381
Sandhi and Domaine de la Cote 249
Sandlands winery 82, 107, 108, 133, 135, 211, 262, 266–7, 275, 279
Sanford & Benedict 241
Sanford, Richard 63, 241, 298, 300
Sangiocomo family 183
Santa Barbara (Central Coast) 103, 127, 166, 202–3, 225–6, 230, 240–50, 298
  AVAs 244–5, 384
  Funk Zone 242
  growing conditions 242–4
  wineries
    early 241–2
    list 245–50
Santa Clara Valley (Central Coast) 26, 193, 194, 206–7, 211–12, 221, 381
Santa Cruz Mountains 135, 188–200
  AVA 198, 380
  geography, geology, climate 101, 105, 188–90
  history 192–4
  vineyards 190, 197
  viticulturists and winemakers 191, 194–8
  wines and wineries 60, 198–200
Santa Lucia Highlands 102–3, 183, 382
  wineries 216, 217–19, 238
Santa Margarita District 236, 384
Santa Maria Valley 203, 244, 384
  wineries 227, 241, 246
Santa Ynez Valley 103, 203, 244, 384
  wineries *206,* 243, 245, 247–8
Sauvignon Blanc 96, 107, 129, 155, 161, 180, 183, 227, 244, 250, 264
Saxum winery 239
Scar of the Sea and Lady of the Sunrise winery 227
Scharffenberger 119
Scheid winery 219–20
Schoener, Abe 160
Scholium Project winery 160
Schramsberg winery 183
Screaming Eagle 160, 183
Seghesio winery 344
Seiad Valley 286, 288, 388
Selyem, William 121, 131
September 11 attacks 77, 304
Servicemen's Readjustment Act 37, 38
Shafer winery 68, 184
Shake Ridge Ranch 279–80
Shannon Family 127–8, 130
Shasta County 288, 388

Shenandoah Valley 279, 280, 384
Sideways movie 78, 83
Sierra Foothills 101, 273–6
AVAs 387
history 28, 115, 206–7, 273–5
map 272
wineries listed 282–4
Sierra Pelona Valley 292, 294, 388
Siletto, Ron 221
Silicon Valley research 94–5
Silver Oak winery 67, 88, 156, 335, 338, 343,
Sine Qua Non 249–50
Sinskey, Robert 65, 139
Siskiyou County 287, 288, 388
Sixteen600 winery 156–7
60 Minutes 78, 167, 305, 317
Skywalker Ranch 112
Sloughhouse 260, 264, 385
Smith, Andy 80–81
Smith, Justin 239
Snowden-Seysses, Diana 340
social media 73, 82–5, 146, 300, 312
Solano County (North Coast) 112, 186–7, 380, 385
Solorio, Arnulfo 86
Solorzano, Ruben 243
Somm movie 75, 83
Sonoma Coast 63, 98, 143–4, 378
wineries 108, 214
Sonoma County (North Coast) 130–42, 337
AVAs 103, 142–4, 377–8
farming 59–60, 141
Monarch Challenge 215
vines 113, 130–31
wineries 65, 66, 67, 80–81, 98, 127, 131–42
listed 144–58
Sonoma Mountain 65, 99, 117, 130–32, 143, 377
wineries 66, 141
Sonoma Valley 132, 134, 138, 141, 142, 145, 377
Soto, Rodrigo 139, 142
South Coast 294, 388
Southern California 289–96, 388–9
known details of 289
Los Angeles producers 110
Spanish Las Californias 14, 255, 390
Spanish Arrival 15–16, 398
vineyards and winemaking 16–18
sparkling wine 119, 157, 188, 193, 309
Spencer, Tim and Barbara 261
Spock, Marjorie 48–9
Spottswoode 161, 184
Spring Mountain District 168, 170, 330, 379
Spurrier, Stephen 68, 164
Squaw Valley-Miramonte 271, 386
St Helena 161, 170, 173, 213, 284, 379
St. Amant winery 261
St. Supery 161
Sta Rita Hills 242, 244, 298, 384
Staglin winery 160
Stag's Leap Wine Cellars 67–8, 159, 164–5, 185–6, 298
Stags Leap District 168, 170, 179, 184, 379
Stanislaus County (Central Valley) 386
State of California 20–22
Steele, Jed 126, 127, 360
Steinbeck, John 215–16
Stewart, Lee 34
Stone, Larry 249
Stony Hill 108, 127, 159, 163, 208, 213
Storybook Mountain winery 161
Suisun Valley 187, 380
Sullivan, Charles 8
Sunnyside winery 42
Sussman, Eric 154
sustainability 159, 332–44
adapting to 332–3
building design 336–7, 342–3, 366
certification 65, 365–6, 368–9
collaboration and protocols 334–6
habitat restoration 343–4
innovative practices 340–41
logging 118
packaging 339–40
pollinators/Monarch Challenge 214–15
regenerative farming 302–3, 344–9, 354–60
renewables 342–3
waste management 338–9
water management 336–8, 345–6
see also California climate; climate change; California farming
Swan, Joseph 60, 131, 149–50
sweet and bulk table wines 41–2

Taber, George 164
Tablas Creek winery 204, 205, 232, 233, 239–40, 306, 354
Talbott winery 220
Talley Vineyards 206, 225
Taribó, Mireia 246
tariffs 89, 310
Taylor, Jack and Mary 163
Tchelistcheff, André 33, 163
Tehachapi Mountains 251, 271
Temecula Valley (Southern California) 389
Templeton District 236
Terah winery 222, 224–5, 280
Thackery, Sean 233
Theopolis Vineyards 4, 85, *115*, 124

three-tier system 373–4
TikTok 85
To Kalon vineyard 161, 162, 166, 179, 360
topography and geology of California *92*, 93, 97–101
Torres family 150, 335
Touriga Nacional grape 108, 128
Tracy Hills 271, 388
Trefethen (was Eschole) 162
Tres Sabores winery 4, 125, 160, 169, 176
A Tribute to Grace and Folded Hills 245
Trinchero family 261
Trinity County (Far North) 286, 288, 387–8
Trinity Lake 286, 288, 388
Trout Gulch Vineyards 191
Truett Hurst winery 344
Trump, Donald 88–90, 309–10
Tuolumne and Mariposa 281
Turley 110, 121, 161, 210–11, 231, 274, 284, 402
  see also Passalacqua, Tegan
Twenty-first Amendment 373–4
Twitter (X) 83, 85, 209
Two Eighty Project 208, *209*, 300
Two Wolves winery 250

Ultramarine winery 157
undocumented workers 301
United Farm Workers 40, 45–6, 50–53, 270
United States Department of Agriculture 49
United States, history timelines
  1873–1967 25
  1920s–1975 40
  1960s–1991 54
  1991–2025 74
University of California, Berkeley 58–9, 63
University of California, Davis 55, 104, 208, 231
  initiatives 6, 159, 208, 336, 337, 340, 401
  research 2, 96–7, 110, 148, 212–13, 233, 268, 352
  teaching winery 94, 95
University of California, Santa Cruz 63
Upper Lake Valley 129, 377
US Department of Agriculture 28, 49, 357, 368

Valdiguié grape 3, 238, 277–8
Valesco, Pete 45
Van Lente, Katrina 64
Vandendriessche, Christopher 81
Vera Cruz, Philip 46
Verdad and Lindquist Family winery 227–8
Verité 157
Vice winery 186
Vietnam War 56, 62, 69
Viña Madre 110, 290, 290
vineyards
  global acreage 301
  wildfire damage 330–31
  see also grapes, most planted; planting; individual AVAs; individual wineries
Vitaevino 299
Volk, Kenneth 232
Volstead Act 23–4, 28

Wade Cellars 88
Wagner, Charlie and Chuck 67, 68
Walker, Andy 110
Wallace, Hardy 82, 142, 146, *147*, 161, 280
WarRoom Cellars 119, 205
water management 336–8, 345–6
  diversion systems 43
Waters, Alice 58–61, 64, 84
Webb, Brad 140, *195*
Welch, Celia 160
West Sonoma Coast 63, 108, 132, 142, 144, 378
  wineries 108, 132, 142, 152, 154, 157, 178, 214
Westbrook Hinds, Chad and Michelle 288
White Americans, rise of middle class 37–9
White Rock 81
Wild Horse Valley 168, 171, 187, 380
wildfires (2017 and 2020) 85–6, 325–9
Williams Selyem winery 157–8
Willow Creek 229, 231, 234, 236, 286, 287, 383
Wine Advisory Board 33–4, 351,
Wine Advocate, The 71–72, 84–5
Wine Berserkers 83
wine clubs 79–80
wine descriptors 75–6
wine discussion boards 83
Wine Disorder 83
wine fairs 86–8
WINefare 87
Wine Folly 83
Wine Institute *see* California Wine Institute
wine investors 308–9
winemaking *see* California wine, historical events; cultural significance of wine
wineries *see* individual AVAs; individual wineries
Wine Twitter 83, 85
Winiarski, Warren 76, 163, 164, 165, 298, 300
Winkler Index 104–5
Winkler, A. J. 216
Winters Highlands 187, 257, 385
Wirz vineyard 221
Woodbridge, Coast Selections 166

World Health Organization 316
World War II 34–7, 41

Yarrow, Alder 85
Year 2000 (Y2K) 75, 76–7, 304
Yellow Dog Vineyard 344
Yolo County (Central Valley) 257, 385
York Mountain 228–30, *229*, 230, 383
York, Alan 64–5, 120, 139–42
Yorkville Highlands 122, 123, 375
Yount, George 161–2
Yount, Hilary and Anthony *229*, 230
Yountville 105, 161–2, 168, 170, 174, 379
Yuba County (Sierra Foothills) 275, 276, 387
Yucaipa Valley (Southern California) 389

Zaca Mesa winery 241–2
Zinfandel
  field blend 107, 136–7
  history of 2, 28, 78, 107, 115, 182
  plantings *see* individual AVAs; individual wineries
  wine 60, 161, 277, 279, 282
Zinfandel Advocates & Producers (ZAP) 171

More from The Classic Wine Library

The Classic Wine Library is a premium source of information for students of wine, sommeliers and others who work in the wine industry, but can easily be enjoyed by anybody with an enthusiasm for wine. All authors are expert in their subject, with years of experience in the wine industry, and many are Masters of Wine. The series is curated by an editorial board made up of Sarah Jane Evans MW, Richard Mayson and James Tidwell MS. Continue expanding your knowledge with the titles below.

The wines of southwest U.S.A.

Few would associate America's frontier states with wine. And yet it is here, in 1629, that the country's first *Vitis vinifera* vines are said to have been planted. Since the 1970s the wine industries of New Mexico, Texas, Arizona and Colorado have evolved. This book takes each state in turn, focusing on growing conditions, key issues and prominent producers.

Wines of the Rhône

This guide to one of the great French wine regions covers all the appellations of the Rhône, featuring interviews with some of the most respected winemakers of the region, and tackles the issues facing the Rhône's wines with clarity and authority. An ideal introduction for those new to the Rhône, while providing fresh insights for long-time admirers of the wines.

The wines of Australia

In *The wines of Australia*, sommelier Mark Davidson tastes his way round the new Australian wine world. He delves in to the country's history, culture, growing environment and grape varieties before exploring the wine regions, with key producers introduced and their wines assessed. *The wines of Australia* captures the character of one of the most exciting wine-producing countries on the planet.

Browse the full list and buy online at
academieduvinlibrary.com

Also published by Académie du Vin Library

Our list is small – and will never be large – because we value quality over quantity. We choose our books with care – above all for their readability, but also because we genuinely believe they have something important to say about the world of fine wine that will enhance your drinking pleasure.

On California: From Napa to Nebbiolo ... Wine Tales from the Golden State

"California has taught me how to write about wine. But California has also taught me about natural beauty so grand I have found myself weeping just driving over the Golden Gate Bridge ... California, a kind of Camelot, knows how to endure. California will always give us – as it has given us in this book – many stories to write." – Karen MacNeil.

Wine Confident: There's No Wrong Way to Enjoy Wine

Kelli A White

Kelli White loves every aspect of wine and wants to ignite the same kind of passion in her readers. For those who have fallen in love with wine already, this book is an imaginative and practical guide to embarking on its greatest adventures. For those who haven't yet, it is the spark to light the fuse.

Behind the Glass: The chemical and sensorial terroir of wine tasting

Gus Zhu MW

What makes red wine red? Do genetic differences, culture and life experience change our perception of wine? And is there science behind the obscure language of tasting notes? The answers to all of these questions and more are explored in *Behind the Glass*, a smart, accessible investigation into the science behind a glass of wine – and our appreciation of it.